DATE DUE

GRAZING MANAGEMENT

nutrient
requirements
15, 39
43
42
43
diet
attributes affecting
36, 80
livestock prod.
167-8, 193

GRAZING MANAGEMENT

An Ecological Perspective

Edited by
Rodney K. Heitschmidt
and
Jerry W. Stuth

TIMBER PRESS
Portland, Oregon

© 1991 by Timber Press, Inc.
All rights reserved.

ISBN 0-88192-190-4
Printed in Hong Kong

TIMBER PRESS, INC.
9999 S.W. Wilshire
Portland, Oregon 97225

Library of Congress Cataloging-in-Publication Data

Grazing management : an ecological perspective / edited by Rod K.
  Heitschmidt and Jerry W. Stuth.
      p.   cm.
    Includes bibliographical references (p.   ) and index.
    ISBN 0-88192-190-4
    1. Grazing--Management. 2. Range management. 3. Grazing-
  -Environmental aspects. 4. Range ecology. I. Heitschmidt, Rodney
  K. (Rodney Keith) II. Stuth, Jerry W.
  SF85.G73 1991
  636.01--dc20   90-19900
                                                                    CIP

CONTENTS

Foreword 7
Acknowledgments 8

CHAPTER 1
An Ecological Perspective
   D. D. Briske and R. K. Heitschmidt 11

CHAPTER 2
Range Animal Nutrition
   J. E. Huston and W. E. Pinchak 27

CHAPTER 3
Foraging Behavior
   Jerry W. Stuth 65

CHAPTER 4
Developmental Morphology and Physiology of Grasses
   D. D. Briske 85

CHAPTER 5
Ecosystem-level Processes
   Steve Archer and Fred E. Smeins 109

CHAPTER 6
Hydrology and Erosion
   Thomas L. Thurow 141

CHAPTER 7
Livestock Production
   R. K. Heitschmidt and C. A. Taylor, Jr. 161

CHAPTER 8
Wildlife
   T. G. Barnes, R. K. Heitschmidt, and L. W. Varner 179

CHAPTER 9
Social and Economic Influences on Grazing Management
   J. R. Conner 191

CHAPTER 10
The Decision-Making Environment and Planning Paradigm
   J. W. Stuth, J. R. Conner, and R. K. Heitschmidt 201

Appendix A
Animal Species List 225

Appendix B
Supplemental Plant Species List 226

Literature Cited 229

Index 255

FOREWORD

This book was written to help resource managers broaden their perspective on management of grazing animals and to heighten managers' awareness of their role in maintaining the integrity of ecological systems. A conceptual synthesis of a rapidly expanding literature base addresses the fundamental ecological concepts and managerial principles pertaining to grazing management. We focus on ecological and managerial constraints common to all grazed systems; the application of these concepts serve only as specific examples. The concepts and principles discussed are applicable to a diverse array of vegetation types and geographical regions even though rangelands have been specifically referenced. Further, the preponderance of examples from Texas rangelands simply reflects the professional backgrounds of the authors.

We address a broad readership including students, educators, scientists, agency personnel, and informed practitioners in the field. Each author has written his respective chapter with the assumption that the reader possesses some familiarity with fundamental ecological principles and grazing management concepts. The chapters have been written to present an independent treatment of a specific topic but have been cross-referenced to provide an expanded discussion of specific concepts. This book represents the culmination of six years of involvement by the various contributors in an integrated approach addressing grazing management research.

Chapter 1 contains a general discussion of the basic principles of grazing management and is written to provide a conceptual framework to integrate the remaining nine chapters. Chapter 2 provides an overview of the nutritional aspects of grazing animals with specific attention to the animals' various morphological and physiological adaptations utilized to garner sufficient resources to meet a specific set of nutrient requirements. Chapter 3 addresses the foraging behavior of grazing animals in relation to the adjustments required to facilitate nutrient acquisition in various plant communities. The physiological and morphological consequences of defoliation on individual plants and their implications for population and community structure are considered in Chapter 4. Chapter 5 expands upon these concepts and focuses on the long-term impacts of herbivory on ecological succession and ecosystem stability and function. The consequences of biomass removal, shifts in species composition, and the physical impact of herbivores on the hydrological characteristic of plant communities are the subject of Chapter 6. Chapter 7 defines how livestock production is influenced by animal numbers, kind and class, and temporal and spatial distribution of animals in a wide array of management regimes. Indigenous wildlife populations occupy many rangelands so Chapter 8 contrasts needs of wildlife and livestock with emphasis on the concept of critical habitat requirements. Chapter 9 evaluates the socio-economic aspects of grazing management decision processes. Chapter 10 discusses the management planning process for developing effective grazing management strategies.

ACKNOWLEDGMENTS

The authors wish to thank Ms. Lili Lyddon for illustrating all the figures in the book. Our special thanks to our reviewers:

CHAPTER 1 J. W. Walker, W. K. Lauenroth, W. A. Laycock, F. E. Smeins, R. H. Hart, F. S. Guthery.
CHAPTER 2 Mike Galyean, John Walker, Wayne Greene, Don Adams, Chris Krysl.
CHAPTER 3 Fritz Senft, Mark Stafford Smith.
CHAPTER 4 J. H. Richards, Carol Bilbrough, Steve Ekblad, Rhett Johnson, Johanna Pate, Lee Thornhill, Jeff Murphy.
CHAPTER 5 Guy McPherson, Allen McGinty, Joe Trlica, Neil West, Joel Brown.
CHAPTER 6 Will Blackburn, Joel Brown, Jim Dobrowolski, Bob Knight.
CHAPTER 7 Neil Tainton, Allen Wilson, Dick Hart, Jerry Holechek, Harvey Blackburn, Clinton Owensby, David Bransby.
CHAPTER 8 John Kie, Fred Guthery, David Swift.
CHAPTER 9 Wayne Hamilton.
CHAPTER 10 Wayne Hamilton, Dennis Sheehy.

We also wish to express our appreciation to Drs. Joe Schuster, Earl Gilmore, and Carl Menzies for their administrative encouragement throughout the writing process. Particular thanks are given to Richard Abel, editor, Timber Press, for his editorial guidance. Special thanks also to Sylvia Dudash and Marcella Smith for typing and managing the many drafts of this book. Finally, we express our sincere gratitude to the Texas A&M University System for providing us the professional environment in which to create this book.

GRAZING MANAGEMENT

CHAPTER 1

AN ECOLOGICAL PERSPECTIVE

D. D. Briske and R. K. Heitschmidt

INTRODUCTION

Grazing or **herbivory** is the process by which animals consume plants to acquire energy and nutrients. Grazing management involves the regulation of this consumptive process by humans, primarily through the manipulation of livestock, to meet specific, predetermined production goals. Both the grazing process and associated managerial activities occur within ecological systems and are therefore subject to an identical set of ecological principles which govern system function. These ecological principles impose an upper limit on animal production which cannot be overcome by management. The fact that both the grazing process and efforts to manage it are influenced by a common set of ecological principles justifies the evaluation of grazing management in an ecological context.

Humankind has historically fostered and relied upon livestock grazing for a substantial portion of its livelihood because it is the only process capable of converting the energy in grassland vegetation into an energy source directly consumable by humans. Biochemical constraints determine that herbivores, function as "energy brokers" between the solar energy captured by plants in the photosynthetic process and its subsequent use by humans (Southwood 1985). The inability of humans to directly derive caloric value from the 19 billion metric tons of vegetation produced annually in tropical and temperate grasslands and savannas (24 millions km^2; Leith 1978) provides the ultimate justification for evaluating grazing as an ecological process.

Ecological processes associated with grazing have probably not changed appreciably since the initial appearance of grasses and grazers in the fossil record some 45 million years ago (Stebbins 1981). However, a rapidly expanding human population, escalating degradation of natural resources, and increasing socio-economic pressures have all increased the complexity associated with the management of grazed systems. Hundreds of experiments have been conducted and thousands of pages written addressing the ecological processes and plant and animal responses within grazed systems. Unfortunately, little attention has been directed toward an interpretative synthesis of this information. The diverse subject matter encompassing several disciplines (e.g., ecology, animal science, hydrology, economics, and systems science), and the difficulty associated with identifying an organizational scheme capable of encompassing this large body of information, have undoubtedly contributed to the lack of subject matter synthesis. The absence of a unified conceptual framework has impeded both the study and the management of grazed systems.

The objective of this chapter is to identify and evaluate the fundamental processes associated with grazing within the context of ecological systems. We begin by reviewing the structure and function of ecological systems; follow with an evaluation of the major effects of grazing on energy flow and nutrient cycling within ecological systems; and conclude with a discussion of the ability of humans to regulate the ecological processes affecting plant and animal production within grazed systems. The topic is treated in a broad, conceptual manner to provide the organizational framework necessary to evaluate the relative relationships among specific components and processes within grazed systems. Specific subject matter areas are treated in greater detail in subsequent chapters.

ECOLOGICAL SYSTEMS

Structure of Ecological Systems

Ecological systems or **ecosystems** are defined as assemblages of living organisms in association with their physical and chemical environment. The ecosystem concept is intended to demonstrate the interrelationship or interdependence among the various components within a system, rather than to delineate a specific set of organisms within a geographic area (Odum 1971; Begon et al. 1986). Consequently, ecosystems are arbitrarily defined depending upon the interest of the investigator.

The living (**biotic**) component of ecological systems is classified according to the strategy organisms use to acquire energy and nutrients from the nonliving (**abiotic**) component. The two most basic strategies are **autotrophic** (self-nourishing), and **heterotrophic** (other-nourishing) (Odum 1971; Begon et al. 1986). Autotrophs acquire energy from solar radiation by photosynthesis, while heterotrophs acquire energy by ingesting other organisms. **Autotrophs** or **producers** include all species of green plants, while **heterotrophs** or **consumers** encompass all animal species including microorganisms (Fig. 1.1). The abiotic component defines the physical and chemical components of the system. The ultimate energy source, solar energy, and raw materials (e.g., CO_2, H_2O and nutrients) necessary to convert solar energy into chemical energy are also present within the abiotic component.

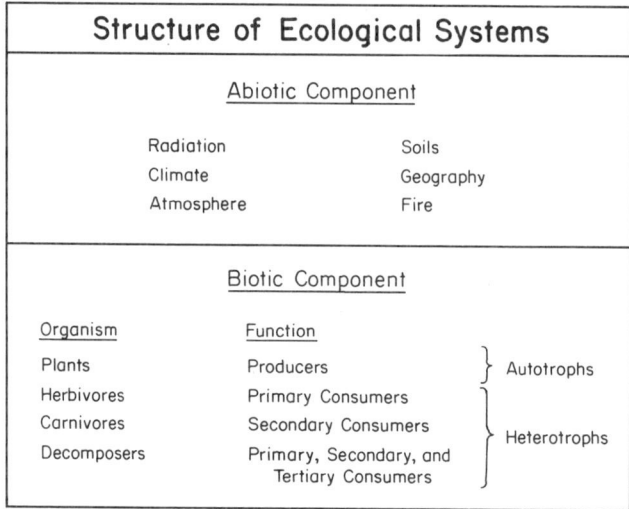

Figure 1.1. Generalized description of the structure of ecological systems. The abiotic (nonliving) component comprises the physical and chemical environment of the biotic (living) component.

Function of Ecological Systems

Energy Flow. Energy flow within ecological systems may be viewed in terms of an economic analogy (Gosz et al. 1978). Economics, like ecology, is concerned with the movement of valuable commodities through a series of producers and consumers. A viable ecological system depends on the flow of energy just as a viable economy depends on the exchange of currency. An analysis of energy flow provides a balance sheet for monitoring energy inputs and outputs within a system not unlike a ledger of credits and debits. The magnitude and efficiency of energy flow between various feeding levels within systems can also be evaluated. The initial capture of solar radiation by vegetation, the efficiency of vegetation utilization by herbivores, and the efficiency with which ingested energy is converted into animal growth comprise the major energy transfer processes in grazed systems. Equally important, but less obvious, is the influence energy flow exerts on the managerial components of systems including resource availability, supply:demand ratios, and price-market structure.

Solar energy is initially converted into chemical energy by photosynthesis within the chlorophyll-containing cells of plants. The energy captured within plants is subsequently transferred to one of two general categories of heterotrophic organisms (Fig. 1.2). In the absence of herbivores, energy is transferred directly into litter following plant senescence. A series of microorganisms, primarily bacteria and fungi within the soil (**decomposers**), utilize this organic matter as an energy source eventually releasing heat energy in association with microbial respiration. This pattern of energy flow defines the **detrital food chain** (Golley 1960; Odum 1971). In the presence of herbivores, a portion of the energy initially captured by plants is consumed and converted into animal tissue. Herbivores, in turn, may be ingested by other consumers at higher feeding levels (i.e., carnivores and humans). Heat energy is released by consumer respiration at each feeding level. This pattern of energy flow through the system defines the **grazing food chain**. Energy is transferred from the grazing to the detrital food chain in the form of feces and animal tissue following death.

Two thermodynamic laws govern the flow of energy within ecological systems (Golley 1960; Odum 1971). The first law states that energy can be transformed from one form to another (e.g., conversion of solar energy to chemical energy by photosynthesis), but cannot be created or destroyed. The second law establishes that

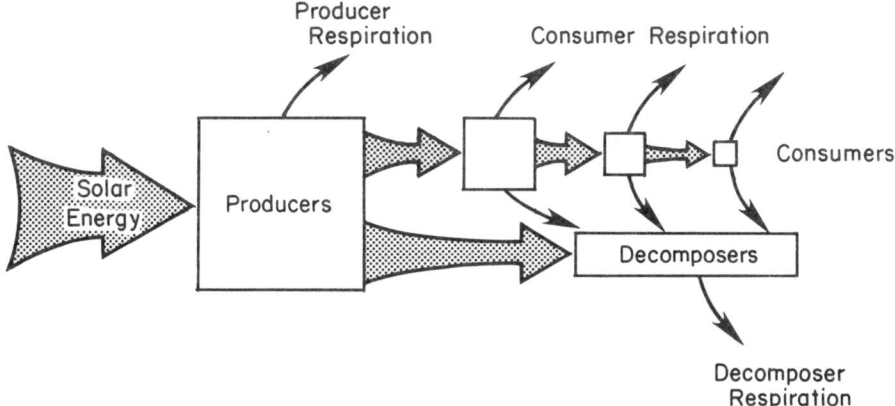

Figure 1.2. Simplified illustration of energy flow through ecological systems. Solar energy is initially captured by primary producers and transferred through at least one consumer feeding level to form the grazing food chain, or directly into the decomposer compartment to form the detrital food chain (after Whittaker 1972).

energy transformation processes are not 100% efficient. These laws dictate that a large proportion of the chemical energy, approximately 90%, transferred between **feeding (trophic) levels** within a system is converted to heat energy of limited value to the biotic portion of the system. The energy "loss" between feeding levels results from an inefficient transfer of organic matter (e.g., gaseous, urinary, and fecal losses) and the energy required for internal maintenance of organisms (i.e., maintenance energy). For example, only a portion of the solar energy converted into chemical energy by photosynthesis is realized as growth because a portion is utilized in respiration. Similarly, animals utilize a large portion of the total energy ingested for basal metabolism, thereby diminishing the amount of energy available for growth or transfer to subsequent feeding levels within the system (see Chapter 2).

The capacity of ecological systems to produce biomass would initially appear limitless given the large, continuous supply of solar energy. However, above-ground **primary productivity** (plant growth/area/time) is less than 1000 kg/ha/yr in many grasslands (Sala et al. 1988). Primary productivity is limited by two general categories of **ecological constraints**. The first constraint involves the quality of solar radiation available at the earth's surface. Only about 45% of the solar energy is within the appropriate region of the spectrum to be effective in photosynthesis (Gosz et al. 1978; Begon et al. 1986). The remaining 55% of the spectrum is primarily composed of long-wave thermal energy unavailable for conversion into chemical energy. However, this energy affects ecological systems, in that it is absorbed as heat energy by the atmosphere, soil, and vegetation to generate the thermal environment and power the hydrological and nutrient cycles (e.g., energy required to evaporate water).

The second category of ecological constraints limiting primary productivity involves the occurrence of rigorous abiotic factors which prevent solar energy capture from being maximized. Water, temperature, and nutrient limitations frequently prevent a sufficient leaf canopy from developing to intercept the available photosynthetically active radiation (Lewis 1969; Begon et al. 1986). For example, plant canopies may be nonexistent for several months of the year in temperate or arid and semi-arid regions. Similarly, abiotic limitations mean that maximum photosynthetic efficiencies are seldom, if ever, attained when a plant canopy is present. It is generally estimated that less than 1% of the solar energy at the earth's surface is converted into chemical energy by terrestrial vegetation (Lewis 1969; Begon et al. 1986). This is despite the fact that a conversion efficiency of approximately 20% can be realized for individual leaves under ideal environmental conditions before a biochemical limitation is encountered within the photosynthetic process (Lawlor 1987). It is important to note that the amount of solar energy captured in primary production represents the *total* amount of energy available for utilization by heterotrophic organisms within the system.

Secondary productivity (animal growth/area/time) is also limited by two broad categories of ecological constraints in addition to the availability of primary production. The first constraint involves the inability of herbivores to consistently consume the majority of the primary production produced. Primary production varies widely through time and space, making it difficult to balance herbivore density with the fluctuating food resource. That portion of the herbaceous biomass available in excess of current animal demand senesces within a period of weeks following its production (Parsons et al. 1983; Chapman et al. 1984). Eventually this material enters the decomposer compartment as litter (detrital food chain). In addition, most primary production in grasslands is located below-ground as roots and crowns making it inaccessible to large herbivores (Sims and Singh 1978b; Stanton 1988).

The second category of constraints limiting secondary productivity is related to the quality of primary production (nutritional value; see Chapter 2) ingested by herbivores. A substantial portion of the total energy ingested by herbivores is lost as methane (in ruminants), urine, or feces, and a large portion of the **metabolizable energy** is utilized in basal metabolism (see Chapter 2). It is only the remaining energy, approximately 10% of the total ingested, which is available for animal growth (Dean et al. 1975; Rode et al. 1986).

Nutrient Cycling. The availability and transfer of nutrients constitutes a second indispensable function of ecological systems. **Essential nutrients** (carbon, nitrogen, phosphorus, etc.) form an integral component of biochemical processes and metabolic pathways within organisms which directly influence the initial capture and flow of energy through the system (Odum 1971; Wilkinson and Lowery 1973). For example, photosynthesis increases linearly over a range of leaf nitrogen concentrations (Field and Mooney 1986), and animal growth frequently increases with increasing nitrogen availability in the diet (Mattson 1980). However, unlike energy flow, nutrients cycle from their reservoir within the soil or atmosphere into the biotic component of the system (i.e., producers and consumers) and then back into soil or atmospheric reservoirs within the system (Fig. 1.3).

Plants initially assimilate many of the essential nutrients from the abiotic environment subsequently used by animals. For example, herbivores can only acquire nitrogen by consuming plants, even though the atmosphere contains approximately 80% nitrogen by volume (Odum 1971; Wilkinson and Lowery 1973). However, as with energy, nutrients may be transferred through either the grazing or detrital food chain within the system (Fig. 1.3). Regardless of the food chain in which they are incorporated, nutrients are eventually returned to their inorganic form following organic matter decomposition by microorganisms within the decomposer compartment (Stanton 1988).

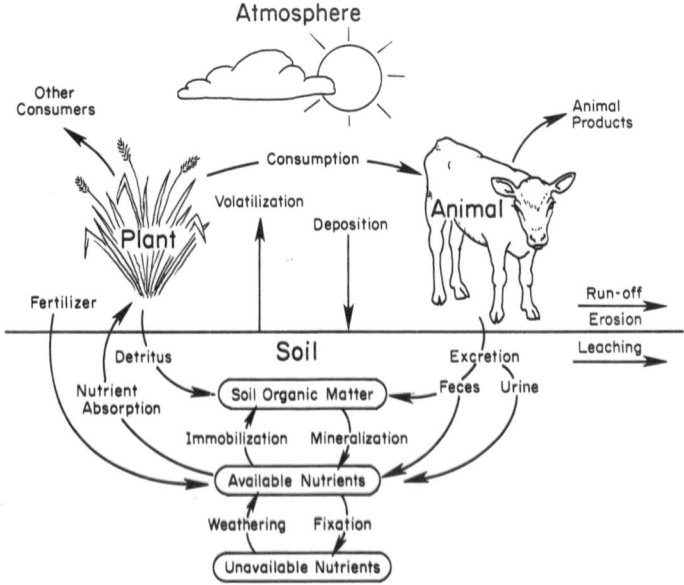

Figure 1.3. Simplified illustration of nutrient cycling within ecological systems. Nutrients move from their respective reservoirs within the abiotic component of the system, into the biotic component, and back into the environment in a cyclic pattern (after Wilkinson and Lowery 1973).

GRAZING AND ECOLOGICAL SYSTEMS

Plant consumption by herbivores (i.e., grazing food chain) introduces an additional feeding level between the primary producers and the decomposers. The objective of this section is to examine how grazing influences energy flow and nutrient cycling within ecological systems. A case study is presented to quantitatively illustrate the influence of grazing on energy flow and demonstrate the utility of the ecosystem concept to the investigation of grazed systems.

Energy Flow

An Ecological Dilemma. The percentage of annual above-ground primary production utilized by herbivores varies greatly, but estimates generally range between 20% and 50% (Scott et al. 1979; Detling 1988). Although much higher levels of utilization can occur, in excess of 90%, they are generally restricted to specific regions or years. Insects and small mammals may consume as much as 10–15% of the annual aboveground production. An even smaller portion of the total annual primary production is utilized by domestic herbivores because approximately 60–90% of the production occurs below-ground in grassland systems (Sims and Singh 1978; Stanton 1988). The portion of annual production not utilized by herbivores is eventually consumed by microorganisms in the decomposer compartment while a small portion is stored as organic compounds within a long-term carbon pool in the soil (Clark 1977).

The **fundamental ecological dilemma** encountered in grazed systems is the inability to simultaneously optimize the interception and conversion of solar energy into primary production and the efficient harvest of primary production by herbivores (Parsons et al. 1983). Severe grazing ensures that available production is efficiently harvested, but eventually reduces production by minimizing the subsequent capture of solar energy. Alternatively, lenient grazing maximizes primary production, but a large percentage of the production is incorporated into the decomposer compartment without being consumed by herbivores. Contrasting patterns of energy flow between the grazing and detrital food chains are clearly illustrated in an experiment conducted with two perennial ryegrass pastures grazed by sheep (Fig. 1.4). One of the pastures was grazed leniently (24 sheep/ha) to maintain a leaf area index of approximately three (leaf area:ground area ratio of three), and the other was grazed severely (47 sheep/ha) to maintain a leaf area index of one. The total amount of energy captured by photosynthesis was 43% greater in the leniently grazed pasture as opposed to the severely grazed pasture because a greater percentage of the available

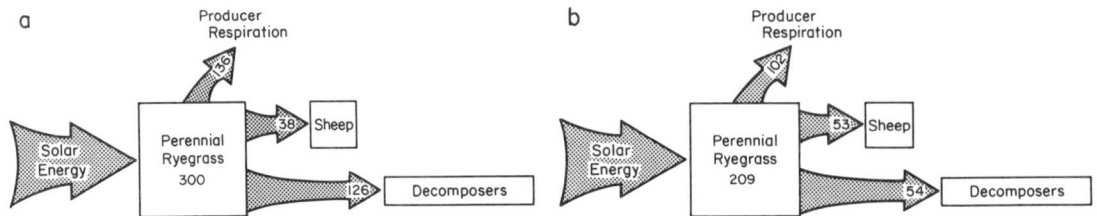

Figure 1.4. Energy capture and flow (kg carbon/ha/day) within (a) leniently and (b) severely grazed perennial ryegrass pasture. A greater amount of solar energy is converted into ryegrass production in the leniently grazed pasture, but this grazing regime reduces the relative amount of energy consumed by livestock and increases the relative amount of energy transferred into the decomposer compartment in comparison with the severely grazed pasture (after Parsons et al. 1983).

solar radiation was intercepted by the plant canopy. However, animal consumption was 40% greater in the severely grazed pasture than in the leniently grazed pasture, even though production was less, because of greater livestock numbers per hectare. The end result was that 42% of the energy captured by ryegrass in the leniently grazed pasture entered the detrital food chain, while only 13% was consumed by livestock. By contrast, only 26% of the energy captured by ryegrass in the severely grazed pasture entered the detrital food chain, while 25% entered the grazing food chain. The remainder of the energy captured by photosynthesis (49%) was either allocated to root growth or used in plant respiration. These data illustrate that primary production and efficient biomass utilization cannot be **maximized** simultaneously because of the contribution of leaf area to both processes.

A Case Study. The influence of grazing on energy flow within ecological systems can best be illustrated quantitatively by evaluating a simple case study from the Texas Experimental Ranch near Throckmorton, Texas. This mixed-grass prairie site was grazed continuously throughout the year at a stocking rate considered severe for the region. The system yielded a mean above-ground herbaceous production of 3183 kg/ha/yr (Heitschmidt et al. 1987a), individual animal gains of 200 kg/cow, and livestock gains of 53.5/kg/ha/yr (Heitschmidt et al. 1990).

These production values can easily be converted into energy values by virtue of the fact that 1 kg of plant and animal tissue contains approximately 19.7 and 23.5 MJ (mega [million] joules) respectively (Golley 1961; Odum 1971). (A joule [J] is the expression for energy in the International System of Units; 1 calorie = 4.19 J; 1 J = 0.24 calories.) The total amount of solar energy received annually at the site provides a convenient reference point with which to compare energy values at various locations within the system. The total input of solar energy at the latitude of the Experimental Ranch (33° 20'N) is approximately 63,000,000 MJ/ha/yr (Cinguemani et al. 1978). In this system, approximately 62,705 MJ/ha/yr is captured in herbaceous above-ground vegetation; 313,525 MJ/ha/yr is captured in total (above- and belowground) herbaceous vegetation (assuming 80% of the primary production is belowground; Stanton 1988); and 1,257 MJ/ha/yr is transferred into livestock gains (Fig. 1.5). Decreasing energy values from the initial input of solar energy to vegetation and finally to livestock gains clearly demonstrate the inefficiency of energy transfer within the system.

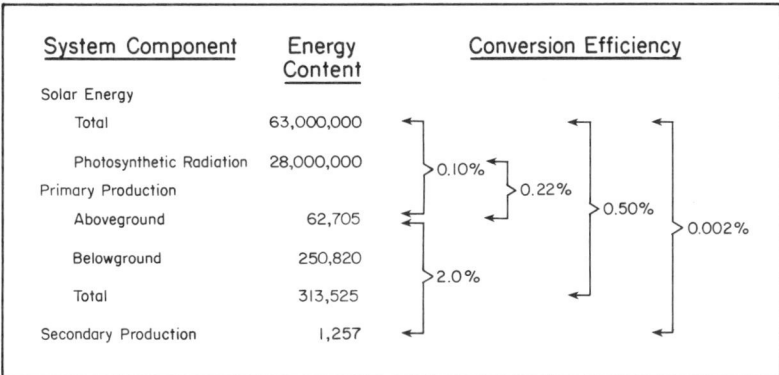

Figure 1.5. Energy content (MJ/ha/yr) and transfer efficiencies (%) for primary and secondary production in relation to total and photosynthetically active solar radiation at the Texas Experimental Ranch (from Heitschmidt et al. 1987a; 1990). Energy values are calculated by multiplying primary and secondary production values by 19.7 and 23.5 MJ/kg, respectively. Conversion efficiencies represent the quotient of two energy values at specified locations within the system multiplied by 100.

An important aspect of energy flow analysis is that **transfer efficiencies** can be calculated by dividing the amount of energy captured within one level of the system by the amount of energy in a preceding level (Golley 1960; Pimm 1988). Aboveground herbaceous vegetation captured 0.10% and 0.22% of the total solar radiation and photosynthetically active radiation/ha/yr respectively (Fig. 1.5). A slightly greater conversion efficiency, 0.50%, is observed when total (above- and belowground) herbaceous annual production is considered. Conversion efficiencies decrease even further when energy transfer into livestock production is calculated because of the energy loss which occurs between feeding levels. Approximately 0.002% of the energy available in total annual solar radiation is transferred into livestock gains in comparison with 2% of the energy available in above-ground production. These minimal conversion efficiencies, which progressively decrease with the incorporation of additional feeding levels, are similar to those estimated in other grassland systems (Macfadyen 1964; Coleman et al. 1976; Snayden 1981; Akiyama et al. 1984).

The intrinsic inefficiencies of energy flow in grazed systems should not be interpreted to imply that these systems possess an insignificant potential for secondary production. Grazing could potentially yield an estimated 7.5 trillion MJ of animal production annually from grasslands and savannas of the world, assuming that large herbivores consume 20% of the above-ground production and possess a conversion efficiency of 2% (Fig. 1.5). These estimates demonstrate the tremendous importance of grazing to the human food supply on a global basis. Substantial increases in secondary production can be attained with only modest increases in ecological efficiencies resulting from effective management strategies. A combined increase of only 0.01% in harvest and conversion efficiencies would potentially increase secondary production by 75 billion MJ.

Nutrient Cycling

Grazing may also modify the rate and pattern of energy flow in ecological systems by influencing nutrient availability. Nutrient availability, in turn, governs the efficiency with which organisms acquire and process energy. Grazing affects nutrient cycling by accelerating the rate of nutrient conversion from an organic form (amino acids and proteins) to an inorganic form (nitrate and ammonium). This process, termed **mineralization**, is critical to grassland production because a large proportion of the essential nutrients are bound in organic matter within the soil (Wilkinson and Lowrey 1973; Woodmansee et al. 1978). However, only those nutrients in specific inorganic forms are available for plant absorption.

Grazing increases mineralization by reducing the particle size of plant material (e.g., chewing and rumination) and providing a favorable environment for microbial activity (e.g., high body temperatures; Wilkinson and Lowrey 1973; Floate 1981). Yet herbivores retain only a small portion of the nutrients consumed, thereby rapidly returning most nutrients to the system in urine and feces. Nutrients excreted in urine—primarily nitrogen, potassium, magnesium, and sulfur—are in the inorganic form and therefore immediately available for plant absorption (Wilkinson and Lowrey 1973). In contrast, a greater proportion of nutrients in fecal material and ungrazed plant material than in urine are bound in organic compounds and must be mineralized by decomposers prior to plant absorption. Consequently, a portion of the nutrients incorporated into primary production become available for reabsorption more rapidly when transferred through the grazing food chain than when transferred directly into the decomposer compartment (Wilkinson and Lowrey 1973;

Floate 1981). Estimates of higher nutrient concentrations in vegetation of grazed than of ungrazed systems support the assumption of increased rates of nutrient cycling (Detling 1988).

Nutrient transfer through the grazing food chain potentially increases the rate of cycling, but in so doing, it also increases the potential for nutrient losses from the system (Woodmansee 1978; Floate 1981). **Nutrient losses** from grazed systems primarily occur as volatilization, leaching, soil erosion, and livestock removal from the system (Wilkinson and Lowrey 1973; Woodmansee et al. 1981). Nutrient losses are highly variable and are influenced by a large number of environmental variables including nutrient solubility, soil morphology and chemistry, climate, and topography (see Chapter 5). Nutrient losses associated with herbivore removal from the system are minimized by the limited productivity of many grasslands and the digestive physiology of herbivores (Floate 1981). Nutrient availability is frequently limited by low above-ground productivity containing low nutrient concentrations, 1.5–2% nitrogen in live and less than 1% in dead grassland vegetation (see Chapter 2). In addition, most nutrients ingested by herbivores are voided as urine or feces, thus leaving only a relatively small proportion to be removed as animal products (Wilkinson and Lowrey 1973).

The limited amount of information available indicates that grazing does not increase nutrient losses from the system thereby creating a negative nutrient balance, but this potential does exist (Wilkinson and Lowrey 1973; Woodmansee 1978; Floate 1981). Atmospheric nutrient inputs, 0.3–1 kg/ha/yr in the case of nitrogen, which represent the largest input into nonfertilized systems, in conjunction with the increased cycling rates, appear sufficient to offset grazing-induced losses from most systems. The large nutrient pool within the organic component of the soil may also buffer nutrient losses in the short-term (Woodmansee 1978). However, additional research is required to more definitively assess the long-term consequences of grazing on the cycling of essential nutrients within grazed systems.

Grazing Optimization Hypothesis

Grazing has traditionally been viewed as having a negative impact on the subsequent rate of energy capture and primary production within grazed systems through a series of direct (see Chapter 4) and indirect affects on plant growth (see Chapters 5 and 6). However, the grazing intensity necessary to induce a decrease in primary production is difficult to establish definitively. The "**grazing optimization hypothesis**" suggests that an optimal grazing intensity can potentially increase primary production over that of an ungrazed system (Fig. 1.6). A limited amount of evidence exists to support the grazing optimization hypothesis (Dyer and Bokhari 1976; McNaughton 1979; Hart and Balla 1982; Paige and Whitham 1987), but it does not appear to be a significant ecological process operating on a regular basis in grassland systems (Belsky 1986; Heitschmidt 1990). Illustrations of the grazing optimization hypothesis tend to exaggerate the potential increases in primary production resulting from an optimal level of grazing relative to the potential decreases which may occur in response to severe grazing, that is, the potential increase in production is shown to be equivalent to the potential decrease.

It is important to recognize that much of the data collected in support of the grazing optimization hypothesis were derived from grazed systems where herbivore density and movement were not directly regulated by humans. In these systems, primary production and herbivore density fluctuate widely in a series of continuous feedback loops in response to climatic variation (Sinclair 1975; Walker et al.

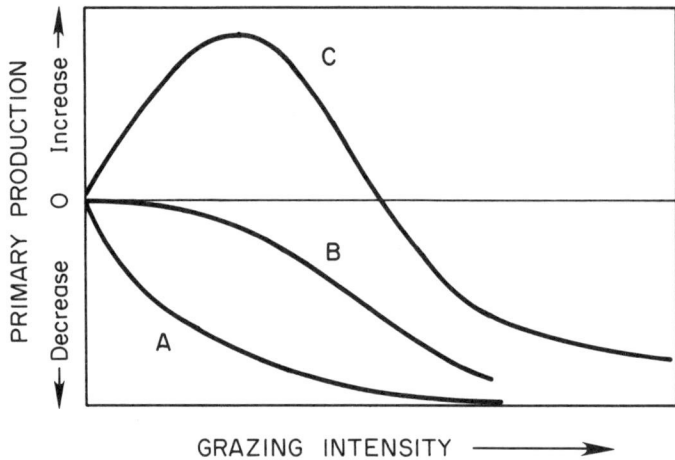

Figure 1.6. Three potential responses of primary production to increasing grazing intensity as indicated by the grazing optimization hypothesis. Primary production may: (A) decrease with increasing grazing intensity, (B) remain unaffected until intermediate levels of grazing intensity are attained and then decrease, or (C) increase with increasing grazing intensity to an optimal level and then decrease (from Detling 1988).

1987). Conversely, herbivore density and movement are rigidly restricted in managed systems, and precautions are taken to minimize deleterious consequences on animal production. Consequently, the grazing intensity in many, if not all, managed systems may frequently exceed the intensity required to consistently stimulate primary production as indicated by this hypothesis (Heitschmidt 1990). This difference likely explains why the hypothesis originated with researchers working in natural rather than managed systems and why the hypothesis receives limited support from natural resource managers.

HUMANS, GRAZING, AND ECOLOGICAL SYSTEMS

Grazed systems are manipulated by humans to meet a diverse set of personal- and/or firm-level production goals (see Chapter 9). The most pervasive of these goals is the maximization of livestock production or profitability on a sustainable basis. The strategies used to attain the desired production goals vary along a continuum of managerial involvement that can be categorized as either extensive or intensive. However, regardless of the managerial strategies employed, livestock production is limited by several constraints intrinsic to ecological systems. The objective of this section is to briefly examine the degree to which managerial strategies can influence the function of ecological systems, thereby increasing livestock production within grazed systems.

Livestock Production

The inverse relationship between animal production per individual and production per unit land area with increasing grazing intensity is a fundamental production response within all grazed systems (see Chapter 7). This response originates from the combined effects of the following processes: 1) decreasing efficiency of solar energy capture, 2) increasing efficiency of forage harvest, and 3) decreasing conversion efficiency (i.e., the efficiency with which ingested energy is converted into animal products) as grazing intensity increases (Fig. 1.7). Primary production decreases because of a reduction in the availability of leaf area to intercept solar energy. Harvest efficiency increases as an increasing number of animals per unit land area consume plant material before it senesces and is transferred to litter. Conversion efficiency decreases as forage intake restrictions per individual animal limit nutrient and energy availability for growth (Van Soest 1982). The end result is that production per animal decreases as grazing intensity increases while production per unit land area increases. Livestock production per unit area continues to increase with grazing intensity because it is dependent upon both individual animal performance and the total number of animals. Eventually, production per unit area decreases rapidly with increasing grazing intensity because increasing livestock numbers are no longer able to compensate for the limited production per individual animal. Therefore, the grazing intensity which **maximizes** sustainable animal production per unit area is that which **optimizes** the processes of solar energy capture, harvest efficiency, and conversion efficiency within a system.

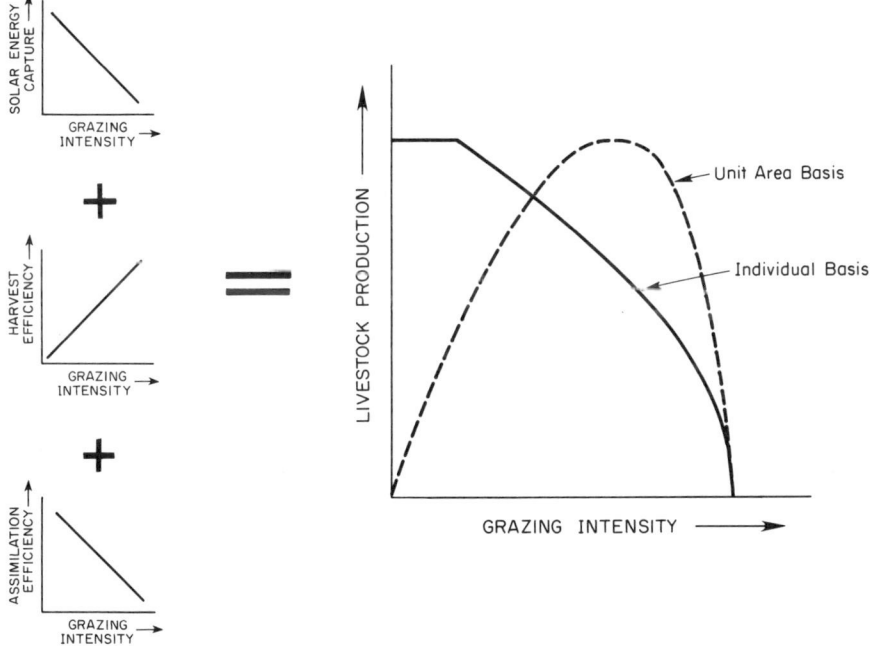

Figure 1.7. Livestock production per individual and per unit area originate from the combined effects of efficient solar energy capture (i.e., primary production), forage harvest efficiency, and conversion efficiency in response to grazing intensity.

The primary ecological constraints limiting the magnitude and efficiency of animal production within grazed systems are summarized as follows:

1. The inefficient capture and conversion of solar energy into primary production, frequently less than 1% per year (Leith 1978; Begon et al. 1986);
2. The limited proportion of total primary production consumed by livestock, less than 20%, considering that 60–90% is below-ground (Stanton 1988) and that approximately 50% of the above-ground proportion is grazed; and
3. The inefficient conversion of ingested energy into animal gains, approximately 10% (Dean et al. 1975; Rode et al. 1986).

These constraints are absolute and defy the best intended and designed managerial strategies. Therefore, managerial strategies must be designed to work within, rather than attempt to overcome or circumvent, these ecological constraints.

The Management Dilemma

Many of the problems encountered in grazing management arise from the attempts of humans to sustain high levels of animal production on a continuous basis. The concepts of overgrazing and undergrazing, for example, address managerial or economic considerations to a greater extent than ecological processes associated with grazing (Crawley 1983). **Overgrazing** refers to situations where improper managerial decisions (e.g., stocking rate) reduce potential livestock production per unit land area by limiting the amount of solar energy captured by species of high nutritive value (e.g., limited leaf area). Similarly, **undergrazing** refers to situations where inappropriate managerial decisions prevent livestock production from being maximized per unit land area because species of high nutritive value are not fully utilized within the limits of sustainable production.

In systems where herbivore movement and survival are not regulated by humans, the concepts of overgrazing and undergrazing are merely points at which primary production and herbivore density show the greatest oscillation within a dynamic equilibrium. Climatically induced periods of limited primary production function as feedback mechanisms to influence herbivore growth, reproduction and survival (Sinclair 1975; Walker et al. 1987). The plant community is provided with a period for recovery during the interval following the return of normal precipitation, but prior to an increase in herbivore density to predrought levels. Herbivore production is determined by climatically induced variation in primary production within these systems, not vice versa as is frequently the case in managed systems (Pieper and Heitschmidt 1988). The period required for vegetation recovery from grazing is frequently eliminated in managed systems by maintaining high livestock density with supplemental feeding during periods of limited primary production (i.e., feed bag syndrome) or the rapid replacement of livestock immediately following the return of favorable environmental conditions (i.e., sale barn syndrome). For this reason, the potential for resource degradation is often greater in managed than in naturally grazed systems.

The managerial intent to maximize livestock production on a sustainable basis magnifies the associated problems of climatic variation and selective grazing. **Climatic variation** determines that the optimal grazing intensity to maximize livestock production is variable in both time and space. For example, in the case study presented previously, above-ground herbaceous production averaged 3183 kg/ha/yr, but ranged from approximately 1500 to 4925 kg/ha/yr within a 4-year period. Similarly, harvest efficiency in the set-stocked treatment averaged 42%, but ranged from 20% to 64%. These data demonstrate that flexible stocking is essential for maximizing livestock production from year to year in most grazed systems. Variability in the magnitude and distribution of precipitation among years is typical of most grazed systems, particularly those in arid and semi-arid environments (Sala et al. 1988).

Selective grazing is displayed to various degrees by all wild (see Chapter 8) and domestic herbivores (see Chapters 2 and 3). Selective utilization of plant species and parts by herbivores in grazed systems frequently decreases harvest efficiency, energy flow within the grazing food chain, and ultimately secondary production. For example, in the case study previously presented, only four of the five most abundant herbaceous species were utilized by cattle, and the most common shrub species, honey mesquite, was never utilized appreciably (Walker et al. 1989b). Consequently, only 17% of herbaceous above-ground production was consumed by livestock, while only about 2% of the total primary production (i.e., honey mesquite plus total below-ground production) was utilized by livestock. Although stocking rate, animal distribution, season of grazing, and mixed-species grazing may minimize selective grazing in some systems, it is impossible to eliminate this problem within most multi-species systems.

Climatic variation and selective grazing frequently interact to affect the magnitude and efficiency of livestock production on both a short- and long-term basis. In the short-term, the amount and relative proportion of primary production by species at a particular location is dependent upon the prevailing environmental conditions and unique physiological requirements of the species (e.g., warm- versus cool-season species). Consequently, the availability and utilization of species varies both seasonally and annually in response to climatic variation and animal preference. Therefore, the appropriate stocking rate to attain optimal harvest and conversion efficiencies also varies within and among years. Stocking rate decisions are difficult to make in a timely manner because the optimal grazing intensity at any given time is dependent upon the occurrence of future climatic conditions. In the long-term, the interaction between climatic variation and selective grazing has a pronounced effect on the rate and direction of ecological succession in grazed systems (see Chapters 4, 5, and 6). Grazing-induced modifications of species composition can greatly affect livestock production depending on the specific plant community and management goals.

Management Strategies

Grazing managers frequently overlook the fundamental ecological basis for implementing various management strategies. Grazing management strategies are intended to increase livestock production by minimizing the detrimental consequences of inherent ecological constraints on the magnitude and efficiency of energy flow within the system (Williams 1966; Lewis 1969). The managerial strategies used to affect the magnitude and efficiency of energy flow can be categorized as extensive or intensive. **Extensive management strategies** are primarily implemented on rangelands characterized by low and/or highly variable production. These strategies focus on the temporal and spatial distribution of various species and numbers of herbivores (see Chapter 7). For example, increased stocking rate and mixed species grazing are strategies commonly employed to increase harvest efficiency and forage quality, thereby increasing the amount of energy and nutrients transferred into the grazing food chain.

Intensive managerial strategies rely upon the direct incorporation of energy inputs into a system beyond those associated with extensive management. Examples of energy-intensive inputs include irrigation, fertilization, introduction of improved forage species, and a variety of vegetation manipulation procedures (Pimentel et al. 1980; Klopatek and Risser 1982). This level of managerial involvement exceeds that associated with grazing management *per se* which is described as the manipulation of livestock in time and space. Although intensive management can substantially increase primary production over that of extensively managed systems, intensive management does not overcome the ecological constraints limiting energy flow efficiencies previously described (Klopatek and Risser 1982).

The introduction of plant species into a system frequently increases primary production by eliminating competing vegetation and replacing it with genetic material selected for high productivity (Pimentel et al. 1980). Species introduction may also minimize selective grazing by replacing multi-species systems with monocultures. Therefore, species introduction affects the function of ecological systems by potentially increasing the efficiency of solar energy capture and the amount and proportion of energy transferred through the grazing food chain. This example demonstrates that management strategies must affect the magnitude and/or efficiency of energy flow if they are to increase livestock production in ecological systems.

Similarly, vegetation manipulation (e.g., noxious plant control through chemical, mechanical, or biological means) may increase the grazeable proportion, but not necessarily the total amount of primary production. An increase in total aboveground production following removal of undesirable species requires that productivity of the existing desirable species exceed that of the undesirable species. It is unlikely that this will occur in most systems without the use of additional cultural practices (seedbed preparation, plant species introduction, etc.). Yet, if total primary production is not increased by the removal of undesirable species, the proportion available for livestock may be greater, thereby increasing the magnitude of energy flow through the grazing food chain.

Intensive management strategies directly incorporate energy into the system in the form of fossil fuels (Pimentel and Burgess 1980; Klopatek and Risser 1982). For example, large amounts of fossil fuels are required for the industrial production of nitrogen fertilizer and its subsequent application. This is also the case for other intensive management practices, including vegetation manipulation and species introduction. An accurate assessment of net energy output from intensively managed systems requires that the fossil fuel input be subtracted from the total system output. When intensive management strategies are evaluated in this manner, they generally display lower ratios of energy output per unit of energy input than do extensively managed systems (Table 1.1). Evaluation in this manner demonstrates that increased production does not result from increased production efficiency within intensively managed systems, but rather from the direct incorporation of **energy subsidies** into the system (Odum 1971; Klopatek and Risser 1982). Energy subsidies also increase the effi-

Table 1.1. Comparison of primary production, energy inputs (excluding solar energy), and outputs for extensive and intensive management strategies. Intensive management increases production, energy output (19.7 MJ $\times$ kg production), and efficiency of solar energy conversion, but substantially decreases the energy output:input ratio which is an indication of production efficiency. Primary production and energy outputs are from Williams (1966) while energy inputs are estimated from Pimentel and Burgess (1980).

System Parameter	Extensive Management	Intensive Management		
	Native Range	Sulfur Fertilization	Clover Introduction	Fertilization and Clover
Primary Production (kg/ha/yr)	887	936	1,823	2,686
Energy Output (MJ/ha/yr)	17,474	18,439	35,913	52,914
Energy Input (MJ/ha/yr)	419	1,722	3,130	4,433
Output:Input Ratio	41.7	10.7	11.5	11.9
Solar Energy Conversion (%)	0.027	0.029	0.057	0.084

ciency of solar energy capture and conversion into primary production by partially overcoming biotic and abiotic limitations to the development of plant canopies and the occurrence of optimal photosynthetic rates (Table 1.1). Many techniques associated with industrialized agriculture are based upon the practice of trading calories of fossil fuel energy for calories of food energy.

The absolute increase in plant and animal production realized from energy incorporated into intensively managed systems minimizes the effect of energy transfer efficiencies on the human food supply. A greater amount of energy is available for human consumption from digestible plant products than from animal products (Odum 1971; Pimentel et al. 1980). For example, if 1000 MJ of energy are available to humans functioning as herbivores by consuming grain, then only approximately 100 MJ are available to humans functioning as carnivores by consuming grain-fed animals. The amount of available energy decreases by a factor of approximately 10 because of the inherent inefficiency of energy transfer between feeding levels, as described by the second law of thermodynamics. In industrial agriculture the conversion of fossil fuel energy to food energy has proven and will prove economically sound, as long as a relatively inexpensive energy supply is available for its continuation. On the other hand, it must be recognized that a large portion of the primary production consumed by livestock in grazed systems is neither digestible by humans nor dependent upon large energy subsidies (Pimentel et al. 1980). Therefore, grazing functions as an essential intermediary process between solar energy capture by vegetation and energy consumption by humans.

SUMMARY

Grazing management can be realistically and profitably evaluated within the context of an ecological system because both the grazing process and efforts to manage it are influenced by a common set of ecological principles. An ecological perspective requires that the ecological processes associated with grazing be identified and organized within the structure and function of ecological systems. Grazing management is intended to minimize the detrimental consequences of several intrinsic ecological constraints on animal and, to a lesser extent, plant production within grazed systems. Management strategies must affect the magnitude and/or efficiency of energy flow if they are to increase livestock production within ecological systems.

The primary constraints limiting production efficiency in grazed systems are summarized as follows:

1. The inefficient capture of solar energy in primary production, frequently less than 1% per year;
2. The limited proportion of total primary production consumed by livestock, less than 20%; and
3. The relatively inefficient conversion of ingested energy in secondary production, approximately 10% of the consumed energy.

These constraints are absolute and defy even well intended and effectively designed managerial strategies. Managerial strategies must be designed, therefore, to work within the limits of these constraints, rather than attempt to overcome or circumvent them. Modest increases in the efficiency of energy flow within the limits established by these intrinsic ecological constraints can substantially increase secondary production. An increase in the efficiency of energy transfer from primary to secondary production of only 0.01% (Fig. 1.5) could potentially increase secondary production by 75 billion MJ in grasslands and savannas globally.

The fundamental ecological dilemma encountered in grazing management is the inability to simultaneously optimize the interception and conversion of solar energy into primary production and the efficient harvest of primary production by herbivores. Severe grazing ensures that available production is efficiently harvested, but may eventually reduce production by minimizing leaf area for the subsequent capture of solar energy. Alternatively, lenient grazing maximizes primary production, but a large percentage of the production is incorporated into litter without being consumed by livestock. Grazing management involves the manipulation of kinds and classes of livestock, stocking rate, grazing season and grazing intensity, as implemented through grazing systems, to optimize these two opposing processes and so maximize livestock production per unit land area on a sustainable basis. The managerial task of optimizing primary production and efficient forage harvest is further complicated by climatically induced variation in plant production and the selective grazing typical of various herbivore species.

A thorough evaluation of the components and processes within grazed systems requires a multidisciplinary effort integrating information from several disciplines. The nutritional requirements of grazing animals (see Chapter 2) and foraging behavior employed to acquire energy and nutrients (see Chapter 3) must be understood to accurately evaluate animal production. Insight into the effects of grazing on individual plant growth and function, plant population dynamics (see Chapter 4), structure and function of communities and ecosystems (see Chapter 5), and hydrological considerations (see Chapter 6) is necessary to evaluate the influence of grazing on system integrity and sustainable production. Finally, an understanding of the integrated effect of these processes on livestock and wildlife production (see Chapters 7 and 8) and economic considerations (see Chapter 9) is essential for the development of ecologically sound management strategies within a complex decision environment (see Chapter 10).

CHAPTER 2

RANGE ANIMAL NUTRITION

J. E. Huston and W. E. Pinchak

INTRODUCTION

Forage production, defined in Chapter 1 as the integrated end-product of conversion of solar energy into plant biomass, is the foundation of range animal production systems. Because plant biomass is of limited caloric value to humans as primary consumers, the value of this renewable resource is in the production of secondary and tertiary products through grazing animals. Temporal distribution of forage production sets boundaries on the opportunities for directly or indirectly utilizing rangeland resources.

Our purpose in this chapter is to depict the role of grazing animals in converting chemicals fixed in plants into animal products (food and fiber). To do so, we trace physiological processes and interactions within the herbivore and describe how these relate to the diets that are consumed. We conclude with a discussion of the implications of these interactions in the nutritional management of range herbivores, primarily domestic livestock.

CHARACTER OF FORAGE

Forage includes browse and herbage which can be consumed by or harvested and fed to animals (Soc. Range Manage. 1989). The structural characteristics of forage are described in various ways and with nomenclature appropriate to the context in which they are considered. Botanists and agronomists approach plant cellular structure from the standpoint of biosynthesis. At what sites or in what organelles do certain chemical reactions occur that result in processes such as photosynthesis, protein synthesis, and nutrient translocation? By contrast, animal nutritionists emphasize attributes of cells and tissues that enhance bio-degradation (Van Soest 1982) and liberation of nutrients. The nutritionist asks what cellular configuration affects the digestibility of protein in the plant leaf. Differences and commonalities in the nomenclature of cell/tissue anatomy and biochemistry employed by botanists and animal nutritionists are illustrated in Figure 2.1.

Strictly for illustrative purposes, consider a teleological comparison of the plant and animal perspectives relating to the plant cell. Cells of young plant tissue are biochemically active, capturing and storing energy, synthesizing proteins and fats, etc. (Fig. 2.1a). These are cytoplasmic activities. Cells of older tissue are comparatively low in biochemical activity. Much of the photosynthate and other synthesized compounds have been translocated to the seeds and roots or deposited in other forms in the cell wall. This leaves the cytoplasm comparatively inactive. Similarly, leaves are

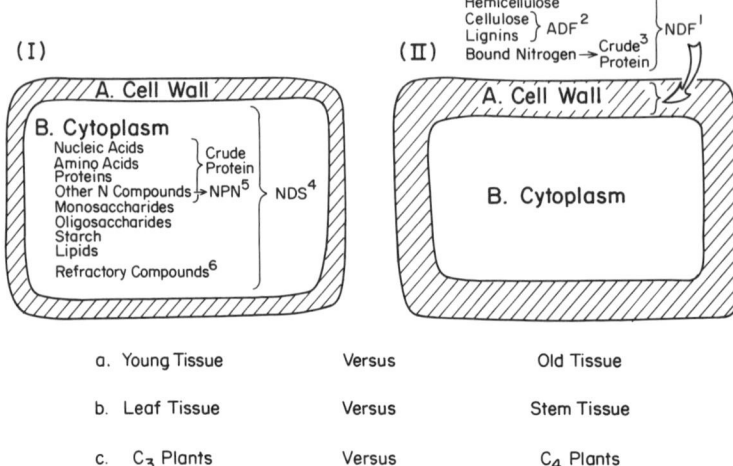

Figure 2.1. Diagrammatic representation of the anatomical and biochemical relationships in plant cell nomenclature between botany and ruminant nutrition. Compounds that are important individually in the physiological processes in plants can be grouped and described for their importance to the consuming ruminant. The relative changes in structural and chemical composition from cell I to cell II depict generalized differences which (a) develop as a plant matures, (b) exist between leaf and stem tissues, and (c) exist between C_3 (cool-season) and C_4 (warm-season) plants.
1. NDF—neutral detergent fiber residue, primarily structural carbohydrates (hemicellulose and cellulose) and phenylpropanoids (lignin). Synonymous with cell wall (CW) and cell wall content (CWC).
2. ADF—acid detergent fiber residue, primarily cellulose and lignin.
3. Crude protein—total nitrogen content × 6.25.
4. NDS—neutral detergent solubles, cell contents + pectin.
5. NPN—non-protein nitrogen.
6. Refractory compounds—A diverse group of primarily secondary plant compounds, in addition to lignin, which affects the digestibility and/or nutritive value of a plant tissue. Included are tannins, flavones, essential oils, steroids, saponins, waxes, and alkaloids. These types of compounds have been variously isolated from the NDS and NDF fractions.

biochemically more active compared with stems that contribute structure and resilience in the overall plant function (Fig. 2.1b). Cool-season (C_3) plants have relatively greater cytoplasm compared with warm-season (C_4) plants that are higher in cell wall (Fig. 2.1c).

The animal measures plant chemicals in terms of availability and nutritional worth, irrespective of their phytochemical functions. Which chemicals are easily accessible? Which are difficultly accessible or inaccessible? The two perspectives relate, in that cell structure and function in plant metabolism closely align with nutrient availability and worth to the consuming animal.

Forage contains fixed energy largely in the form of complex carbohydrates, waxes, terpenes (essential oils, saponins, etc.), and phenylpropanoids (lignins, tannins, etc.). Plant biomass is a virtually infinite number of combinations of these biochemicals determined by plant species and phenological stages. The structure and form of these biochemicals, to a large extent, determine a plant species' capacity to survive (resilience), which is related to the general inverse relationship between nutritional value to grazing animals and plant resilience. The complex carbohydrates, etc., are generally impervious to mammalian gastric and intestinal digestive enzymes. Readily digested proteins and soluble carbohydrates, including simple sugars and starches, on the other hand, usually exist either in lesser proportions (< 40% of dry matter) or are complexed (rendered insoluble and poorly available) with insoluble compounds such as lignins and tannins. Cellulose, the most widely distributed organic compound in nature, is a glucose polymer, differing from starch in the isomeric arrangement of the bonds between the glucose monomers (Fig. 2.2). Intestinal hydrolytic enzymes can cleave alpha linkages in starch, whereas the beta linkages of cellulose are resistant to these enzymes. Cellulose is of nutritive value only to herbivores that have incorporated anaerobic microbial fermentation in the digestive process (Hungate 1966). In the presence of cellulolytic microorganisms, exposed cellulose is broken down with relative ease. However, in many plant species, especially the warm-season perennial grasses, the cellulose is complexed or "encapsulated" by lignin as plants mature (lignification). Therefore, diet selection, to be discussed in Chapter 3, is nutritionally the important element of grazing animal behavior. This is true for plant species as well as plant structural part (leaf, stem, mast) and physiologic age of the plant tissue (new or old growth) consumed.

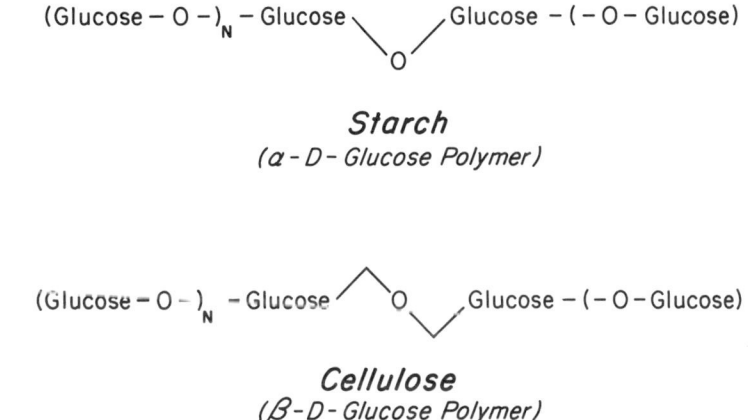

Figure 2.2. The isomeric arrangements (α, β) of the glycoside bonds in the complex carbohydrate polymers starch and cellulose.

THE ANIMAL PERSPECTIVE

Range animals rely on vegetation for the nutrients needed to support bodily processes. The term "quality" is often used to ascribe worth to the components of diet; worth in turn is defined by the chemical composition (e.g., protein content) of the plants selected for consumption. We propose that the proper concept is "nutritional value" because it includes consideration of both the chemical composition of the dietary components and their adequacy for supporting the physiological functions of the consuming animal. For example, a forage species containing 20% protein is considered of higher quality than a similar forage species containing only 10% protein; yet both may be equal in nutritional value to an animal having a relatively low protein requirement. Indeed, the lower protein forage may offer greater overall value to the production system if it also possesses a greater tolerance to grazing, higher production of dry matter, or a longer growing season.

Thus, a proper perspective of the plant:animal interface requires a dual focus to balance short- and long-term production goals. Practices promoting maximum production of animal food and fiber will eventually reduce long-term secondary production by decreasing the stability of the forage resource. On the other hand, an approach which is overly protective from an ecological point of view is economically and sociologically insupportable (see Chapter 9). Hence, both the animal's needs and the consequences that result when these needs are adequately, marginally, or inadequately met determine the proper balance between short-term and long-term productivity.

RANGE HERBIVORES

Foraging animals possessing microbial fermentation capabilities, whether pre-gastric (foregut) or post-gastric (hindgut), are the principal producers of food and fiber from rangelands (McNaughton et al. 1982; Belovsky 1984a). Most pre-gastric fermenters belong to the orders Ruminantia (Bovidae, Cervidae, etc.) and Tylopoda (Camelidae). Most range livestock and big-game species belong to Ruminantia. Grazing animals relying upon symbiotic pre-gastric fermentation and "ruminants" are considered synonymous in this chapter. In terms of economic importance, these foregut fermenters—including cattle, sheep, goats, cervids, and big-game animals—are the most common. However, the suborder, Hippomorpha, which includes the horse (Equidae), is important in some range settings.

Post-gastric vs. Pre-gastric Fermentation

Evolution of microbial fermentation in mammals has been the subject of extensive reviews (Hungate et al. 1959; Janis 1976; Hume and Warner 1980). Figure 2.3 illustrates the comparative digestive anatomy of the non-ruminant (minimal post-gastric fermentation) and post- and pre-gastric fermenting herbivores.

Fiber Digestion. In non-ruminants (a) and post-gastric fermenters (b) in Figure 2.3, foods are exposed to digestion by hydrolytic proteinases (trypsin, pepsin, chymotrypsin, etc.) and carbohydrases (amylase, maltase, lactase, etc.) in the gastric (5) and intestinal regions (6) prior to active fermentation in the large intestine [colon (8) and cecum (7)]. However, because cellulase, the enzyme lysing cellulose, is not present in gastric, pancreatic, or intestinal secretions, cellulose passes through the digestive tract essentially unaltered and provides no direct nutrition to the animal. In

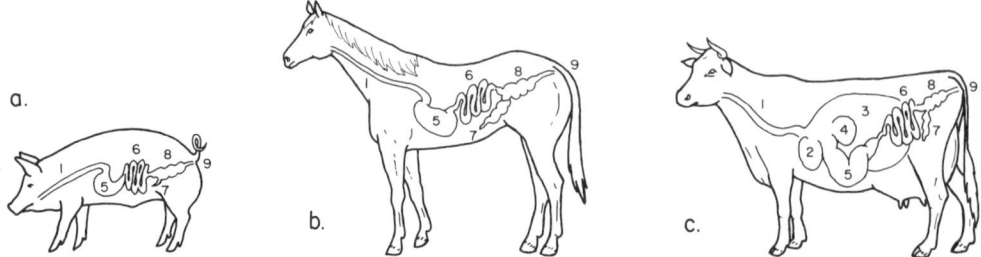

Figure 2.3. Stylized representation of the digestive anatomy and arrangement of (a) non-ruminant, (b) post-gastric fermenter, and (c) ruminant herbivores. 1, Esophagus; 2, Reticulum; 3, Rumen fermentation compartment; 4, Omasum; 5, Stomach (abomasum); 6, Small Intestine; 7, Cecum fermentation compartment; 8, Large Intestine; 9, Anus.

the colon (8) and/or cecum (7), structural carbohydrates, including cellulose and undigested and endogenous residues that have escaped hydrolytic digestion are exposed to microbial fermentation. Fermentation results in the growth and accumulation of microbial cells (primarily bacteria) high in protein. However, there is limited microbial protein catabolism in and amino acid absorption from the colon/cecum (Janis 1976). Hence, the major by-products of fermentation in these herbivores are short-chain organic acids (volatile fatty acids: VFAs), ammonia, carbon dioxide, and methane. A major portion of the VFAs are absorbed and used by the host animal for energy, as discussed later.

By comparison the food consumed by ruminants (Fig. 2.3c) is subjected to microbial fermentation prior (2,3,4) to digestion by hydrolytic enzymes (proteinases and carbohydrases) in the gastric and intestinal segments (5,6). The microbial population is established in the rumen (3) and reticulum (2) (referred to as "rumen", "ruminoreticulum", or "reticulorumen") into which the food enters via the esophagus (1). Consumed material is mixed with existing ruminal microbial populations, portions of previously consumed meals, and both transient and end products of fermentation. After a variable delay period (rumen retention time: RT), particles move into the omasum, and then sequentially to the abomasum (5), small intestine (6), large intestine (7), and rectum (8), from which the remaining residue is excreted as feces (Phillipson and Ash 1965).

Flow dynamics of the ruminal compartment resemble a modified continuous flow system with periodic additions to, and frequent outflow from, the constantly mixed ruminal pool of materials (Fig. 2.4). Rumen retention time of an individual

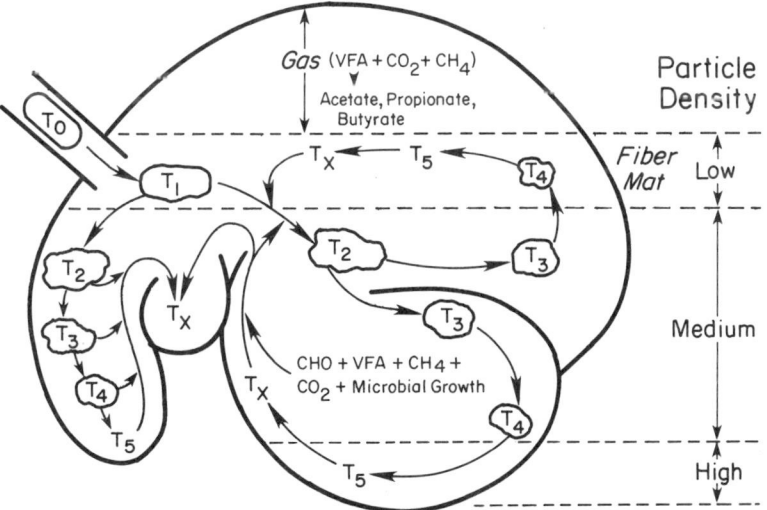

Figure 2.4. Stylized paths of consumed materials within the reticulorumen of ruminants. Individual boluses of dietary material are swallowed (time of swallowing, T_0) and disassociate over time ($T_1, T_2, T_3 \ldots T_x$) into a network of pathways within the reticulorumen. Migration of the particulate matter from these boluses is influenced by particle size and density, frequency and intensity of muscular contraction, and chance. The less dense particles tend to become incorporated in a floating "mat" or "raft" until the size is reduced and density is increased by rumination and fermentation. Dense particles fall to the bottom of the rumen and become transported to the reticulo-omasal orifice area by a fluid gush that occurs as a result of strong lower ruminal contractions. Individual particles may remain in the reticulorumen only a short time (< 1 hr) or an extended period (> 100 hr). The average residence time is between 20 and 60 hr depending on type of diet and species and physiological stage of the ruminant.

particle may be as short as a few minutes or as long as several days, depending on size of the compartment, levels of dry matter and water intake, particle size and reduction rate, particle density, ruminal motility, and chance (Pond et al. 1987). Post-ruminal hydrolytic digestion is similar in the ruminant to that of the non-ruminant (pig) and the post-gastric fermenting animal (horse). The microbial activity in the lower tract (i.e., colon/cecum) of grazing ruminants is quantitatively of less importance than in post-gastric fermenters. However, VFAs and ammonia produced in the cecum may be important to animal status. Krysl et al. (1987a) and Caton et al. (1988b) suggest hindgut VFAs and ammonia production make a significant contribution to the respective pools of these compounds in sheep.

Comparative Protein and Vitamin Nutrition. Proteins are large molecular compounds comprised of approximately 20 individual amino acids bonded together in linear, coiled, or branching chain forms. The relative number of each of these amino acids and the sequence in which they are bonded together determine the character of the particular protein in the tissue (muscle, hair, hoof, enzyme, etc.). Protein in the diet must be broken down to the individual amino acids within the gastrointestinal tract and absorbed as such since the large protein molecule cannot be transported through the intestinal wall. These absorbed amino acids are then used to resynthesize proteins that fit the needs of the animal. Some of the amino acids can be formed within the tissue from materials such as other amino acids that are present in excess. On the other hand, some of the amino acids must be absorbed from the gastrointestinal tract preformed and are referred to as **essential amino acids**. If absorbed amounts of essential amino acids meet or exceed the animal's physiological requirements, protein synthesis can proceed at a normal rate. If one or more are not absorbed in sufficient amounts, tissue protein synthesis is restricted and the associated maintenance or production function is impeded.

Vitamins are "cofactors" or catalysts in metabolic reactions, in that they do not appear in the products of reactions, but must be present for reactions to occur. All vitamins or their precursors must be absorbed from the digestive tract as they cannot be synthesized by mammalian tissue. If vitamins are not absorbed in adequate amounts, metabolic activity is restricted.

Protein and vitamin nutrition are both influenced by microbial fermentation and its location within the digestive tract. Pathways of protein synthesis in microorganisms are similar to those of mammalian tissue except that amino acid requirements are much less specific. The microorganisms, as a mixed population, have no absolute amino acid requirements. Ammonia, derived from most nitrogen-containing compounds, including urea, can be used in the synthesis of "microbial protein" (Fig. 2.5). Likewise, most vitamins are synthesized by populations of microorganisms, except for vitamins A, D, and E.

Ruminant animals are insulated against essential amino acid and most vitamin deficiencies because these compounds are synthesized by symbiotic microbial populations in the rumen and subsequently presented for hydrolytic digestion in the gastric-intestinal region. Once microbial protein passes from the rumen to the gastric-intestinal region, it is hydrolyzed to the individual amino acids which are absorbed for use at the tissue level. Therefore, ruminants can survive on a protein-free diet as long as the diet contains a form of nitrogen to yield ammonia under anaerobic fermentation (Virtanen 1968). Additional insulation against protein deficiency is conferred by ammonia nitrogen recycling (Weston and Hogan 1967). However, over the longer term, a base supply of amino acids in the form of dietary protein may be necessary for maximal fiber digestion and ruminal protein synthesis (Petersen et al. 1985).

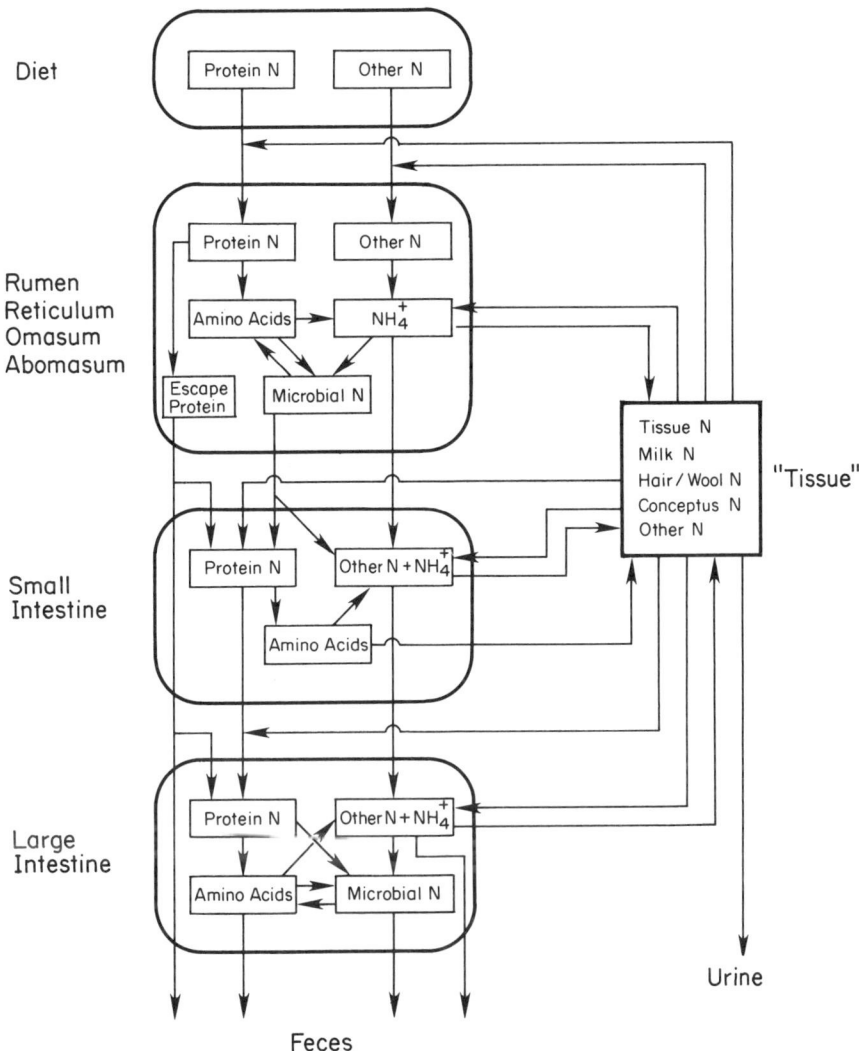

Figure 2.5. Diagram of nitrogen flow in ruminants (adapted from NRC 1985b).

Vitamins synthesized by the microorganisms in the rumen are likewise digested in the lower tract.

The essential amino acids necessary to achieve and sustain maximum production—defined as rapid growth, successful reproduction, and heavy lactation in domestic ruminants—cannot be met solely through microbial protein synthesis (Burroughs et al. 1975). Microbial growth is limited by the maximum level of fermentation which can be supported by a given diet (substrate). Obviously, complete fermentation of a substrate in the rumen can yield only a finite amount of microbial protein. Even at maximum fermentation, microbial synthesis is unable to provide sufficient quantities of amino acids to fully satisfy the physiologic requirements for maximum productivity (**genetic potential**) of some particular animals in a highly productive state (e.g., rapidly growing). Maximum productivity can be achieved only by the addition of escape protein (Fig. 2.5), with a favorable amino acid profile, to augment microbial protein production (Anderson et al. 1988).

Digestion and Flow Dynamics in Ruminants

The chemical components of the diets of ruminants can be separated into two structural fractions of nutritional significance (Van Soest 1967). The first, cell contents (**neutral detergent solubles**: NDS), are those substances found inside plant cells. These organic molecules are soluble and so are readily digestible in the intestine. These substances also tend to be rapidly and extensively fermented before reaching the gastric-intestinal region. The second fraction, cell wall components (**neutral detergent fiber**: NDF), are digested more slowly and less completely. Digestion of the cell wall fraction is performed almost exclusively by microbial hydrolysis and fermentation. Volatile fatty acids produced during fermentation are absorbed through the rumen wall and subsequently metabolized for use at the tissue level as energy. Products of fermentation not absorbed through the rumen wall, including microbial cells, pass to the lower tract together with unfermented dietary residues. These modified (or synthesized) and original dietary and endogenous fractions are exposed to hydrolytic digestion in the gastric-intestinal region.

The total amount and quality of nutrients derived from a grazing animal's diet are determined by the type and amount of forage consumed and the proportioning of the material among five possible fates:

Fate 1. Degraded to products absorbed directly from the ruminal compartment (VFAs);

Fate 2. Modified during fermentation in the rumen and subsequently digested in the lower tract (microbial protein);

Fate 3. Escapes fermentation in the rumen and undergoes hydrolytic digestion in the gastric-intestinal region (bypass or escape protein);

Fate 4. Either modified by or escapes fermentation in the rumen and/or hydrolytic digestion in the gastric-intestinal region to be fermented and absorbed in the colon/cecum;

Fate 5. Bypasses or escapes digestion completely and excreted in the feces.

Some plant components such as cellulose are of greater nutritive value to the animal if they remain in the rumen over an extended period and are either degraded to absorbable end products (e.g., VFAs; Fate 1), or their fermentation contributes to the formation of other substances (e.g., microbial cells) that are subsequently digested (Fate 2). Otherwise such components are destined to either minimal nutrient yield from colon/cecal fermentation (Fate 4) or to be excreted (Fate 5). Other components such as non-fiber bound protein and soluble sugars have greater net value if they escape ruminal fermentation and undergo hydrolytic digestion in the abomasum and small intestine (Fate 3) because respiration losses associated with anaerobic microbial fermentation are avoided. Therein lies the reason ruminants can survive, and are indeed productive, on fibrous forage diets, but are comparatively inefficient, compared with the chicken or pig, at converting feeds high in soluble carbohydrates and protein to animal products.

Thus, inherent species differences in gastrointestinal flow dynamics ultimately influence which species are adapted to particular components of the vegetation on rangeland. Cattle and bison, which have a relatively large-capacity rumen compartment in relation to both body size and nutrient requirements (Demment and Van Soest 1985), also have a long rumen retention time (RT) (Fig. 2.6). These anatomical factors permit cattle to extract a large amount of nutritional value from fibrous materials, often in amounts adequate to satisfy all their nutrient requirements. Conversely, small ruminants, e.g., sheep and goats which possess a relatively small

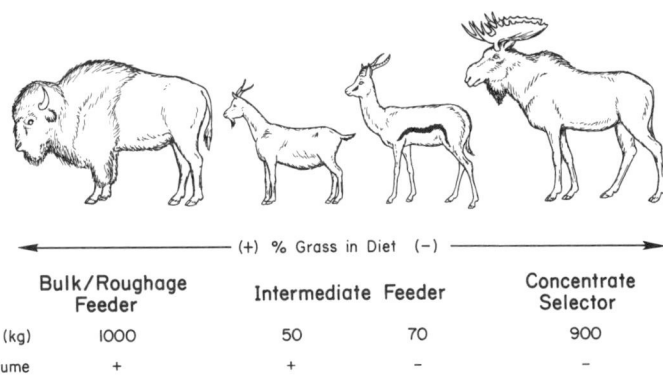

	Bulk/Roughage Feeder	Intermediate Feeder		Concentrate Selector
Body Size (kg)	1000	50	70	900
Rumen Volume	+	+	−	−
Prehensile Strength	−	+	+	+
Parotid Salivary Glands (Presence)	+	+	−	−
Retention Time	+	+	−	−
Fermentation Time	−	+	+	+
Intake (% BWT)	−	+	+	+
Cellulolytic Bacteria	+	+	+	+
Protozoa	+	+	−	−

Figure 2.6. Relationships between ruminant bulk, intermediate and concentrate feeders.
Body size—approximate weight of the mature adult of the species.
Rumen volume—volume capacity of the rumen and reticulum relative to body weight.
Prehensile strength—the physical ability of the species to sort and discriminate among available dietary options.
Parotid salivary glands—the presence of parotid salivary glands related to total volume of secretion of saliva.
Retention time—length of time that consumed particles are exposed to fermentive digestion in the rumen.
Intake—daily consumption of diet relative to body weight.
Cellulolytic bacteria—presence of bacteria in the rumen capable of degrading cellulose.
Protozoa—presence of protozoa in addition to bacteria in the rumen.

rumen compartment in relation to body size and nutrient requirements, cannot extract comparable levels of nutrients from the same fibrous forages. Even though nutrient requirements are greater per unit body weight in small ruminants, rumen capacity is significantly less, retention time significantly shorter, and flow rate significantly faster than in large ruminants (Table 2.1). Hence for relatively equivalent intake levels, fibrous diets are of less nutritional value to small ruminants.

Foraging Strategies of Ruminants

Smaller ruminants have evolved two strategies to overcome the metabolic dilemma described above. The first strategy is reduced RT (Van Soest 1982) which allows a slight shift in the site of digestion of the highly digestible components out of rumen fermentation (Fate 1 and/or Fate 2) and into the gastric-intestinal region (Fate 3), thereby decreasing respiration losses associated with fermentation. Also, the shorter RT is associated with a greater level of intake and a slightly depressed fiber digestibility. Taken together these result in a greater level of intake, a slightly lower digestibility compared to larger ruminants, but an opportunity to equal or exceed total digested nutrient intake (Huston 1978). This strategy is important in survival but

Table 2.1. Comparative nutritional dynamics in livestock species.

Measurement[a]	Animal Species	Reference
Intake, % of body weight		
1.7–2.1	Cattle[b]	NRC (1984)
1.7–4.8	Sheep[c]	NRC (1985a)
1.5–5.9	Goats[d]	NRC (1981b)
Rumen retention time, hr[e]		
33–40	Cattle	Huston et al. (1986)
26–40	Sheep	Huston et al. (1986)
26–29	Goats	Huston et al. (1986)
Dry matter digestibility, %[e]		
48–58	Cattle	Huston et al. (1986)
44–59	Sheep	Huston et al. (1986)
36–52	Goats	Huston et al. (1986)
Diet quality, % IVDMD[e,f]		
50–62	Cattle	Huston et al. (1986)
52–64	Sheep	Huston et al. (1986)
50–62	Goats	Huston et al. (1986)

[a] Ranges include values for dry, pregnant, and lactating animals.
[b] Cows weighing 400 to 600 kg.
[c] Ewes weighing 50 to 70 kg.
[d] Goats weighing 30 to 60 kg.
[e] Values for free-ranging cattle, sheep, and goats grazing in common over a 12-month period on mixed vegetation in the Edwards Plateau region of Texas.
[f] In vitro dry matter digestibility percentages. These are standard estimates of digestibility considering only the influences of plant characteristics. The differences between these values and those for actual dry matter digestibility (above) are related to animal influences.

is seldom effective in allowing the small ruminants to match the productivity of large ruminants when both are limited to high-fiber diets. The second strategy is to consume a high quality diet which necessitates a greater degree of discrimination in diet selection. Size and prehensile agility of the lips, teeth, and tongue ultimately determine an animal's ability to selectively consume plant species, individual plants on offer within a species, and even discrete plant parts, all from a heterogeneous assemblage of plant biomass. Significant differences in the morphological structure of mouth parts exist in pre-gastric fermenters and post-gastric fermenters (Fig. 2.7) which reflect the types of forages consumed. Generally, increased pliability of the lips and manipulative capacity of the tongue denote greater levels of selectivity.

Range herbivores have been variously classified into as many as six classes based upon the types of foods eaten (Langer 1984). Figure 2.6 is a modified form of the system described by Hofmann and Stewart (1972) applied to ruminants.

Bulk/roughage grazers (cattle, bison, cape buffalo, etc.) graze comparatively indiscriminately on the herbaceous fraction of vegetation by wrapping their tongue around individual clumps of plant growth and, with a short jerking motion of the head, breaking the clump loose then drawing it into their mouths. Once in the mouth, the material is wetted with salivary secretions, chewed slightly, formed into a cylindrical "bolus" with the teeth and tongue, then swallowed (Fig. 2.4). Later, when the animal is at rest, swallowed material is regurgitated, chewed extensively, then reswallowed (rumination).

Concentrate selectors (white-tail deer, mule deer, dik-dik, etc.) characteristically have pliable and often split lips, soft muzzles, and agile tongues (Fig. 2.7). Hence,

Figure 2.7. Prehensile mouth parts of (a) cattle, (b) sheep, (c) goats, and (d) deer reflecting the degree of selectivity and harvest efficiency expressed by these herbivores.

these animals can select plants or plant parts high in cell contents (protein and other soluble fractions; NDS) and low in cell wall (cellulose and fibrous fractions; NDF). Bite sizes are smaller and more discrete, even consisting of single leaves, leaf tips, fruits, seeds, or fallen mast.

Intermediate feeders are a diverse group characterized by dietary plasticity not found in either bulk/roughage feeders or concentrate selectors. Diet is characterized by variety and frequent compositional changes. The domestic sheep is classified as an intermediate feeder, but its diet often approximates the bulk/roughage group. The goat is a true intermediate feeder, and its diet selections clearly overlap the entire array of forages.

Such a classification of the feeding behavior of grazing animals is useful to better understand species adaptability to specific forage conditions but should not lead the reader to believe these are rigid relationships because "crossover" in feeding habits regularly occurs. Especially within sympatric ruminant populations, all species select diets from an array of available plant materials which vary in space and time (see Chapter 3). Availability is the first and most important determinant of what a grazing animal consumes. When the opportunity is presented for selection among types, species and morphological parts of plants, ruminant populations regularly exhibit "preferences" in the materials selected. This ability to discriminate between available materials is sufficiently pronounced that in vegetatively productive periods the diets of ruminant species grazing in common are almost completely different. Conversely, during periods when the amount and diversity of forage are limited, dietary overlap between sympatric species is very high.

Summary of Comparative Nutritional Physiology of Herbivores

The distinction of ruminants relative to their adaptability to forage-based animal production systems stems from three characteristics unique to this group of animals. First, by virtue of the evolution of a pre-gastric fermentation chamber, ruminants can more effectively utilize structural carbohydrates (NDF) than either non-ruminants or post-gastric fermenters of comparable size. Increased retention time under conditions of anaerobic fermentation leads to more complete digestion and utilization of forage. It must again be noted that ruminant species vary widely both in RT and the extent of fermentive degradation of forage components.

Secondly, whereas non-ruminants depend on preformed amino acids and vitamins in their diets, ruminants are comparatively free of these requirements. Simple forms of dietary or endogenous nitrogen (ammonia releasing compounds, i.e., urea, proteins, amino acids, etc.) can be used by ruminants in the microbial synthesis of protein which subsequently is digested in the gastric-intestinal region. This adaptation is further enhanced by the ability to recycle urea via salivary and ruminal mucosal secretions. Microbial protein generally fulfills the minimal amino acid requirements of ruminants for maintenance and moderate levels of production. Genetically possible levels of production in animals in stages of high productivity cannot be achieved without the addition of escape protein to increase the supply of essential amino acids.

Lastly, dietary overlap of sympatric animal species can be very high or low depending upon forage diversity and availability, environmental conditions and management. The net effect of these three physiological and behavioral characteristics is that ruminants, as a group, are well adapted to production systems on rangeland.

NUTRITIONAL REQUIREMENTS OF GRAZING ANIMALS

The nutrients required by animals are energy, protein, vitamins, and minerals. The **concept of requirements** is generally seen as the amounts necessary to support "normal" metabolic activity. That is, the animal's requirements are thought to be met when it gives evidence of normal health and vigor, normal rate of growth, normal reproduction, and/or normal lactation levels. Obviously, "normal" is not identical in all members of the same species at all times, so these requirements should be seen as a set of ranges.

Nutrients as limiting factors, while an important concept, should not be thought of as a rigid one-to-one relationship. Generally, nutrients are utilized in the hierarchical order of maintenance, reproduction, lactation, and storage (Fig. 2.8). However, across a population of animals, reproduction and lactation can occur when the diet does not provide the "required" levels for these functions. Within that same population, a certain proportion of animals can even reproduce or lactate at nutrient levels well below maintenance "requirements." Despite the absence of rigor, the concepts of nutrient requirements and priority of use are fundamental to an understanding of animal nutrition and management.

The overview of nutrient requirements which follows is a general outline. Refer to the National Research Council Series on nutrient requirements (NRC 1981b, 1984, 1985a) for greater detail.

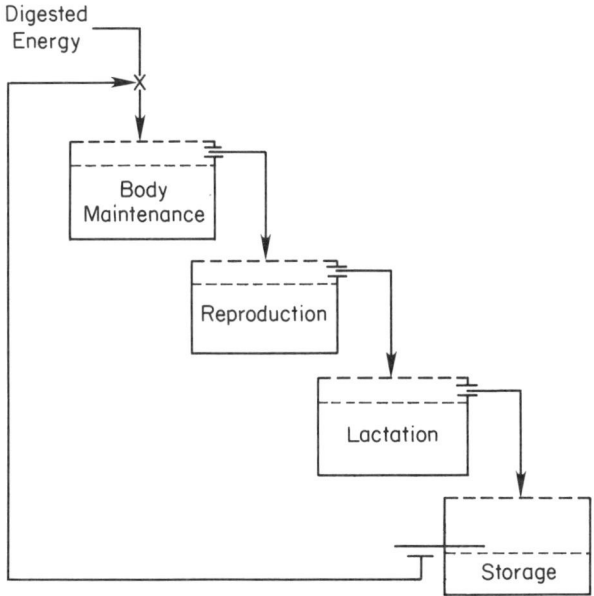

Figure 2.8. Generalized diagram indicating the prioritization of energy use by ruminants.

Energy Requirements

Energy is required primarily in making (anabolism), but sometimes in breaking (catabolism), chemical bonds during animal metabolism. Metabolic processes requiring energy include muscle contraction, nerve impulses, and tissue synthesis.

An example of energy being expended to synthesize protein from amino acids (AA) to form tissue is shown in Equation 1.

(1) $AA_1 + AA_2 \xrightarrow{ATP \to ADP + P} AA_1 - AA_2$

Amino acids are bonded together in peptide sequences during protein synthesis. The energy necessary for this bonding comes from a **coupled reaction** during which a high-energy phosphate bond in adenosine triphosphate (ATP) is cleaved yielding adenosine diphosphate (ADP) and a free phosphate radical. Formation of these high-energy bonds occurs as a result of **respiration** (Equation 2).

(2) $Glucose + Oxygen \xrightarrow{ADP + P \to ATP} Carbon\ dioxide + Water$

In most animal systems, glucose is broken down (oxidized) during respiration to carbon dioxide and water. During this chemical change, energy is captured in the formation of a high energy phosphate bond, which is then available for tissue protein synthesis (Equation 1) or another energy-requiring metabolic process. In ruminants, energy is captured primarily during respiration of VFAs that are produced during fermentation in the rumen (Fig. 2.4), then absorbed into the bloodstream in the rumen wall. These VFAs are metabolized through a network of pathways (simplified in Fig. 2.9) and ultimately yield carbon dioxide (CO_2), water (H_2O), and captured energy in

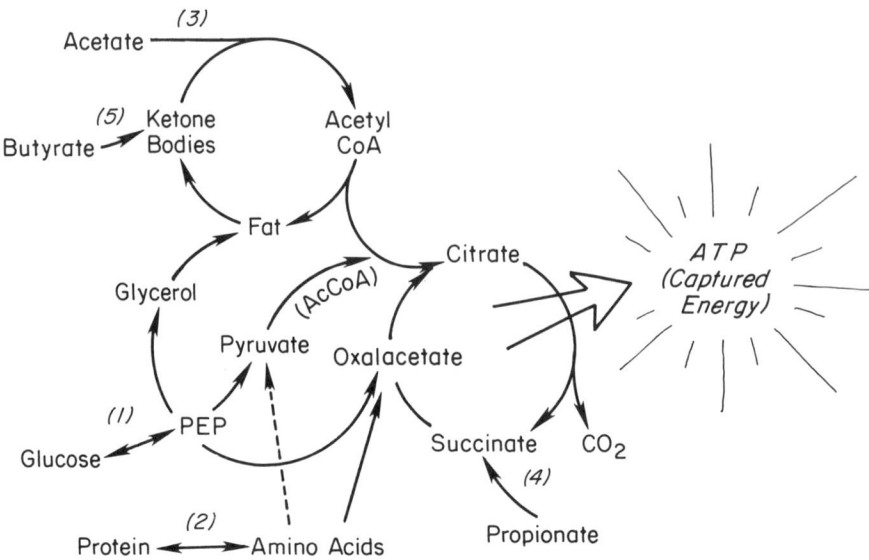

Figure 2.9. Major pathways of respiration for absorbed metabolites that yield energy in ruminants.

the form of high-energy bonds (ATP). Although ruminant tissue can metabolize glucose (1) and protein (2), most captured energy arises from either acetate (3), propionate (4), or butyrate (5), the main VFAs produced during rumen microbial fermentation.

Grazing ruminants derive energy primarily from plant carbohydrates, lipids, and proteins, but not all consumed energy is captured in a form usable to the animal. **Total dietary energy** includes all combustible energy of the diet measured in calories (cal), kilocalories (kcal: 1000 cal), or megacalories (Mcal: 1000 kcal), but not all dietary energy is captured in a form utilizable by the animal. That is, if a cow consumes 20 pounds of hay which if burned would give off 50,000 kcal of heat, then the cow would have eaten 50 Mcal total energy. This total or gross energy (GE) is partitioned (Fig. 2.10) into digestible energy which is DE = GE−fecal energy; metabolizable energy which is ME = DE−Urinary and methane energy; and finally net energy which is NE = ME−heat increment. Net energy is the amount of energy available for **maintenance** (energy required to maintain normal health and vigor) and **production** (energy required for growth, reproduction, lactation, etc.). The metabolizability of digestible energy, ME/DE, is rather constant at approximately 82% (NRC 1984). However, the digestibility of gross energy, DE/GE, and the net availability of metabolizable energy, NE/ME, vary with the chemical composition of the diet and the metabolic function for which the net energy is used.

Expressions of the energy value of feeds and forages are defined in Table 2.2. Components of the diets of grazing animals can have dry matter digestibility (DMD) values from 14–85% depending on the amount of cell contents (NDS) and cell wall constituents (NDF) in the dry matter. The net availability of metabolizable energy (NE/ME) in a forage varies from about 90% when used for maintenance down to less than 20% for an incremental increase in intake high on the productivity curve (Fig. 2.11; Van Soest 1982; Fox et al. 1988). Therefore, the energy value of a quantity of forage varies as a function of its digestibility and its ability to meet the energy required to support a desired metabolic process or productivity level.

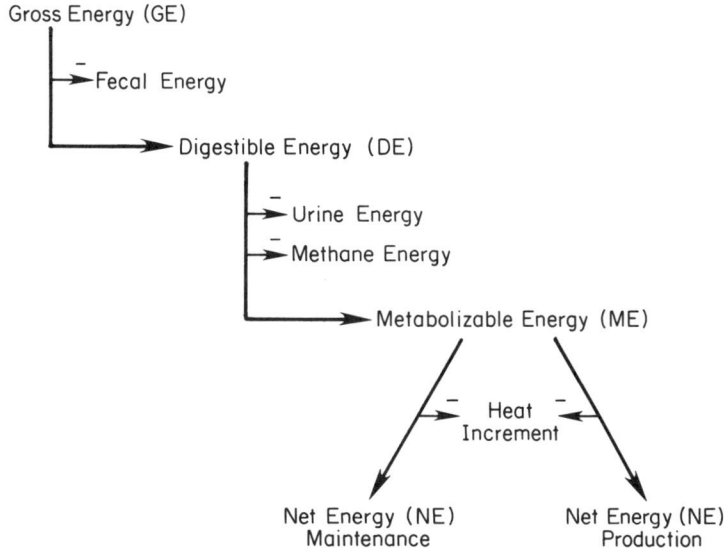

Figure 2.10. Catabolism of dietary energy depicting the energetic losses involved with digestive and metabolic processes in the ruminant.

Table 2.2. Expressions of energy and protein in forages.[a]

Energy

Dry Matter Digestibility or
Digestible Dry Matter (DDM):
$$\text{DMD (\%)} = \frac{(\text{DMI} - \text{FDM})}{\text{DMI}} \times 100$$

Organic Matter Digestibility:
$$\text{OMD (\%)} = \frac{(\text{OMI} - \text{FOM})}{\text{OMI}} \times 100$$

Digestible Organic Matter:
$$\text{DOM (\%)} = \frac{(\text{OMI} - \text{FOM})}{\text{DMI}} \times 100$$

Total Digestible Nutrients: $\text{TDN (\%)} = [\text{DOM (\%)} \times 1.05]$

Digestible Energy:

$$\text{DE (Mcal/lb)} = \frac{[\text{TDN (\%)} \times 2]}{100} \quad \text{or} \quad \frac{[\text{DOM (\%)} \times 2.1]}{100}$$

$$\text{DE (Mcal/kg)} = \frac{[\text{TDN (\%)} \times 4.4]}{100} \quad \text{or} \quad \frac{[\text{DOM (\%)} \times 4.6]}{100}$$

$\text{DE (Mcal)} = [\text{TDN (lb)} \times 2] \quad \text{or} \quad [\text{DOM (lb)} \times 2.1]$

$\text{DE (Mcal)} = [\text{TDN (kg)} \times 4.4] \quad \text{or} \quad [\text{DOM (kg)} \times 4.6]$

Metabolizable Energy: $\text{ME (Mcal)} = [\text{DE (Mcal)} \times .82]$

Net Energy: $\text{NE (Mcal)} = [\text{ME (Mcal)} - \text{HI}]$

Protein

Crude Protein: $\text{CP (\%)} = \text{N (\%)} \times 6.25$

Digestible Protein:
$$\text{DP (\%)} = \frac{(\text{CPI} - \text{FCP})}{(\text{DMI})} \times 100$$

[a] DMI—Dry matter intake   HI—Heat increment
FDM—Fecal dry matter   N—Nitrogen
OMI—Organic matter intake   CPI—Crude protein intake
FOM—Fecal organic matter   FCP—Fecal crude protein

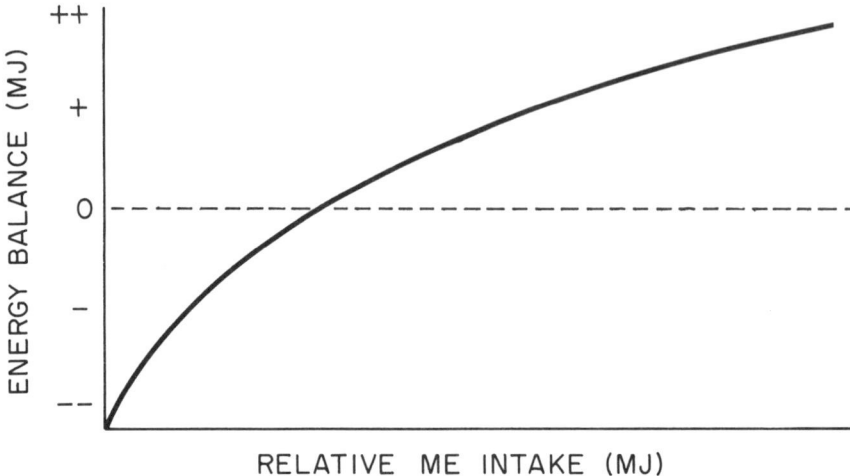

Figure 2.11. Efficiency of metabolizable energy utilization with increasing levels of energy intake. When animals are in a negative energy balance (below maintenance), energy is used very efficiently. Above energy balance in the production portion of the curve, increasing energy intake results in greater productivity but at a decreasing marginal efficiency. The flattening of the curve indicates it takes more ME intake for a net increase in energy balance.

Protein Requirements

Ruminant animals require protein in the diet to supply nitrogen (ammonia) and amino acids for intraruminal microbial activity and amino acids for cellular-level tissue metabolism. Protein expressions are defined in Table 2.2. Suboptimal protein supply to the microbial population in the rumen results in a lowered fermentation rate, decreased digestibility of food consumed, and decreased voluntary intake (Kempton and Leng 1979). Protein requirements in ruminants include protein and/or nitrogen requirements of the ruminal microbial population. Generally, microbial requirements are met at 6–8% crude protein in the diet. Animal requirements range from 7–20% in the diet depending upon species, sex, and physiologic state. Normally animal protein requirements are satisfied by a combination of microbial and dietary escape protein (Fig. 2.5). As animal protein requirements increase, the animal becomes more dependent on dietary escape protein.

Priority of protein use can be expressed in the same fashion as priority of energy use (Fig. 2.8). Maintenance requirements are met first and include repair and replacement of body tissue. After maintenance requirements are met, absorbed amino acids are used for productive functions until one of three limitations are encountered:

1. The supply of amino acids in the correct proportion is depleted. That is, the synthesizing system literally runs out of one or more of the necessary amino acids to build the protein;
2. One or more of the other necessary nutrients required in coupled reactions become limiting. This is easily understood for limited energy by reviewing the coupled equation, Equation 1. Alternatively, other nutrients, particularly vitamins or minerals, are not present in the proper proportion and limit protein synthesis;
3. The animal's genetic capability for performing a particular function has been

reached. Genetic potential should be viewed as a variable range in a manner similar to nutrient requirements, but generally as a point on the production curve beyond which additional nutrients produce no practical response. Thus a beef cow's requirements for protein or energy are met at a lower level of protein or energy intake than that required by a dairy cow in high lactation.

Vitamin Requirements

Vitamins are organic compounds that must be present at the cellular level to act as catalysts in metabolic processes. As noted earlier, many of the vitamins are synthesized by the ruminal bacteria and subsequently absorbed from the intestinal tract. With few exceptions, vitamin A is the only vitamin that is likely to limit the productivity of grazing ruminants. Vitamin A does not occur in plant tissue, but is synthesized by the animal from chemical precursors in plants, mainly beta carotene, but other plant pigments as well. Vitamin A deficiency is most likely to develop during an extended period of low temperature and/or drought when green plants are unavailable to the animal. The second most likely deficient vitamin in grazing ruminants is vitamin E. This condition can become especially severe when combined with low selenium in the diet.

Mineral Requirements

Minerals required by animals are classified as either macro-minerals or micro-minerals according to the amounts required. Those required in relatively large amounts, the **macro-** or **major elements**, are sodium, chlorine, calcium, phosphorus, magnesium, potassium, and sulfur. In each case these elements are either a constituent of animal tissue or are required in large amounts to carry on metabolic functions. Mineral elements required in small amounts, **micro-** or **trace elements**, include iodine, iron, copper, zinc, manganese, cobalt, molybdenum, and selenium. These generally have special functions as either low-level components of certain tissues, or as cofactors for certain metabolic reactions.

Macro-minerals. All of the major elements are potentially problematic in the range setting. Those most likely deficient in range forages are sodium, chlorine and phosphorus. Deficiencies of salt (sodium chloride) and/or phosphorus can result in perverted animal behavior such as indiscriminate eating of rocks, sticks, bones, etc. and reduced forage intake and productivity. Deficiencies of the remaining four are unlikely under normal range conditions, but where deficiencies occur, the effects can be as devastating as in the cases of the more common deficiencies. A magnesium deficiency, for example, is associated with grass tetany that occurs during lush plant growth periods that appear to provide the opportunity for high production. Reduced potassium can also depress animal productivity, by reducing the appetite and so the food intake. Particular attention should be given to the macro-mineral status of animals grazing on drought or winter dormant forages for extended periods of time.

Micro-minerals. The trace elements, although needed in only minute amounts, are crucial to normal animal metabolism. Iodine is a component of the hormone thyroxine, iron equips blood cells to carry oxygen, and cobalt is required by microorganisms to synthesize vitamin B_{12}. Many of the minor elements are cofactors in the enzyme systems involved in energy and protein metabolism. Therefore, "minor" or "trace" should not be interpreted as meaning of less qualitative importance. Animals cannot function normally without an adequate supply of any of the required elements, major or minor.

It is not possible at this writing to make definitive predictions about micro-mineral deficiencies and toxicities due to the wide disparities in the amounts required as compared to the macro-minerals. Trace element deficiencies are less widespread, less predictable, more difficult to recognize, and probably quantitatively less important than major element deficiencies. Exceptions to this general statement include those regions deficient in selenium, iodine, or cobalt. It should be noted that rangelands deficient in these micro-elements are extensive throughout the world.

Toxicities resulting from consuming excessive amounts of micro-elements also occur in natural settings. An example is peat scours, a high-molybdenum-induced copper deficiency, on high organic matter soils. Yet, the importance and extent of trace element imbalances on rangelands remains largely undetermined.

NUTRITIVE VALUE OF FORAGES

Nutritive value is an inclusive expression used to encompass all nutritional attributes of a forage in relation to its overall value to the consuming animal. However, the term is often used in the more restrictive sense of **forage quality**, including protein content, digestibility, or simply palatability. The reader is encouraged to develop the broad view of quality which includes consideration of usefulness of forage constituents (nutritive value) for particular productive purposes in animals as proposed above. This section discusses systems of nutritional description of forages and the classification of forage types for application in grazing management.

Nutritional Description of Forages

A useful description of forages must somehow relate to the nutrient groups required by animals. These groups were enumerated as energy, protein, vitamins, and minerals. The **Proximate Analysis System** was developed over 100 years ago in an attempt to use chemical determinations to describe the value of feeds for animals. The proximate factors used as components are crude fiber (CF); crude protein (CP); crude fat, often stated as ether extract (EE); nitrogen-free extract (NFE); and ash. The most widely used proximate component analysis has been for crude protein.

$$(3) \; CP \, (\%) = \% \, \text{Nitrogen} \times 6.25$$

The protein contained in a wide array of forages averages about 16% nitrogen. So the standard procedure is to determine the nitrogen content of a forage, multiply that value by 6.25 (100/16) and refer to the product as **crude protein**. Crude fiber (CF) and NFE fractions were intended to estimate the less and more easily digested portions of feeds, respectively. When applied to forages this arbitrary partitioning does not adequately differentiate the digestibilities of these fractions.

The adoption of the proximate analysis system to describe feed fractions led to the development of Total Digestible Nutrients (TDN) approach. The latter was an attempt to more adequately describe the energy value in feeds. **Total digestible nutrients** are defined as the sum of the digestible portion (% composition × coefficient of digestibility, COD) of each of the proximate organic components with an adjustment factor of 2.25 for EE. Ash is not included because it contains no energy, while EE is increased because fat contains about 2.25 times the energy per unit weight compared with carbohydrates. The TDN system has been very useful over a long period of time in assigning values to feedstuffs that are relatively constant in composition, but is less adequate for forages, especially range forages which vary widely in chemical components within the proximate fraction.

The detergent fiber analysis system (Van Soest 1967) was a major improvement in the evaluation of the nutritional characteristics of forages. Partitioning cell content (NDS) from cell wall (NDF) distinguishes that portion that is essentially totally digestible from that which is partially and variably digestible, respectively. Further fractionation of the NDF into its components including acid detergent fiber (ADF), acid insoluble ash (AIA), lignin, and silica has refined the analysis of the fibrous portion. A very useful adjunct to this system of analysis was the development of a two-stage, micro-digestion technique (Van Soest et al. 1966). This technique, in vitro digestion of dry matter (IVDDM), provides an approximation of the digestibility of plants and plant parts. Further computorial correction to an organic matter basis provides an estimate of digestible energy content in megacalories.

However, IVDDM does not take into account the variable effects of rate of fermentation, digesta flow rate, and retention time on digestive efficiency (Huston et al. 1986). These factors vary among animal species and in response to **associative effects** of companion dietary constituents. That is, the nutritional value of a dietary constituent can be enhanced by the addition of another dietary constituent which supplies a limiting nutrient.

Determinants of Nutritive Value

Forage quality is determined by various combinations of micro-and macro- scale biotic and abiotic factors (Morley 1981; Wheeler and Mochrie 1981; Van Soest 1982). The inherent morphological, anatomical, physiological, and chemical characteristics of each plant species determine its potential nutritive value. Abiotic and temporal factors modify this potential.

Examples of biotic factors can be found in the differences in quality between grasses utilizing three-carbon (C_3) versus four-carbon (C_4) photosynthetic pathways and between monocotyledonous (monocots) and dicotyledonous (dicots) plants (Table 2.3). In the first example the C_4 plants, commonly termed warm-season species, contain less mesophyll and greater proportions of schlerenchyma, epidermis, and vascular tissue than C_3 plants, cool-season species (Fig. 2.12). Vascular bundles are densely packed and parenchyma bundle sheaths thick-walled in C_4 grasses (high NDF), therefore inhibiting microbial digestion in the rumen, while reduced mesophyll (low NDS) provides less protein and soluble carbohydrates. Lignin concentra-

Table 2.3. Expected range of crude protein (CP), digestibility (DMD), and phosphorus (P) content in warm and cool season forages from native and improved pastures throughout the world.

Forage Class	Pasture Type	Growing Season	Life Form	CP%	DMD%	P%
Grass	Native	Warm	Annual	—	50–73	—
			Perennial	2–15	20–65	.08–.28
		Cool	Annual	2–25	60–95	.03–.48
			Perennial	3–25	42–94	.05–.35
	Improved	Warm	Annual	4–18	46–69	—
			Perennial	2–25	36–68	.05–.35
		Cool	Annual	3–30	50–91	—
			Perennial	5–30	30–76	.08–.28
Forbs				4–23	42–91	.10–.46
Browse (shrubs)				4–32	14–74	.08–.54

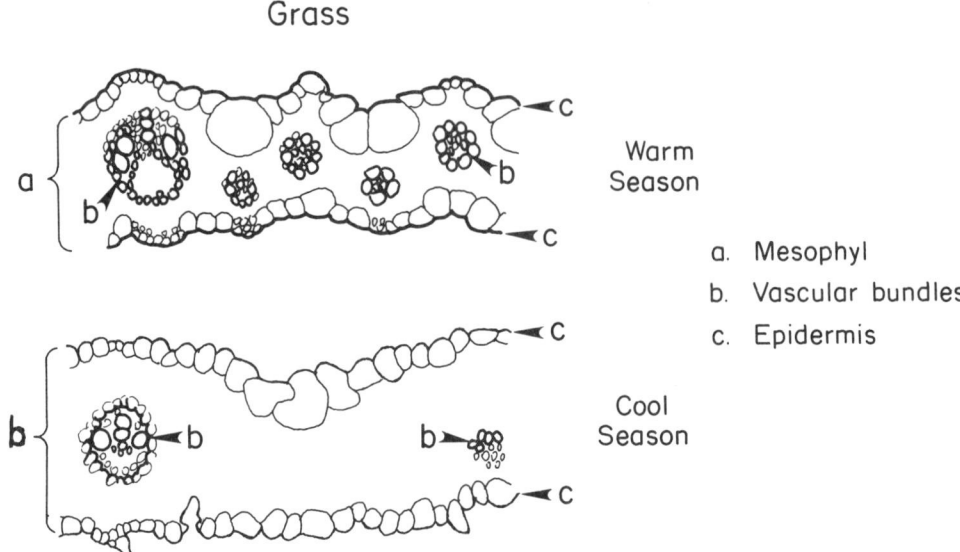

Figure 2.12. Cross sections of leaves of (a) warm-season and (b) cool-season grasses depicting the anatomical differences which lead to differences in nutritive value between these types of forages.

tions are higher and leaf:stem ratios lower in warm-season grasses than in cool-season grasses. Stems have significantly greater proportions of structural carbohydrates and lignin (high NDF) in all forages, while leaves have greater proportions of cell contents (high NDS) and crude protein than stems.

Shrubs and most forbs are dicots, and their leaf biomass is generally of higher nutritive value than that of grasses (monocots) (Table 2.3). Non-woody plant parts of dicots have greater quantities of cell solubles than monocots and lower levels of structural carbohydrate and lignin. This apparent advantage is often offset, however, by biologically significant proportions of secondary compounds (tannins, volatile oils, alkaloids etc.) in a number of shrub and forb species. Many of these secondary compounds produce inhibitory and/or toxic effects on the microbial fermentation (Hegarty 1982). Hence, even if the quality of a particular plant species is comparatively high, inhibitory factors may reduce the utilization of the metabolizable nutrients (Burns 1978).

Food materials of the highest quality are found in metabolically active tissues (live leaves, stems, flowers, etc.) or storage tissue (seeds, fruits, and roots). Live plant tissue is of higher quality than dead. Similarly, younger live tissue by virtue of its greater metabolic activity is of higher quality than older live tissue. Generally, live leaf is of higher quality than live stem because of its greater photosynthetic activity. Nutrient quality declines as the rate of development or recruitment of new leaf tissue decreases and the rate of senescence increases (see Chapter 4). While the overall quality of live leaf material may not change drastically with age, increasing amounts of senescent material dilute nutrient density (Greene et al. 1987). Concurrent changes in the leaf:stem ratio also occur as a plant matures. In terms of the energy flow (see Chapter 1) and standing crop (g/m^2), available gross energy ($Kcal/m^2$) usually peaks when stems have elongated in mid-anthesis. However, maximum available net energy (NE $Kcal/m^2$) occurs earlier in the late vegetative and early anthesis stages before significant reproductive culm elongation occurs (see Chapter 4).

Turning to the abiotic factors which affect forage quality, the most important are air temperature and soil moisture. These environmental conditions modify the rates at which live material is accumulated and senescence occurs. Generally, the leaf and stem tissue of grasses grown at high temperatures is lower in both digestibility and crude protein content. Lignification and the formation of structural carbohydrates (NDF) occur rapidly at elevated temperatures, causing a concomitant reduction in the cell soluble fraction. Shrubs and forbs usually exhibit little change in leaf quality until senescence; however, stems of forbs and juvenile leaders of shrubs exhibit exaggerated declines in quality with advancing age (Petersen et al. 1987). Below-normal ambient temperatures that occur during the growing period frequently reduce growth rate and respiration rate, thereby reducing the rates of senescence, stem elongation, and lignification. These reduced rates effectively extend vegetative growth further into the growing season, so the resultant standing crop maintains greater proportions of digestible dry matter and protein than the same forage crop under normal temperature conditions.

Restricted soil moisture can either increase or decrease forage quality. If moisture is restricted during the vegetative growth stage creating slowed growth, but not senescence, delayed maturation maintains forage quality in a manner similar to lower ambient temperature. However, if restriction progresses to severe water stress, forage quality decreases in response to nutrient translocation and senescence of plant parts. In most rangeland environments, drought is often accompanied by above-normal ambient temperatures which exacerbate the plant's growing conditions by increasing the rate of evapotranspiration.

In summary, the primary factors influencing the quality of forage are the plant species present and their level of metabolic activity. The more active a particular tissue is the greater its quality. Environmental conditions in turn modify this activity by affecting the rate at which it occurs.

Practical Classification of Range Forages

A variety of plant communities, each having a unique assemblage of plant species, occurs in rangeland ecosystems. Intra- and interspecific competition among plants for resources and interactions with prevailing climatic conditions lead to formation of plant communities (see Chapter 5). Animals, however, are neither plant taxonomists nor community ecologists and consume plants according to availability and preference (see Chapter 3). Whether a plant is an increaser, decreaser, or invader (see Chapters 4 and 5) is immaterial to the animal. Instead, the amount of live-to-dead and leaf-to-stem material available, presence or absence of inhibitory factors, etc., in various species or species groups are the only matters of concern (see Chapter 3).

Generally, animals select from the highest quality components of the available forage pool first. Some plant species are highly nutritious but available only in limited quantities, while more readily available species are less nutritious. As the pool of highest quality plants is depleted, increasing quantities of the next highest quality components are consumed. These selection and consumption processes are integrated through space and time (see Chapter 3). Although each rangeland environment is composed of a unique agglomeration of plant communities, each with particular vegetational characteristics, the following general classification of their functional nutritional components has been proposed (Huston et al. 1981).

Semiarid and arid rangelands are usually dominated by a particular forage type that is relatively high in quality during early vegetative growth but quickly declines in quality as the forage accumulates and matures. This forage type provides the majority of organic matter consumed by grazing animals on rangeland and is termed the **production component**. On temperate and tropical rangeland, this component is comprised of perennial grasses. Characteristics limiting the nutritional value of these plants are the very same as those ensuring their availability for consumption. Their content of structural carbohydrates is quite high, and they enter dormancy during unfavorable periods and reinitiate growth during favorable periods. Adult bulk feeders can maintain acceptable levels of productivity when grazing these forages, provided their reproductive cycle conforms closely to the temporal nutrient profile of the vegetation. In other ecosystems, the production component may be annual grasses, as for example California annual grassland or shrubs in salt desert shrub ecosystems, but the common characteristic of the production component is that it ultimately determines the sustained animal yield potential because it is the principal stable component under existing grazing conditions.

Other plant species provide a **quality component** to diets of ruminants on rangeland. These species, which differ from one ecosystem to the next, provide only a minor amount of forage, but that forage is significantly higher in nutrients CP, DE, etc., than the production component. Certain perennial forbs, shrub leaf buds and tips, mast, fruits, seeds, etc., contribute disproportionately to the productivity of bulk feeders both by raising the overall diet quality and preventing nutrient deficiencies, for example vitamin A and phosphorus. Perhaps more importantly, quality components provide a suitable diet for grazing and browsing small ruminants having higher nutritional requirements, thereby increasing the overall production potential of a specific rangeland. The quality component is also important to big-game populations.

The plant species making up the **level component** of forage materials in this classification system can be characterized as those which remain green throughout the grazable portion of the year. These species rarely produce forage that is either exceptionally high nor ruinously low in nutrient content, but offer fair- to good-quality forage during all seasons. The level component competes with the production component in a plant community for space, moisture, and nutrients, but substitutes for the quality component during periods of dormancy and can significantly reduce reliance on supplemental feed. Examples of plants in the level component include elk sedge in the Intermountain area of North America and Texas wintergrass in north and central Texas. Leaves of evergreen browse species, such as fourwinged saltbush of western North America, belong to this component.

Plant species that are of exceptionally high quality and available episodically make up the **bonus component**. These species are the antitheses of the production component species, being neither stable nor predictable. In continental climates, annual forbs and grasses commonly form this component. When present, these plants contribute significantly to the live standing crop and offer a substantial short-term opportunity for enhanced animal production. Sufficient management flexibility must exist to exploit their presence. Animals having high nutritional requirements, such as growing or heavily lactating animals, make the most efficient use of this component. This component is also particularly important to upland and non-game birds and big-game animals.

In this classification system, **null component** plant species are those not used unless the availabilities of the other components, particularly the production component, are severely restricted. Significant animal consumption of these forages indicates a badly depleted forage resource. In Texas these plants include prickly pear, creosote bush, tarbush, honey mesquite, Texas persimmon, broomweed, and croton. These species are of limited value to grazing animals yet may be an extremely important part of the diets of sympatric mammals and bird species. The presence of these plants is often mistakenly considered desirable by stockmen because they are viewed as emergency forage. But this view is incorrect. Cyclic utilization of this component is an indicator of unstable nutrient intake where nutrient demand grossly exceeds nutrient availability from alternative components.

The **toxic component** includes all species poisonous or injurious to grazing animals. Many of these species serve dual roles. They are of some value in other components but are harmful when consumed in excess or at a particular stage of growth. Examples of these dual-role plants in Texas include kleingrass, peavine, sacahuista, oaks, johnsongrass, and pricklypear. Acute effects of toxicity are obvious and can be dealt with promptly. Conversely, chronic effects often go undetected and may even be more costly by virtue of reducing production efficiency.

NUTRIENT INTAKE AND UTILIZATION

Ruminants optimize forage consumption to meet their nutrient requirements if no physical or metabolic restrictions are imposed (Weston and Poppi 1987). **Voluntary intake** of forage is the amount consumed by the animal when its accessibility to forage is unrestricted. In such a case, regulation of intake is dependent only on endogenous mechanisms triggered either within the animal or by some characteristic(s) of the forage (Baile and Forbes 1974; Forbes 1980; Van Soest 1982; Grovum 1986). **Forage (nutrient) intake** under grazing conditions is a modified expression of voluntary intake and is influenced by forage quality (Table 2.4), forage availability (Table 2.5), forage harvestability, environmental stress, and management (Chacon and Stobbs 1976; Hodgson 1977; Arnold and Dudzinski 1978; Finch 1984; Allison 1985; Young 1986, 1987a). We group environmental stress with nutrient intake in this discussion because nutrient demand for travel, diurnal and seasonal thermal fluctuations, and predator avoidance are more pronounced under free-grazing than controlled feeding conditions.

Forage intake of grazing ruminants is usually controlled by distension of the reticulum and cranial sac of the rumen (Grovum 1986). Distension of this sensory region is decreased by digesta passage to the lower tract and/or by reducing ingesta volume and mass through mastication and fermentation. Mastication, primary and secondary, is the major means of particle size reduction (McLeod and Minson 1988) resulting in more dense, less bulky digesta and more rapid fermentation and passage.

Table 2.4. A summary of relationships between forage intake and diet quality.

rence	Location	Pasture Type	Animal	Forage Species	Season	Intake	CP (%)	Digestibility (%)
ila et al. 87	New Mexico	Desert grassland	Steers	Mesquite	Spring	(a).9–1.4	14–16	48–55
				Broom snakeweed	Summer	1.3–2.0	14	50–63
					Fall	1.4	13	48
				Mesa dropseed	Winter	1.3	7	41
l et al. 87b	New Mexico	Native range	Steers	Blue grama	Spring	(a) 2.4–2.45	14–18	47–73
					Summer	2.14–2.6	13	61–65
					Fall	1.9–1.96	8–11	52–62
					Winter	1.91	7	53

Table 2.4. Continued.

Reference	Location	Pasture Type	Animal	Forage Species	Season	Intake	CP (%)	Digest (%)
Pfister and Malechek 1986	Northeastern Brazil	Native caatinga	Wethers	Deciduous trees with herbaceous understory	Spring	[a] 1.2	17–19	61
					Summer	2.8	12–14	52
					Fall	2.5	13	48
					Winter	2.0–2.6	17–24	50–69
						[a] 1.2	18–19	53
						2.6	12–15	55–57
						2.0	13	46
						2.0–2.5	12–25	46–59
Adams et al. 1987a	Northern Great Plains	Native shortgrass prairie	Steers	Western wheatgrass Blue grama Needle-and-thread grass Threadleaf sedge	Spring	[a] 1.9–2.1	14–15	72
					Summer	1.9–2.0	9–13	67–74
					Fall	1.9	7	63–64
Horn et al. 1979	Oklahoma	Bermudagrass pasture	Cow	Bermudagrass	Spring	[b] 117	20	53
					Summer	101	16	53
			Calf		Spring	—	20	53
					Summer	41	18	53
Wanyoike and Holmes 1981	England	Perennial ryegrass	Yearling steers and heifers at low (L) and high (H) stocking rates	Ryegrass	Spring[H]	[a] 1.85	11	
					Spring[L]	2.28	11	
					Summer[H]	1.41	15	
					Summer[L]	1.64	15	
Holechek and Vavra 1983	Northeastern Oregon	70% forested 30% grassland	Heifers	Idaho fescue Bluebunch wheatgrass Sandberg bluegrass Elk sedge Snowberry	Late spring	[a] 2.20	14–15	69–70
					Early summer	2.17	11	49–55
					Late summer	1.97	8–14	57–60
					Fall	2.14	8–11	54–55
Holechek and Vavra 1982 Holechek et al. 1982	Oregon	Grassland	Steers	Bluebunch wheatgrass Idaho fescue Sandberg bluegrass Junegrass Snowberry	Late spring	[a] 2.28–2.39		
					Early summer	1.73–2.04	7–15	40–65
					Late summer Fall	1.89–2.39		
		Forest		Elk sedge Kentucky bluegrass Ninebark Spiraea Idaho fescue	Late spring	[a] 2.05–2.36		
					Early summer	1.73–2.06	8–15	39–67
					Late summer	1.50–2.05		
					Fall	1.89–2.39		
Powell et al. 1982	Nebraska	Sandhill range site	Steers	Blue grama Lead plant Little bluestem Needle-and-thread Prairie sand reed Land switchgrass	Early June	[a] 1.94	14	67
					Early July	1.92	9	62
					Late July	1.99	8	63
					Late August	1.75	8	52

Table 2.4. Continued.

ence	Location	Pasture Type	Animal	Forage Species	Season	Intake	CP (%)	Digestibility (%)
et al. 3	Colorado	Grass prairie	Yearling steers and heifers at High (H) and Low (L) stocking rates	Blue grama	June 1969(H)	(c) 3.1	13	41
					1970(H)	2.7	12	62
					1969(L)	4.2	14	51
					1970(L)	2.4	13	59
					July 1969(H)	4.2	12	48
					1970(H)	2.7	10	56
					1969(L)	4.9	12	51
					1970(L)	3.9	11	62
					Aug. 1969(H)	5.7	11	50
					1970(H)	3.6	10	54
					1969(L)	6.4	10	57
					1970(L)	4.5	9	60
e et al. 0	New Mexico	Plains grassland	Cows Cows Cows Heifers	Blue grama Wolftail Sand dropseed Red Threeawn	June August June August	(a) 2.05–3.39 .47–2.17 1.06–2.14 .37–1.73		71–79 77–79 73–79 76–80
r et al. 7	Peru	Fescue/ Muhly	Alpacas	grass 95% sedges and reeds 4.5% forbs .5% grass 93% sedges and reeds 7%	Wet season Dry season	(a) 1.8 (a) 1.6	8 13	53 64
and venza 6	Utah	Crested wheatgrass Crested wheatgrass shrub	Ewes		Winter Winter	(a) .93–1.0 (a) .98–1.34	5–7 7–9	32–46 24–48
s et al. 3	Utah	Intermountain Region	Mule deer	Big sagebrush Crested wheatgrass Douglas rabbitbush	Winter	(a) 1.73	30	76
et al. 0, 1981	Australia	Pangola grass Rhodes grass	Steers Wethers	Stem Leaf Stem Leaf		(a) 1.00–1.31 1.38–1.64 .68–1.87 1.85–2.04		49–59 48–60 43–55 44–56
et al. 9 phalaris tuberosa harvested at 3 nt stages of maturity	Australia		Wethers	Phalaris tuberosa forage	Stage I Stage II Stage III	(c) .93 .86 .69	26 20 11	80 74 54
e et al. 2	Canada	Hays	Wethers	Iroquois alfalfa Saratoga bromegrass Timfor timothy Champ timothy Timothy hay (weathered)		(a) 3.89 3.52 2.81 2.53 2.20	19 11 11 9 6	

(a) % BW
(b) g/kg .75 BW
(c) kg/day

Table 2.5. Relationships between forage availability and intake restriction across forage types.

Reference	Forage Type	Animal	Standing Crop or Forage Allowance at Which Intake was Limited
Allden and Whittaker 1970	Wimmera ryegrass, Brome, Sub-clover	Wethers	<3000 kg/ha green herbage <1800 kg/ha dry forage <6000 kg/ha forage <5000 kg/ha forage
Arnold and Dudzinski 1966	Phalaris-Sub-clover pasture	Liecester × Merino ewes	<2500 kg/ha
		Corriedale ewes	<1600 kg/ha dry forage
	Native pastures	Young Merino sheep	<1300 kg/ha green DM
	Ryegrass		<1000 kg/ha green DM
	Cocksfoot		< 900 kg/ha green DM
	Phalaris-Sub-clover		<1300 kg/ha green DM
Hodgson 1976		Lambs	< 140 g OM/kg LW/day
		Calves	45–65 g OM/kg LW/day
		Ewes	<1200 kg/ha
Hodgson 1977	Temperate grasses and clover	Cattle	<1100–2500 kg/ha DM
	Subtropical improved pasture	Sheep	1100–4000 kg/ha DM
Handl and Rittenhouse 1972	Crested wheatgrass pasture	Steers	< 135 kg/ha
Fox 1987		Lambs, calves and cows	<1700 kg/ha < 80 g OM/kg LW/day
Baker et al. 1981	Improved perennial ryegrass	Cows, calves	17–51 g DM/kg cow + calf
Havstad et al. 1983	Crested wheatgrass	Heifers	140–920 kg/ha
Chacon and Stobbs 1976	Improved Setaria pasture	Cows	1000 kg/ha DM leaf material
Arnold and Dudzinski 1967	Phalaris-Sub-clover pasture	1962: ewes	
		Lactating	<2023 kg/ha
		Dry	<1345 kg/ha
		Pregnant	<1441 kg/ha
		1963: ewes	
		Pregnant	< 826 kg/ha
		Lactating	< 943 kg/ha
		Dry	< 738 kg/ha
Forbes and Coleman 1987	Plains and Caucasian bluestem pasture	Steers	< 4.4 T OM/ha
Black and Kenney 1984	Wimmera ryegrass Kikuyu grass	Ewes	< 1 T/ha
Pinchak et al. 1990	Southern mixed grass prairie	Cattle	< 700 kg/ha OM

Animal Factors Affecting Nutrient Intake

Voluntary intake may decrease before, and increase after, parturition in both sheep and cattle (Jordan et al. 1973; Weston 1982; Warrington et al. 1988). Decreased intake during late gestation is attributed to decreased reticulorumen capacity caused by a combination of rapid fetal growth and/or increased deposition of abdominal fat and hormonal mechanisms (Forbes 1971; Baile and Della-Fera 1981). The extent to which these mechanisms ultimately control voluntary intake is not known. Voluntary intake increases post partum, but lags behind increased energy requirements for lactation by 2–6 weeks, apparently because of the time required for the rumen to increase in size and reestablish maximum volume (Weston 1982).

There is no clearly defined relationship between body condition (fatness) and nutrient intake in cattle and sheep (Freer 1981; Weston 1982). The general consensus is that abdominal fat restricts voluntary intake 3–30% (Cowan et al. 1980; Freer 1981; Fox 1987), although various effects of fatness have been reported (Bines et al. 1969; Holloway and Butts 1983; Adams et al. 1987b). Conversely, animals in a depleted state consume greater quantities of moderate- to high-quality forages (**compensatory intake**).

Beef cattle and sheep of different genetic backgrounds exhibit markedly different voluntary intakes (Arnold and Dudzinski 1966; Table 2.5) and efficiencies of production. Maintenance requirements of beef cattle account for 70–75% of the ME requirements through a production cycle, under pen-fed conditions (Ferrell and Jenkins 1987). While limited quantitative data are available (Osuji 1974; Havstad and Malechek 1982), the maintenance energy costs of free-ranging cattle are estimated to be 20–50% greater than under pen-fed conditions (Cook 1970). Therefore, the mature size and milk production capability of cows could have a marked effect on their efficiency of production under grazing conditions. Metabolizable energy intake increases as mature size and milk production increases. Similarly, Havstad and Doornbos (1987) reported voluntary intake of ¾ Simmental cows was greater than Hereford cattle under free-ranging conditions. Under conditions of low forage quantity and/or quality the production potential of ¾ Simmental cattle was not achieved.

Animal genotype and phenotype can have marked effects on voluntary intake and efficiency of production. Dairy cattle breeds have higher maintenance (Solis et al. 1988) and lactation (NRC 1978, 1984) energy requirements and intake per unit weight than beef breeds. These are attributed to differences in physiological prioritization of tissue growth and maintenance (Solis et al. 1988). Dairy breeds have a higher proportion of soft tissue organ mass having high maintenance requirements. Additionally, dairy breeds store a larger proportion of fat internally than beef breeds, thereby decreasing insulatory capacity. *Bos indicus* cattle (Brahman type) have been found to exhibit lower maximum intakes of moderate quality diets, under minimal stress, than *B. taurus* (Hunter and Siebert 1985a, 1985b). Lower intake may be the result of *B. indicus* having a smaller digestive tract; however, on poor quality tropical grasses, *B. indicus* digests forages more completely and still exhibits greater voluntary intake than *B. taurus* (Hunter and Siebert 1985a, 1985b). Voluntary intake of moderate- to high-quality forages is greater for *B. taurus* than for *B. indicus*. When low-quality tropical grass diets are supplemented with nitrogen, voluntary intake of *B. taurus* is greater than *B. indicus*, indicating *B. indicus* may have a greater capacity to recycle nitrogen (Hunter and Siebert 1985b). Adaptability of these cattle species to the thermal environment also influences intake patterns. Based upon these findings for domestic ruminants, selecting genotypes suited to a particular range setting is an important management consideration.

Influence of Environmental Factors on Nutrient Intake

Thermal conditions affect intake more than any other environmental factor (see Chapter 3). The range of temperature and humidity where the ruminant is at relative equilibrium with the environment is the **thermal neutral zone** (TNZ). Beef cattle have a TNZ for intake of 10–25 °C (NRC 1981a; Finch 1984). Below the TNZ, **cold stress**, intake increases in response to heat loss down to −25 °C if fill limitations are not encountered. Above the TNZ, **heat stress**, intake decreases in response to heat loading. Abrupt changes in temperature, i.e., blizzard or sleet, may cause a transitory decrease in intake, even within the TNZ. At sustained temperatures below −25 °C, grazing time and intake may be restricted under free-ranging conditions to minimize energy expenditures for grazing (NRC 1981a; Adams et al. 1986, 1987a; Young 1986). As would be expected from their origin, *B. taurus* are more cold tolerant than *B. indicus* animals. The reverse is true in terms of heat tolerance (Finch 1984). Intake responses follow the same trends as tolerances. Crosses of these cattle types exhibit intermediate intakes across the ranges of heat and cold stress.

Forage Quality and Nutrient Intake

Level of forage intake and associated forage-quality interactions are complex functions that vary through time and across animal and forage types (Table 2.4). Generally, short-term intake responds in positive manner to increasing digestibility up to 80% (Hodgson 1977; Freer 1981). However, because ruminants tend to consume forage in response to physiological requirements, long-term intake regulation is relative to a certain level of homeostasis in body condition. Hence, the treatment which follows is an attempt to blend both short- and long-term intake responses to forage quality relative to physiological requirements.

Long-term voluntary intake patterns are determined by the amount of food needed to meet the physiological requirements but modified by the amount which can be consumed before physical constraints are encountered. Both are affected by forage quality, in that less food is needed if the food items have higher concentrations of nutrients, and more food can be physically consumed if the bulky, indigestible fraction is lower. Figure 2.13 illustrates the relationship between forage digestibility and intake assuming no other restrictions. The descending curve represents forage intake needed for maintenance requirements for digestible dry matter, 4.3 kg (9.5 lb) DDM, for a 500-kg (1100-lb) beef cow (NRC 1984). At 20% digestibility, 21.5 kg (47 lb)/day of forage must be consumed to permit the cow to extract the required 4.3 kg of DDM. However, only 5.4 kg/day of an 80% digestible forage must be consumed to supply the same 4.3 kg DDM. The ascending curve depicts the theoretical maximum consumption of forages within the range of 20–80% digestibility, assuming a constant 1% body weight of feces (Conrad et al. 1964). The two curves intersect at approximately 46% forage digestibility and 9.3 kg/day forage intake. Note that to the left of the point of intersect, maximum intake falls below required intake. In the above example, the cow fed a forage that is less than 46% digestible cannot consume enough to reach the required amount of DDM. In the right-hand portion of the figure, maximum intake rises above required intake, so the cow can consume greater amounts of forage at these digestibility levels than are required to meet DDM maintenance needs. The model proposed by Conrad et al. (1964) postulated that voluntary intake tends to take on the pattern formed by the area below both curves. In which case, voluntary intake of forage increases as the digestibility of the forage increases to the point of intersect. Further increases in digestibility lead to decreased food intake and so no change in DDM intake occurs.

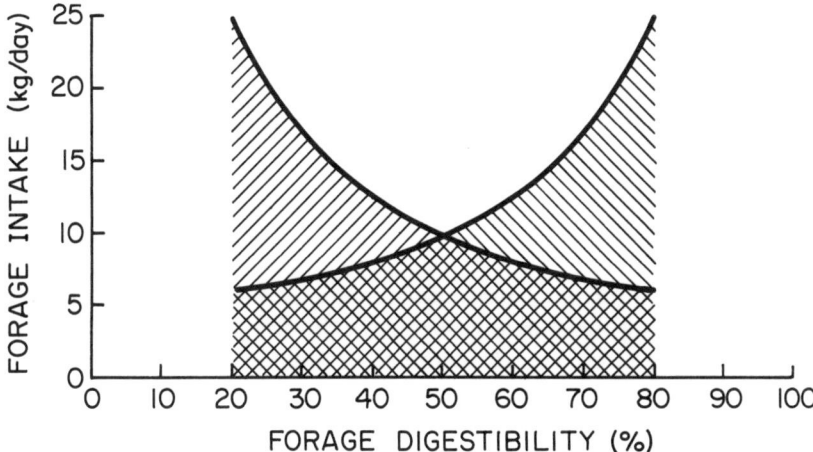

Figure 2.13. Relationship between forage digestibility and long-term forage intake patterns. The descending curve shows the required forage intake to yield 5 kg of digestible dry matter as forage digestibility increases from 20% to 80%. The ascending curve depicts the maximum possible (approximate) forage intake by an adult cow as forage digestibility increases.

This model has been challenged in recent years as being inaccurate and too simplistic (Freer 1981; Grovum 1986), and in some cases with good reason. For example, low digestibility forages are, almost without exception, also low in protein. A small addition of protein to the diet dramatically increases intake of a low-quality forage, indicating that its inherent low digestibility alone did not lead the animal to consume less. On the other side of the scale, grazing animals do not abruptly quit eating the moment their daily nutrient requirements for on-going physiologic processes are met. This fact is easily seen in cows becoming overly fat after the loss of an infant calf or failure to breed. Animals clearly initiate and terminate feeding in response to an array of physical, chemical, and humoral signals (Grovum 1986). In the lower digestibility range, physical factors are most important, although ruminal nitrogen status is certainly involved. At higher diet digestibility, physical factors are less important so internal chemical and humoral factors become more important, in producing hunger and satiety signals. Although the model shown in Figure 2.13 does not account for all factors modifying forage intake, it does depict generalized long-term forage intake patterns of grazing ruminants.

The area to the left of the point of intersect, below 46% digestibility in this example, forms the **zone of response**. As forage quality increases, nutrient intake (Ventura et al. 1975) and productivity increase. If a cow which is not lactating and at mid-pregnancy consumes 9.3 kg of forage, normal growth of fetus and some accumulation of fat for later use after parturition occurs. However, if she consumes forage of lower quality, < 46% digestibility, little or no fat accumulates and fetal development is retarded. Once born, the calf will be smaller and weaker. The cow will produce less milk, wean a lighter calf, and have a reduced probability of rebreeding on schedule. In the extreme case, < 30% digestibility, the cow is malnourished and will eventually die.

The area to the right of the point of intersect, above 46% digestibility in the example, forms the **zone of adequacy**. As forage quality increases above the 46% digestibility level, the model indicates that the cow is correspondingly less stimulated to consume the forage, and thus intake declines. Because requirements are met at a lower level of intake of a more digestible diet, no decline in productivity accompanies the decline in intake. An animal previously restricted by either quantity or quality of diet to the point of nutrient depletion increases intake to a greater level than depicted. Once recovered from the depleted state, voluntary intake is adjusted lower.

Adequate data on forage intake in free-grazing ruminants in different physiological states and over a wide range of forage digestibility is limited due to the difficulty of making such measurements. However, sustained access to forages in the higher range of digestibility is rare under range conditions. If this occurs, animals having higher nutrient requirements (stockers, replacement heifers) should be grazed to make the most efficient use of this resource. The art and science of grazing management is matching the nutrient supply in the forage to the nutrient requirements of the foraging animal to reach sustained optimal productivity. In that spirit we submit the Huston–Pinchak Theorem:

GOOD ENOUGH IS EXCELLENT!

Diets of range animals typically fluctuate above and below the theoretical point of intersect (Fig. 2.13) in a more or less cyclic fashion based upon short-term intake responses to quality of forage consumed (Table 2.4). During periods of high physiological requirements such as early to mid-lactation, the animal may not be capable of consuming adequate amounts of forage to prevent tissue loss. During periods of low physiological requirements and on occasion higher forage quality, nutrient intake may greatly exceed current requirements and result in substantial tissue accretion. We define this spectrum of forage quality as the **normal range** of forage quality and intake. This spectrum is specific for each of the animal species, types, ages, and uses in production systems and is reflected in Table 2.4. The data in this table illustrate a wide array of seasonal trends of intake in response to forage quality over an equally wide array of forage type-animal species combinations, i.e., "a spectrum of normal ranges." A forage or assemblage of forages providing a diet in the normal range is therefore correct for that production system. Good enough is excellent.

Influence of Forage Availability and Structure on Nutrient Intake

An obvious interaction exists between the quantity and quality of available and consumed forage (see Chapter 1). Selective utilization of areas within pastures as well as selective utilization of plants and plant parts within these areas (see Chapter 3) make it difficult to determine which component of available forage is regulating intake. Table 2.5 is an overview of the dynamic interactions between forage and animal type, demonstrating the relationships between forage intake and forage availability. Generally, standing grass crops below 1000 kg/ha restrict forage intake by sheep and cattle on temperate native grasslands of North America. However, on improved pasture, temperate and tropical, standing crops become limiting between 1000 and 4000 kg/ha (Stobbs 1973; Forbes and Coleman 1987). Differences within and between regions are related to forage species or species mix of the pastures. The vertical distribution of leaf and stem biomass and their live and dead fractions ultimately limits intake (Chacon and Stobbs 1976; Poppi et al. 1980; Freer 1981; Forbes and Coleman 1987). Hence, the amount of available live leaf biomass (kg/ha) within an exploitable zone (see Chapter 3) determines the maximum rate of intake. Departure from maximum rate of intake results from the decline of live leaf within this zone below a critical threshold. The point at which this threshold is reached varies with forage species, growing season, length of grazing period, and animal species.

Historically the relationships between forage availability and intake have been described in relation to forage standing crop (Table 2.5). However, overwhelming evidence exists that the amount of leaf and the ratio of leaf:stem within harvest horizons ultimately determine the upper limit of intake, and therefore production, for a given set of forage conditions at a specific point in time. The frequency, severity, and duration of periods of restricted intake determine the sustained animal yield capacity of any land area. Short-term conditions can be overcome through supplemental (substitutional) feeding. Chronic intake restriction can be overcome by destocking and/or increasing forage production and/or increasing of the amount of leaf material. The latter two remedies depend on increased cultural inputs.

SUPPLEMENTAL NUTRITION MANAGEMENT

Supplemental nutrition management is defined as the implementation of practices specifically aimed at improving the nutritional status and/or efficiency of converting available forage into animal products in a given circumstance. Supplemental nutrition is an option when the forage base fails in quantity and/or quality of nutrients to meet the physiological requirements of the grazing animal. Supplemental feeding is targeted at correcting nutrient deficiencies or providing nutrients to stimulate intake, digestion, and/or utilization of forage (Table 2.6). In a broader sense, supplemental nutrition management includes corrective practices to align nutrient supply with nutrient demand. Replacing a quantity of forage that would otherwise have been consumed by feeding an alternate feed supply is called substitution. Supplying a limited nutrient, e.g., protein, to animals having unrestricted forage available of poor quality is called **supplementation**. Huston et al. (1988) (Fig. 2.14) clearly demonstrated the potential stimulation in forage intake of low-quality forages by

Table 2.6. Relationships between type and amount of supplement fed and forage intake.

rence	Animal Type	Forage Type	Forage Quality CP (%)	Dig. (%)	Type of Supplement	Rate	Response
nhouse et al. 70	Heifers	Native range of mixed cool and warm season grasses in winter		39	Protein (P)	1.16, 2.07, 3.0 g/kg BW$^{.75}$/d	CP supplementation had little effect on intake or FDMD.
					Energy (E)	.02, .041, .061 .081 Mcal/kg BW$^{.75}$/d	Energy supplementation > .041 Mcal/kg$^{.75}$ depressed FI but not FDMD. TDMI and TDMD increased with increasing levels of energy.
s et al. 1988	Cows	Bermudagrass hay	12		Energy	.3% BW	Intake of hays declined with cow supplementation and was greater for BG than OG. Supplementation improved OMD.
		Orchardgrass hay	11				
hner 1980	Cows	Native winter range of blue grama, Western wheatgrass Needle-and-thread	6–9	54	Protein (CSM)	1.5 kg	Supplementation decreased FOMI. Forage intake and DMD were lower in grain-fed cows. Protein increased TDMD over FDMD.
					Energy (Barley)	1.4 kg	

Table 2.6. Continued.

Reference	Animal Type	Forage Type	Forage Quality CP (%)	Quality Dig. (%)	Type of Supplement	Rate	Response
Judkins et al. 1987	Steers	Native blue grama rangeland	8–10		Protein (CSC) Alfalfa Control	1.7 kg/hd 3.6 kg/hd	Cattle receiving alfalfa had slightly lower than the Control s or those receiving Total intake increa with supplementa (Control < CSC < alfalfa).
Cook and Harris 1968	Wethers	Utah desert rangeland	8		Protein and energy Protein Energy	.5–.75 lb/day .33 lb/day .33 lb/day	.5 lbs of suppleme decreased FI but .. increased FI. P+E supplement depre DMD but increase protein digestibilit Energy reduced pr digestibility but pr increased protein digestibility, DMD, and MEI.
Adams 1985	Steers	Russian wild rye in the fall	7	62	Energy (Corn)	.3 kg/100 kg BW in morning and afternoon	Supplementation depressed intake i and p.m. DE intak greater in p.m. sup mented steers
Lusby et al. 1976	Cows	Little bluestem winter range		54 63	Protein	Moderate High	Cows receiving m ate level of supple mentation consum more forage.
Scales et al. 1974	Calves	Colorado sandhill shortgrass range	5.2		Protein Energy P+E	Six combinations	Supplementation tended to depress consumption exce the lowest level.
Caton et al. 1988a	Steers	Winter native range Blue grama Sideoats grama Sand dropseed Wolftail Mat muhly Carruth sagewort	7.1–9.2	47–50	Protein (91% CSM) Control	.58 and .83 kg/hd/day	Supplemented ste forage intake (11.6 BW) tended to be greater (P<.12) th control steers (9.3 BW). In vitro digestibilit (OMD) was greate supplemented tha trol diets, 50 and respectively. OM disappearance rate situ and NH$_3$ were greater for supple mented steers.

CP	— Crude protein		DMD	— Dry matter digestibility
FDMD	— Forage dry matter digestibility		FI	— Forage intake
TDMI	— Total dry matter intake		CSC	— Cottonseed cake
TDMD	— Total dry matter digestibility		CSM	— Cottonseed meal
BG	— Bermudagrass		DE	— Digestible energy
OG	— Orchardgrass		MEI	— Metabolizable energy intake
OMD	— Oraganic matter digestibility			

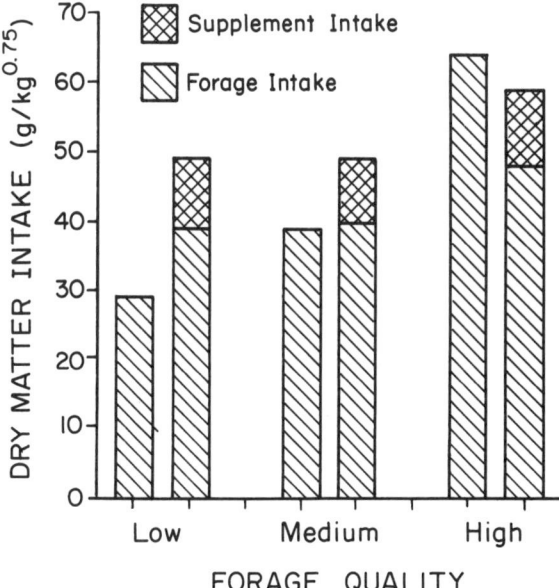

Figure 2.14. Effect of forage quality on intake responses to supplemental feeding. This is the result of an experiment in which sheep were offered forages of different quality ad libitum either with or without supplemental feed (Huston et al. 1988). The low-, medium- and high-quality forages were wheat straw (3.4% CP and 41% IVDMD), sorghum hay (5.9% CP and 54% IVDMD), and oat hay (13.8% CP and 65% IVDMD), respectively. The supplement contained 28% CP and approximately 3.3 Mcal/kg DE. Supplement feeding level provided 60 g CP per sheep per day.

sheep with low levels of protein supplementation. Generally, field experiments have been less conclusive (Table 2.6), although low levels of protein supplementation on poor-quality (< 6% crude protein) diets can stimulate forage intake (Caton et al. 1988a).

Grazing management is a primary means of achieving a balance between animal demand and nutrient supply. Decisions on animal populations (species, breeds, and classes), stocking rates, breeding dates, pasture sizes, rotation schedules, etc. (see Chapter 7) set the degree of match or mismatch between the supply of and demand for nutrients. Supplemental nutrition management in this context is then the fine adjustment in the balance between supply and demand.

The following discussion describes four general categories of mismatches of nutrient supply and demand. The difference between high and low quantities relates an animal's ability or inability to achieve adequate intake of forage in a reasonable length of grazing time (see Chapter 3). High and low quality refers to the normal range defined in the previous section.

Quantity and Quality High

This range condition is seldom found on a sustained basis but often occurs on a short-term basis. Seasonally, such a condition occurs on temperate rangelands during the late spring growth period. Small grain pastures—wheat, oats, rye—provide forage of this type until mid-anthesis. This is an important interval for animals matched to forage within the **normal range** as this "up" period follows and precedes a "down" period. Therefore, it is essential for recovery from a past depletion period and preparation for future depletion.

It is conceivable that under some conditions both diet quantity and quality consistently, or at least frequently, exceed requirements. In such cases, forage quality and nutrient intake rise above the normal range. The expected results are overly fat animals, reduced efficiency in transferring dietary nutrients into animal products, and possibly reduced individual animal performance. The corrective management strategy is to restructure the grazing population. That is, animals having greater productive potential and a greater capability for utilizing the high-quality forage should be selected or the stocking rate of existing animals increased.

Quantity High and Quality Low

This range condition commonly occurs when abundant plant growth is followed by an extended period of temperature- and/or moisture-induced dormancy. This condition is characteristic of the dormant season in the temperate region. The residual forage contains comparatively high proportions of structural carbohydrates, thereby diluting its energy and protein value. The key concern is whether the digestibility of the forage fluctuates within the normal range. Remember that forages that fall in the lower region of the normal range for dry cows are virtually always below the normal range for lactating cows and small ruminants, sheep, goats, deer, etc. A supplemental nutrition program should provide the limiting nutrients (e.g., protein, phosphorus, vitamin A). This supplemental feeding program may stimulate forage consumption if protein and/or phosphorus are critically low, or may decrease forage consumption by substitution (Fig. 2.14). Assuming that protein and/or phosphorus are not limiting in the forage, forage consumption decreases by approximately one-half of the amount of concentrates fed.

Quantity Low and Quality High

The converse of the previous range profile, this condition favors small ruminants, especially goats and deer, which are flexible in their foraging behavior. This condition is characteristic of shrub-dominated landscapes and often results from overgrazing and/or protection from fire (see Chapter 5). Supplemental feeding can be used to increase the stocking rate, but if the range is properly stocked with the correct animal types, supplemental feeding does not improve the productivity of the individual grazing animals.

Special use pastures can also be assigned to this category such as small grain pastures of extremely high quality, especially in protein. Feeding grains to growing animals, lambs and calves, on small grain pastures allows an increase in stocking rate without altering animal performance. In this case, an almost exact substitution occurs—the small grain forage intake is reduced by the amount of the grain fed. Benefit is realized because the high concentration of protein is more efficiently distributed to a larger number of animals resulting in greater net secondary productivity.

Quantity Low and Quality Low

This range condition is best typified by desert or arid landscapes. The limited standing crop typically contains an abundance of structural carbohydrates, lignin, and/or secondary plant chemicals that reduce palatability, intake, and utilization. The proper nutritional strategy in this circumstance is to encourage high plant selectivity by maintaining a low stocking density. Feeding during dormant interim periods provides a balance of nutrients when little or no alternative natural supply is available.

Similar seasonal conditions exist on rangelands overstocked during periods of dormancy. Very little forage is available, and that which remains is low in quality. For best results in the short term, a good quality hay or a complete feed should be provided. Heavy rates of stocking on yearlong range lead to similar nutritional conditions during winter dormancy (Greene et al. 1987; Heitschmidt et al. 1987c), hence establishing a cyclic pattern. The range between the highs and lows in nutritional adequacy is too broad to fit within the nutritional state characterized earlier as the normal range. The lows are too low for adequate recovery during the highs; thus, productivity is substantially reduced. Management alternatives include reduced stocking to increase quantity and quality of diet or liberal feeding which rarely yields economic returns and only prolongs an unsustainable ecological condition.

CONCLUSIONS

So, what have we said? Ruminant animals are placed on rangeland as primary consumers of the vegetation formed by the capture of solar energy. In a natural state, these animals would in turn adapt spatially and in proper numbers for more or less sustained survival. However, the human demand for the offtake of consumable products (food and fiber) imposes a requirement in excess of survival and so creates an equilibrium that is less than a natural balance. Restricted movement, altered numbers, and controlled breeding impose an unnatural match between what is offered by the vegetation and what is required by the grazing animal.

Through an understanding of what nutrients are important, their probable concentrations and fluctuations in forages, and their requirements by animals, management can partially align nutrient supply and demand on rangeland. Supplemental nutrition management is then required to provide a fine adjustment for optimal productivity. Perhaps the most important aspect of management is the recognition of what is involved in grazing behavior and diet selection, the topic of Chapter 3.

PLANT SPECIES CITED

Grasses

COMMON NAME	SCIENTIFIC NAME
Bermudagrass	*Cynodon dactylon* (L.) Pers.
Bluebunch wheatgrass	*Elymus spicatum* (Pursh) Scribn.
Blue grama	*Bouteloua gracilis* (H. B. K.) Lag. ex Steud.
Brome	*Bromus* spp.
Caucasian bluestem	*Bothriochloa caucasica* (Trin.) C. E. Hubb.
Crested wheatgrass	*Agropyron cristatum* (L.) Gaertn.
Elk sedge	*Carex geyeri* Boott.
Fescue	*Festuca* spp.
Idaho fescue	*Festuca idahoensis* Elmer.
Johnsongrass	*Sorghum halepense* (L.) Pers.

COMMON NAME	SCIENTIFIC NAME
Junegrass	*Koeleria cristata* (L.) Pers.
Kentucky bluegrass	*Poa pratensis* L.
Kikuyu grass	*Pennisetum clandestinum* Chiov.
Kleingrass	*Panicum coloratum* L.
Little bluestem	*Schizachyrium scoparium* (Michx.) Nash.
Mat muhly	*Muhlenbergia richardsonis* (Trin.) Rydb.
Mesa dropseed	*Sporobolus flexuosus* (Thurb.) Rydb.
Muhly	*Muhlenbergia* spp.
Needle-and-thread	*Stipa comata* Trin. & Rupr.
Orchardgrass (Cocksfoot)	*Dactylis glomerata* L.
Pangola grass	*Digitaria decumbens* Stent.
Perennial ryegrass	*Lolium perenne* L.
Phalaris (Hardinggrass)	*Phalaris tuberosa* L.
Plains bluestem (King Ranch bluestem)	*Bothriochloa ischaemum* (L.) Keng.
Prairie sandreed	*Calamovilfa longifolia* (Hook.) Scribn.
Red three-awn	*Aristida longiseta* Steud.
Rhodes grass	*Chloris gayana* Kunth.
Russian wildrye	*Elymus junceus*
Sacahuista	*Nolina texana* S. Wats.
Sandberg bluegrass	*Poa sandbergii* Vasey.
Sand dropseed	*Sporobolus cryptandrus* (Torr.) Gray.
Saratoga bromegrass	*Bromus* spp.
Setaria	*Setaria anceps*
Sideoats grama	*Bouteloua curtipendula* (Michx.) Torr.
Switchgrass	*Panicum virgatum* L.
Texas wintergrass	*Stipa leucotricha* Trin. and Rupr.
Threadleaf sedge	*Carex filifolia* Nutt.
Timothy	*Phleum pratense* L.
Western wheatgrass	*Elymus smithii* Rydb.
Wimmera ryegrass	*Lolium rigidum* Gaud.
Wolftail	*Lycurus phleoides* H. B. K.

Forbs

COMMON NAME	SCIENTIFIC NAME
Annual broomweed	*Xanthocephalum dracunculoides* (DC.) Shinners
Broom snakeweed, Perennial broomweed	*Gutierrezia sarothrae* (Pursh) Britt. and Rusby
Carruth sagewort	*Artemisia carruthii* Wood
Croton	*Croton* spp.
Iroquois alfalfa	*Medicago sativa* L.
Peavine	*Astragalus nuttallianus* DC. var. *nuttallianus*
Spiraea	*Spiraea lucida* Dougl.
Sub-clover	*Trifolium subterraneum* L.

Shrubs

COMMON NAME	SCIENTIFIC NAME
Big sagebrush	*Artemisia tridentata* Nutt.
Creosote bush	*Larrea tridentata* (DC.) Cov.
Douglas rabbitbrush	*Chrysothamnus viscidiflorus* (Hook.) Nutt.
Fourwinged saltbush	*Atriplex canescens* (Pursh) Nutt.
Lead plant	*Amorpha canescens* (Nutt.) Pursh
Ninebark	*Physocarpus malvaceus* (Greene) Kuntze
Snowberry	*Symphoricarpos albus* (L.) Blake
Tarbush	*Flourensia cernua* DC.

Succulents

COMMON NAME	SCIENTIFIC NAME
Prickly pear	*Opuntia* spp.

Trees

COMMON NAME	SCIENTIFIC NAME
Honey mesquite	*Prosopis glandulosa* Torr. var. *glandulosa*
Oak	*Quercus* spp.
Texas persimmon	*Diospyros texana* Scheele

CHAPTER 3

FORAGING BEHAVIOR

Jerry W. Stuth

INTRODUCTION

Optimum livestock production on grazing lands is in part a function of the ability of the animal species employed to harvest nutrients in an effective and efficient manner. An understanding of the temporal and spatial dynamics of the grazing process in a complex environment is critical for optimizing livestock production.

Each pastoral situation offers a unique environment from which the animal must garner nutrients, maintain thermal balance, and interact socially with other individuals of the herd both to sustain itself and the species. Each population employs an evolutionary strategy directed toward maintenance of fitness. Reported scientific studies indicate that both domestic livestock and most economically important wild ungulates forage optimally and are energy maximizers. That is, they maintain fitness by feeding optimally to consume the greatest amount of energy and/or other nutrients (Schoener 1969, 1971, 1983; Charnov 1976; Pyke et al. 1977; Belovsky 1978, 1981a, 1981b, 1984b, 1986a, 1986b; Krebs and Davies 1978; Whitman 1980; Hixon 1982; Owen-Smith and Novellie 1982; Black and Kenney 1984; Kenney and Black 1984; Belovsky and Slade 1986; Horner and Staddon 1987). Thus, grazing managers must develop an understanding of the grazing patterns employed by the animals they are managing. The strategic view involves understanding the genetic evolution and predisposition of a given species to forage. Mechanisms employed by the animal have their roots in the history of the evolution of the species (Westoby 1974; Ellis et al. 1976; Rosenzweig 1981) and are largely inherent in the animal. The inherent nutritional aspects are discussed in detail in Chapter 2. The objective of this chapter is to provide an overview of the inherent behavioral aspects of grazing.

FORAGING TACTICS OF ANIMALS—THE PLANT-ANIMAL INTERFACE

Perhaps the single most important aspect of understanding plant-animal interactions is developing an appreciation of the foraging process (Crawley 1983). When an animal grazes a plant, a hierarchy of instinctive responses and behavioral actions have been taken by it that leads to the point of prehension and consumption (McNaughton 1987; Senft et al. 1987; Senft 1989) (Fig. 3.1). Each landscape unit (pasture, paddock, block, allotment) is composed of a complex of different habitats or distinct groupings of plant species in communities (Fig. 3.2). Habitats are delimited by the type of plant species present, their spatial arrangement, and structural configuration, e.g., a grassland community with scattered trees less than 1 m in height in contrast to a shrubland of dense shrubs over 3 m high with some grassland occupying the

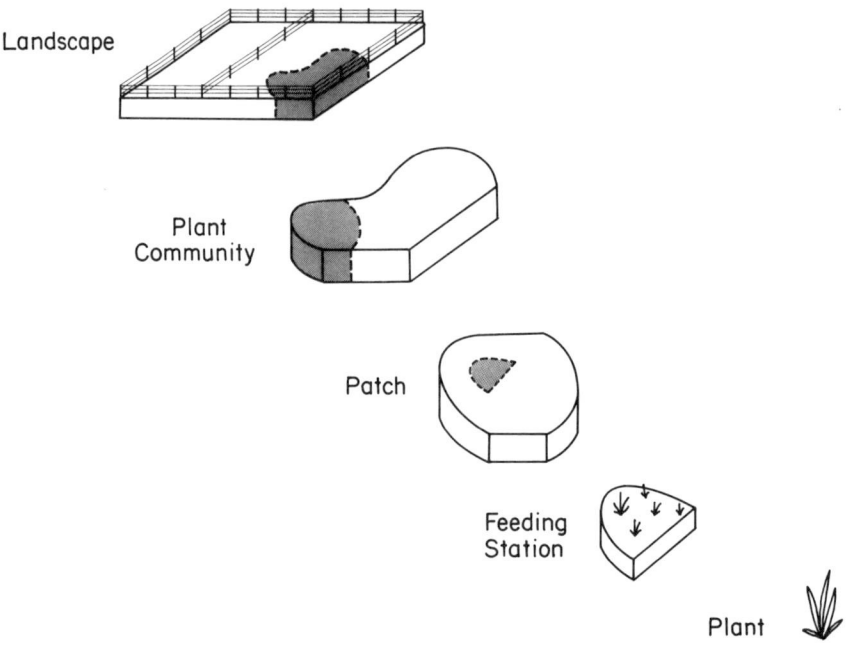

Figure 3.1. Hierarchical view of the diet selection process from the landscape level down to the individual plant.

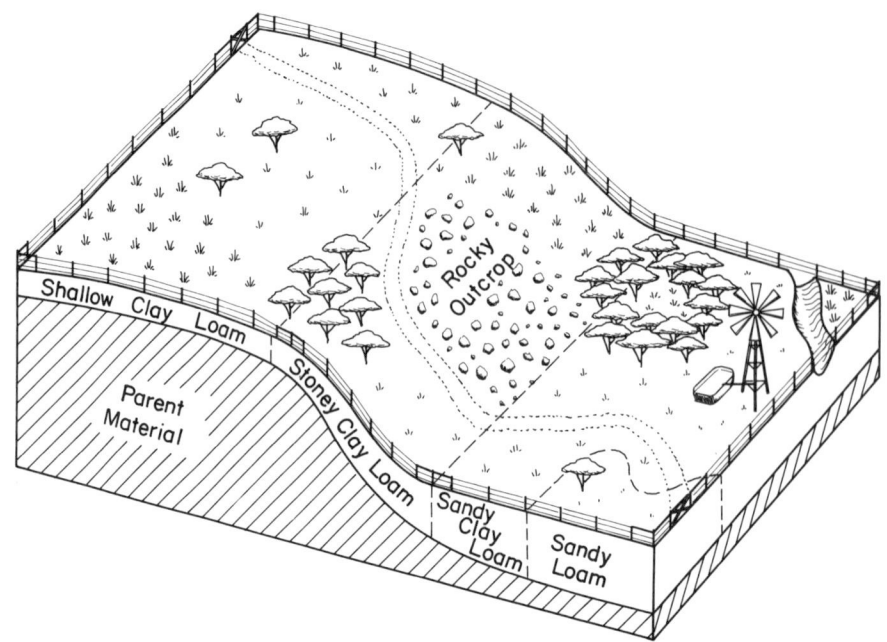

Figure 3.2. Landscape configuration reflecting the unique set of forage resources, water locations, and terrain constraints which effect use patterns by grazing animals.

interspaces. Habitats can be further delineated into patches which contain more homogeneous groupings of species. In the example above, a patch is represented by the shrub aggregation or the grassland interspace.

When the animal has oriented itself in a habitat it must decide when to lower its head and establish a feeding station along its grazing path. Within the feeding station, the animal must then select from among the individual plant species those it will consume, and beyond that which of the plant parts will be eaten. Therefore, the diet selection process has two major levels that must be clearly distinguished, spatial choice and species choice

SPATIAL CHOICE

The Landscape Level of Diet Selection

Spatial choices position the animal in a landscape prior to selecting plant species or parts from among an aggregate of available plants. The landscape level of diet selection is characterized by those physiognomic and thermic features of a management unit which influence animal movement patterns. A given landscape unit (pasture) is characterized by boundaries; distribution of plant communities; degree of assessability, and distribution of water, thermal, and mineral foci (Table 3.1).

The animal must first come to understand the nature of its landscape by locating the boundaries, routes of access and escape, plant communities and contacts, and the seasonality of desirable species. Therefore, the more experience the animal has with the variety of habitats and plant species available to it in a variety of years, the greater its ability to optimize grazing tactics to survive in a management environment (Smith 1984; Senft et al. 1987; Senft 1989). When first introduced into a pasture, animals seek the boundary of their environment (e.g., fence or home range), and once their experience base is sufficiently high, typically 24–72 hours for livestock, they begin the search for the source of the most vital consumption item, water. Free-standing water is the principle focus around which most of the larger ungulates such as cattle, buffalo, and impala orientate their foraging strategies. They seek the most energy-efficient sources of forage as referenced to known water sources (Coleman et al. 1989). Large herbivores are "central place foragers," with the central or home place centered on water. The observed probability distribution of foraging around water (Senft et al. 1985; Smith 1988b) resembles that expected for theoretical central place foragers where foraging activity is universally related to distance from a central point of habitation.

Table 3.1. Characteristics of the landscape level of diet selection which influence animal movement patterns.

Attribute	Components
Boundaries	Fences, home range, migration routes
Distribution of plant communities	Range sites, soils, aspect, elevation, structure, species composition
Accessibility	Slope, gullies, water courses, shrub density, rockiness, roads, trails, fencelines, cut openings, pipeline/utility right-aways
Distribution of foci	Location of water, shade, loafing and bedding sites and other convergent and divergent points in a landscape

Optimum grazing area is defined by an approximate circle whose radius is generally not over 0.8 km from the water source. About 1.6 km is considered the maximum outer limit for a herd of cattle (*Bos* spp.) or flock of sheep (*Ovis* spp.) to balance their forage and water needs (Valentine 1947). However, during drought the effective grazing area is increased as forage supply diminishes (Squires 1982; Walker et al. 1987; Smith 1988b). When calculating stocking rates for a given area, the manager must consider not only the spatial distribution of the animal population in relation to water but also frequency of drought, and reduce stocking rates accordingly (Fig. 3.3). Smith (1988b) has defined the form of this relationship as an inverse log equation, while Andrew (1988) referred to it as a sigmoid logistic curve.

Rough terrain, such as gullies, steep slopes, and/or rocky outcrops, restrict animal movements even when water sources are within otherwise acceptable distances. Heavy concentrations of shrubs have been found to create access problems to potentially grazable areas, especially when the animal encounters high stem densities and overlapping canopies. Animals prefer to use established trails, roads, cut paths/openings, and pipe-line/utility right-of-ways rather than attempt to penetrate thick brushy areas or traverse difficult terrain. Animal movement in rough terrain is determined by the agility of the species in question. Goats (*Capra* spp.) and large ungulates inhabiting mountainous terrain have adapted foraging skills which minimize terrain impact on movement patterns.

Spatial-use patterns of livestock can be regulated by mixing experienced animals with inexperienced animals. This matriarchal system of experience training can be a vital means for reducing the learning curve of replacement animals in a herd, so that if extended drought occurs experienced animals can show others those sites which improve their chances for survival. Large-herd liquidations during a harsh drought have a significant long-term impact on the experience base or living memory within the herd. By contrast, because most domestic animals are creatures of habit they may establish landscape-use patterns that do not optimize animal distributions. Thus, the introduction of new animals may, in certain instances, enhance distribution patterns.

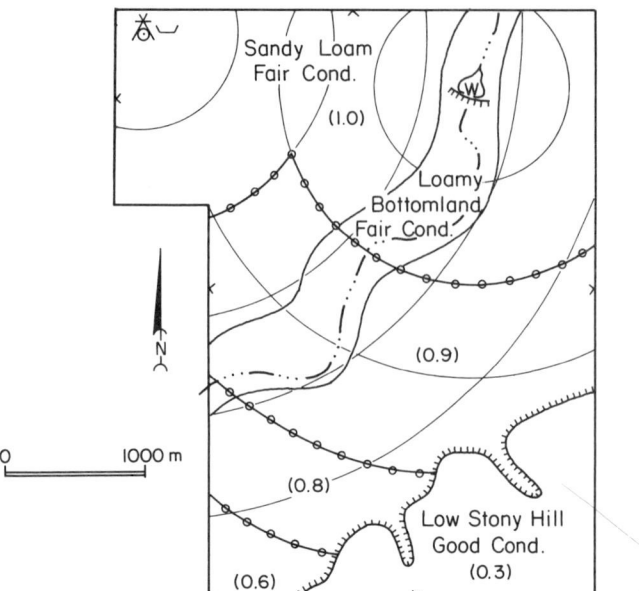

Figure 3.3. Zonal impact of water and shape on adjustments in stocking rate of a grazing unit.

Smith (1988b) proposed that ungulates have a hierarchy of physiological needs which exhibit thresholds for altering activities and subsequent movement within pastures (Figs. 3.4 and 3.5). This hierarchy determines the probability that a given site or patch will be frequented by grazing animals. The greatest need is for water. If visual cues (e.g., windmill, trees) are prominent near water sources, the strength of the water's attraction is increased. Distribution of thermal foci, which allow animals to maintain homoiothermy in a landscape relative to water location, may interact with prevailing winds during the growing season to affect the amount of potential grazing pressure a site will receive. Small domestic stock, particularly goats and sheep, drift against winds, resulting in a noticeably disproportionate use of those portions of pastures where the prevailing winds enter the pasture (Smith 1988b).

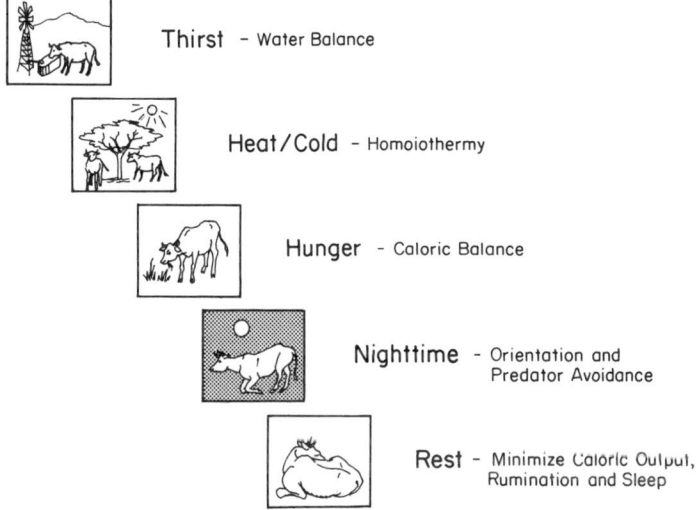

Figure 3.4. Hierarchy of large grazers' physiological and behavioral needs which affect patterns of landscape use. These threshold levels trigger initiation and velocity of movement and frequency of encounter of locals within a landscape.

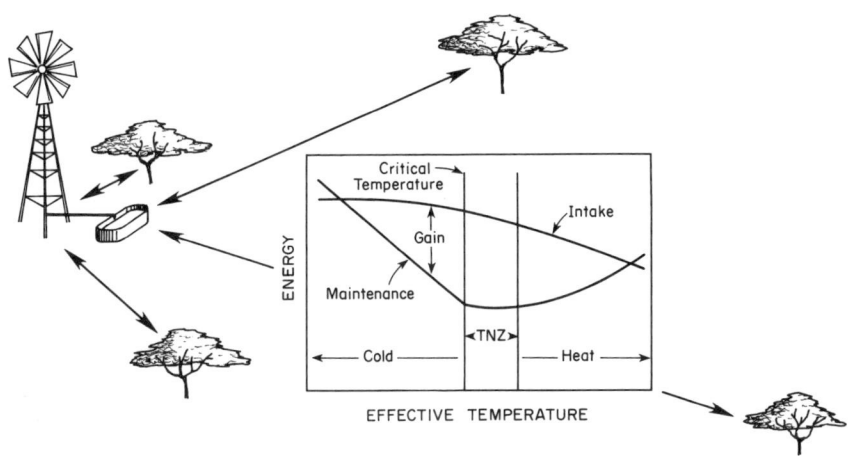

Figure 3.5. Interactive effects of water and thermal foci as they affect energy maintenance and intake of ruminant animals.

The primary orientation to water and thermal balance causes animals to forage away from these foci to meet their nutritional needs. Most ungulates first harvest food, then move to loafing and bedding sites to ruminate and digest the food ingested in a previous grazing bout (meal) and/or to areas for predator avoidance. If the starting point for grazing is a water or thermal focus, the subsequent distance covered by the animal is determined in part by digestive capacity or rate of food passage through the animal and in part by the potential harvest rate of forage encountered, potential grazing velocity, and level of satiety of the animal (Walker et al. 1989a). Once satiated, the animal returns to a thermal, water, or strategic bedding site depending on thresholds of these various needs. The interaction of thermal regulation and digestive capacity is responsible for the noticeable "piospheres" or rings of utilization which diminish in area with distance from water sources.

Grazing time per day is a function of forage quality, thermal balance, and short-term stability of forage supply. Animals reduce daily grazing time as digestibility of forage available declines and retention time of ingesta increases. When daytime temperatures are within the thermal neutral zone of cattle, most grazing, 90%, takes place during daylight hours (Table 3.2). During hot periods cattle reduce afternoon grazing and increase night-time grazing. Cattle have demonstrated little directional grazing after darkness, so as nighttime grazing increases it is in the neighborhood of termination points of grazing at dusk (Walker and Heitschmidt 1988). Evidence is mounting that cattle rely on vision to move about in their environment, so as darkness sets in they loose many of their visual cues, so they do not venture far from night-time bedding areas.

When winter temperatures are below the thermal neutral zone of cattle they limit evening grazing but increase afternoon daylight grazing substantially. Therefore, longer daytime grazing bouts with directed grazing occur during the winter.

When forage supply is restricted, the animal compensates by increasing grazing time. However, if in a severe caloric deficit, the animal tends to give up the search due to the high cost of travel relative to the energy garnered from edible forage located (Coleman et al. 1989).

Plant Community and Patch Level of Diet Selection

Evidence suggests that the animal's selection of a given plant community is largely related to those attributes of a site which influence its ability to harvest nutrients. Table 3.3 provides a comprehensive summary of community attributes and the way they impact animal use of a site.

There have been several studies to isolate those community attributes which affect the selection of communities by a grazer. Senft et al. (1987) established that forage quantity and quality are closely related to the ratio of amount of time spent grazing in the community relative to the area the forage occupies within the landscape. The abundance of seasonally preferred plant species has also been shown to influence the patterns of plant community use (Senft et al. 1985).

Preference for communities is usually measured either by determining the ratio of percent grazing time: percent of land area or percent of grazing capacity of given management unit or landscape of the animal. Implicit in this measurement is that as animals increase time in a site, the greater the quantity of nutrients harvested from the site. This assumption implies that communities which afford an animal species high harvest rates per unit of grazing time are preferred by that animal. Put in another way, plant community profitability can be valued by measuring the potential ingestive rates (g/min) of forage by the animals (Table 3.4). The greater the density of high-

Table 3.2. Diurnal grazing time (hr) of brangus cattle in southeast central Texas (after Stuth et al. 1987).

Period	Time	Spring	Summer	Winter
Morning	06:00–12:00	2.65	2.30	2.50
Afternoon	12:00–18:00	3.35	2.20	4.50
Evening	18:00–24:00	3.30	3.10	0.65
Night	00:00–06:00	0.85	2.35	1.00
Total	—	10.15	9.95	8.65

Table 3.3. Attributes at the plant community and patch level which influence the animal's selection of forage site.

Attribute	Function
Moisture-holding capacity of soil	Forage supply and stability
Species composition	Affects suitability/ stability of the site for general dietary and nutritional needs
Plant frequency	Affects the probability of encounter of plant species by the animal and number of dietary decisions
Abundance	Affects the supply of nutrients
Structure	Affects accessibility and harvestability of plant species and nature of thermal niches provided
Continuity	Affects movement velocity
Size	Affects amount of search area available
Aspect	Affects the thermal characteristics of the site
Orientation in landscape	Position relative to needs focii affects frequency of exposure to grazing

Table 3.4. Relative profitability of plant communities on a mixed brush, clay loam site in south Texas.

Treatment	Bite Weight (g/bite)	Bite Rate (#/min)	Intake Rate (g/min)
		June	
Rootplowed	.61	32.3	14.0
Sprayed	.66	14.4	8.9
Untreated	.87	15.9	11.2
		August	
Rootplowed	.46	35.5	15.4
Sprayed	.49	33.4	15.5
Untreated	.53	24.7	11.3

quality food species, the slower the grazing velocity therefore the greater residence time and intake level attained relative to other communities available to the animal (Senft et al. 1987). If these communities lie between important water and thermal foci,. site preference is magnified.

However, several studies have shown that grazing preference based on occupancy:area ratios can be misleading if assumed to reflect "food value" of a site (Butterfield and Stuth 1991). Figure 3.6 provides a conceptual view of the functional nature of landscape use categories when occupancy:area ratios are contrasted to utilization:herbage mass ratios for the same site. Site preference in this case results in four major preference categories:

1. Grazing preferred;
2. Grazing avoided;
3. Terrain constrained or directed use;
4. High-impact grazing sites.

Grazing preferred sites are those sites with high occupancy:area ratios and high utilization:herbage mass ratios with the major bulk of the animal's forage derived from these sites. Grazing-avoided areas contain low-value food or are inaccessible to the animal(s). Terrain-constrained or directed-use sites are unique in that these sites have high occupancy times yet little utilization relative to herbage mass in the pasture. Examples of preferred directed-use sites include near-satiated grazing of these sites at access points to water sources; sites where normal animal movement causes herd concentration in pasture corners or against gullies, hills, or roads; and prevailing wind-directed grazing. Finally, there are those sites where limited occupancy relative to area in the pastures results in high utilization relative to herbage mass in the pasture high-impact sites. Low potential sites occurring along directional grazing paths where animals are exhibiting high grazing velocities can result in limited occupancy time but high levels of use of available forage.

Generally, pastures in which grazing occupancy time and level of utilization are highly correlated possess few terrain-constrained or high-impact sites. Pasture configurations which result in poor correlations of occupancy time and forage use offer opportunities to manipulate habitat to improve harvest efficiency. Animals apparently establish directed grazing paths which increases the probability of encountering more profitable sites. Highly profitable communities attract ungulates from neighboring foci. Memory of their level of profitability most likely establishes the direction from water, thermal, or resting foci, but directed grazing will occur between foci and preferred sites (Bailey et al. 1988). The duration of grazing in a community or patch relative to another site along a grazing path is largely determined by the relative differences in harvest rates. Low harvest rates (g/min) result in high grazing velocity and so short resident time in the community/patch. The rate of

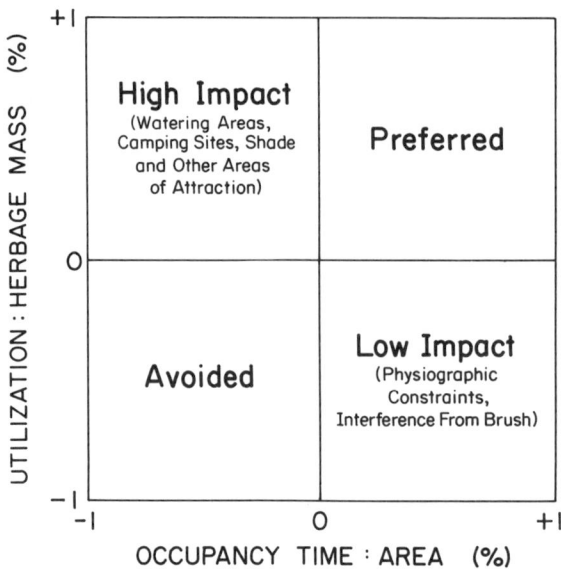

Figure 3.6. Conceptual view of the functional nature of landscape use categories when considering the ratio of occupancy time to percent area occupied contrasted to the ratio of forage utilized relative to herbage mass a plant community contributes to a pasture.

grazing velocity is dependent again on the interaction between level of satiety and distance from thermal/water foci.

Animals who have loafed and ruminated for extended periods leave foci at a higher velocity, not slowing their travel pace until a highly profitable site is encountered (Smith 1988b). As the gut is filled, the rate of ingestion slows. This behavior often leads to high residence time but low harvest rates due to increased selectivity. To understand this phenomena requires a greater understanding of feeding station behavior (Demment and Van Soest 1981).

THE FEEDING STATION LEVEL OF DIET SELECTION

An animal's feeding station is established when it stops walking, lowers its head, and bites a plant. At this point certain sensory cues have caused the animal to stop searching and to select a species or combination of species it perceives as profitable. The pattern of feeding stations is strongly related to the distribution and profitability of patches in a community, the size of the community, and the geographical relationship of the community to the animal's grazing path (Novellie 1978; Ruyle and Dwyer 1985).

Forage behavior at this level can be categorized as search time, time spent travelling between feeding stations, biting rates within feeding stations, and duration of biting while at a feeding station (Stuth and Searcy 1987). Recent studies in foraging behavior of cattle indicate that they have definite seasonal foraging strategies in response to changing plant phenologies and the availability of forage (Stuth et al. 1987). Cattle increase the amount of grazing time allocated to searching between feeding stations when forage conditions are more universally high across species and habitats (Fig. 3.7). The animals appear to select fewer plant species and focus their selection on plant species which offer the maximum amount of green forage per bite (bite size) within the primary food group. This in turn leads to an overall drop in bite rate which is a result of increased search time while actively grazing. If forage becomes limiting during these high-quality periods, animals intensify searching to acquire an adequate daily intake until their preferred primary food group is depleted. However, as the season progresses and the amount of senescent material in the crown or canopy increases, cattle and sheep reduce search time between feeding stations and increase selection time at each feeding station. Each feeding station is more fully exploited during these times. In fact, when the animal stops to graze a feeding station, most of the available green forage is fully consumed before the animal moves on to the next feeding station. During these periods, intraspecific (animal-animal) competi-

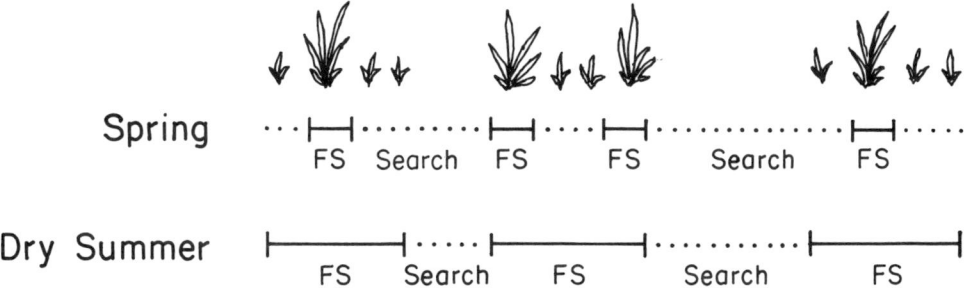

Figure 3.7. Effects of seasonal quality of plant communities on feeding station behavior. High forage quality (e.g., spring) results in short feeding times at stations (FS) with longer search intervals between feeding stations.

tion becomes most critical with respect to the nutritional well being of the individuals. Herding instinct causes the herd to fragment into smaller feeding groups and disperse over a wider area of the landscape when forage supply is low (Smith 1988b). High grazing pressures during periods of high differential palatability between species available often cause nutritional problems for an individual. And as grazing pressure increases the number of unexploited feeding stations diminishes.

Searching between feeding stations occupies 20–30% of the grazing hour and appears to be an adjustment mechanism associated with forage quality. As stated previously, animals reduce overall grazing time as forage quality declines seasonally. However, less of this time is allocated to searching between feeding stations in a community or patch. The net result is reduced seasonal differences in actual grazing time at feeding stations.

Patchiness within communities has its greatest effect on distance between feeding stations (Fig. 3.8). Observed distances travelled between feeding stations can be up to 10 fold greater in distinct patchy communities as compared to communities with dense, continuous swards, 20–25 steps verses 2–3 steps between feeding stations, respectively.

Several studies have focused on the influence of plant communities on ingestion rates (Alden and Whittaker 1970; Chacon and Stobbs 1976; Arnold and Dudzinski 1978; Arnold 1981; Forbes et al. 1985). Most of these studies have analyzed foraging behavior under heavy grazing pressures and rapidly declining monospecific stands of forage. Basically, these studies have found bite size to diminish as herbage supply in a plant community increases while biting rate declines.

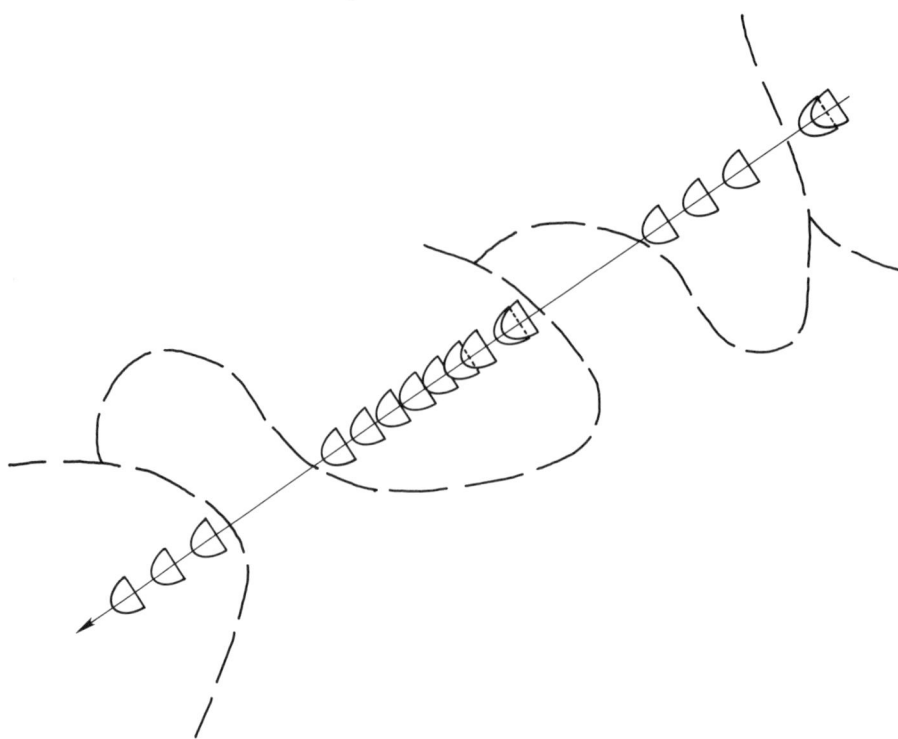

Figure 3.8. Pattern of feeding stations along a directional grazing path as influenced by patch environments which vary in herbage mass (g/m^2) and potential harvest rates (g/min).

PLANT CHOICE

Once an animal establishes a grazing location, its experience with available forage is utilized in a plant species-to-species appraisal and selection process. This process is specific to the animal species (see Chapter 2). Herbivores, as noted earlier, exhibit an evolutionary adaption predisposition to feed on plant species from one or more of their primary food groups, grasses, forbs, and browse (Provenza and Balph 1987a, 1987b). Therefore, the grazing value of a plant depends on the animal species in question (Demment and Van Soest 1981; Hanley 1982; Hanley and Hanley 1982; Owen-Smith and Cooper 1987).

At this point it is essential that the palatability of a plant and the preference for that plant be differentiated (Heady 1964). Palatability refers to those factors inherent to a plant species that elicit a selective response by the animal. Preference involves proportional choice of one plant species from among two or more species and is essentially behavioral. The preference status of a particular plant species is largely dependent upon its inherent abundance, its morpho/phenological characteristics, the array of species on offer, and the species of animal in question. Preference constantly changes as abiotic factors (i.e., season and weather conditions) alter the nature of the plant community. Some species are selected only under specific conditions. Therefore, broad generalizations about species selection and preference should be tempered by the understanding that animal selectivity is a dynamic, situation-specific process. However, recent studies (Colebrook et al. 1987) have indicated that preference can be quantified for an animal species as well as selection order predicted based on the relative rank order of absolute preference values. Implicit in these findings is the concept that specialized or focused grazing on some plant species may relate largely to their relative preference ranking at the time of active growth.

Plants have been generally classified into five general selectivity categories (Table 3.5) which follow the functional categories outlined in Chapter 2. Plant species selected in greater quantities, as a percent of diet, than found in the landscape (percent composition), are referred as preferred or favored species (Fig. 3.9). Such plant species do not generally dominate the diet unless they dominate the community. Instead, preferred species enhance the diet nutritionally, resulting in better-than-normal animal performance. These species have high handling time for the animal but high nutrient concentration and/or are low in floristic composition.

The more abundant species are generally consumed in proportion to their availability and are referred to as proportional or desirable species. When present in high percentages they dominate the diet and usually provide the basis for estimates of

Table 3.5. Preference classification of forage and associated function.

Selectivity Class	Diet:Availability Ratio*	Nutritional Role	Functional Role
Preferred	> +3.5	Performance	Diet enhancer
Proportional	−3.5 to +3.5	Maintenance	Bulk
Forced	< −3.5	Subsistence	Survival
Detrimental	−9	Toxic	Death
Non-consumable	0	+/− Composition	Reduced carrying capacity

*Based on the formula:

$$\text{D:A Ratio} = \frac{\%\text{ diet} - \%\text{ available in field}}{\%\text{ diet} + \%\text{ available in field}} \times 10$$

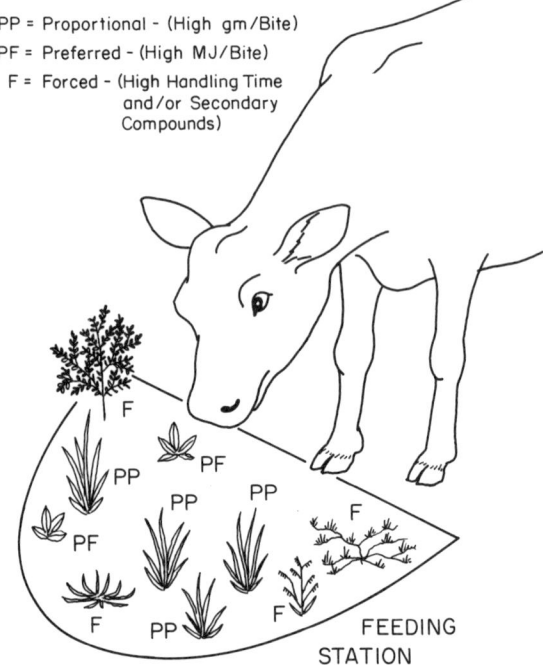

Figure 3.9. Animals are faced with a choice of plants at a feeding station which offer different potential instantaneous intake rates, nutrient density, and secondary compounds.

grazing capacity. These species are not generally as high in nutrients as the preferred species but afford the animal the opportunity to maximize instantaneous intake rates (gm/bite).

Species not readily consumed by animals generally make up a lesser percentage of the diet than the percentage available in the vegetation. Consumption of this undesirable, forced, rejected, or avoided selection group is highly condition specific, i.e., incidental grazing when other preferred and proportional species are abundant, seasonal selection of specific plant parts (mast, pods, fruits, flowers, etc.), or major dietary components when the preferred and desirable species are limited. Species in the lesser selection group allow the animal to survive in a subsistence situation. Incidental consumption is believed to be a response to animal sampling of the environment as conditions change.

Recent studies correlating animal selection ratios (%diet:%available) and a given plant species' inherent abundance in a sward have revealed four types of relationships (Fig. 3.10). Particular plant species are preferred regardless of abundance and the presence of associated species; the preferred species are generally higher successional species. Secondly, there are those species which are consumed proportional to availability and consumption is highly correlated to their inherent abundance. A third group in Figure 3.10, variable, transcends all selection categories, their consumption of which changes from avoidance to preference as herbage mass declines. These plant species are referred to as variable or secondary preference species, and generally exhibit morphological constraints to consumption by animals. Finally, there is the last group or avoided species which is selected at levels below their availability. Selection ratios of avoided species are poorly correlated with their inherent abundance. Those species generally contain undesirable nutritional attributes (see functional group discussion in Chapter 2).

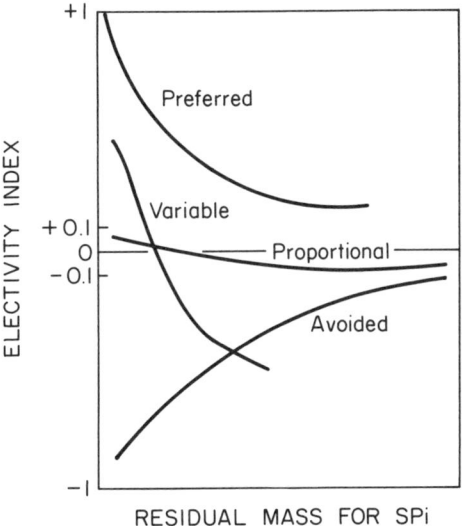

Figure 3.10. Classification of plant species based on their relationship to herbage mass (hg/ha) and associated electivity index (EI):

$$EI = \frac{\%\ diet\ +\ \%\ landscape}{\%\ diet\ -\ \%\ landscape}$$

Nonconsumable species are generally not found in the diet of the animal except in extraordinary situations. Generally, only specific adverse conditions result in any consumption. Exceptions might include pods or fruiting bodies, e.g., mesquite beans or prickly pear cactus fruit. These species generally affect the animal only indirectly by reducing the overall grazing capacity of the range, but can have a positive effect on nutrition. This is particularly true of shrubs when cattle are the primary herbivore, as shrubs create microclimates for certain species which are nutritionally richer than associated species or maintain green material longer into dry or cold periods of the year.

Finally, there are the detrimental or toxic species. When most of the favored species are reduced in the landscape, toxic species express themselves in the diet devastatingly. Cyclic poisonous plant problems in arid regions is a testament to this problem.

If we assume that most ungulates are energy maximizers and feed optimally, we can predict that plant species in the preferred primary food group of an animal which provide high instantaneous intake rates of nutrients without the negative effects of secondary compounds (e.g., phenolic acids) should receive the most selective pressure by the animal. In another words, plant species offering the highest bulk density of unmixed green foliage with the highest nutrient concentration and lowest content of secondary compounds has the greatest probability of being grazed. This conclusion must of course be understood and acted upon in terms of the inherent food group preference and nutritional requirements of the animal species and associated plants in the landscape. For instance, cattle have higher dry matter requirements, lower nutrient requirements, less precise prehensile organs, and larger rumen volume:body volume ratios than goats (Demment and Van Soest 1981). Cattle's buccal/oral cavity is much larger, leading them to form a larger bolus prior to swallowing. Therefore, grasses with their high canopy bulk density of green material are more profitable than a small forb with higher nutrient concentrations but smaller size. Goats on the other hand, can "afford" to select smaller plants that are less profitable for cattle; to a goat a browse leaf presents a proportionally larger bite size than to a cow. Also, the ratio of nutrient concentration per bite:required nutrient concentration must be maintained at higher levels for a goat than for a cow.

Work by Cooper and Owen-Smith (1986) on goats and several African ungulates indicates that plants bearing spines reduce the potential rate of harvest by animals. Shrubs from which large bites could be taken were preferred in this study if secondary compounds were not high.

It appears that ungulates focus their grazing activity on a few highly profitable species when overall forage quality of the landscape is high, with the consequences that search time increases, biting rate declines, and bite size increases (Coleman et al. 1989). When observing grazing strategies through time, one could hypothesize that animals would be attracted to plant communities during rapid growth periods based on the abundance of highly profitable species. As phenologies of plant communities become mixed, animals should reduce species selectivity and focus their attention on communities which offer the greatest harvest rates of green foliage regardless of species. Once herbage is dormant, the animal's only option is to graze on sites with more abundant plant material regardless of greenness.

Plant morphology also influences the probability of being grazed. If grasses elicit a selective response early in their growth cycle, selective pressures increase as relative abundance or phenologies change (Stuth et al. 1987). Therefore, communities with a high proportion of forage utilized in the early growing season have a higher probability of being grazed when environmental conditions are conducive to plant growth.

What are the morphological attributes of primary food groups which influence the grazing decision? In grasses, it appears to be physical presentation of green leaf blade relative to its pattern of senescence and culm development (Fig. 3.11). Grasses with rapid culm development (determinant growth) and strong, midrib leaf structure are selected less frequently if allowed to develop long-standing, senesced leaf material. Degree of sheath development and angle of growth by tillers influence the height and position of leaf blade material relative to the soil surface, so cattle have a much harder time than sheep in selecting short or decumbant species.

Forbs are characterized by two temporal presentations, ephemeral-annual and perennial. Ephemeral-annual forbs grow rapidly and complete their life cycle quickly. Therefore, they present a unique problem for ungulates. They possess high value for short periods in the animal's annual production cycle. The concentration of nutrients in most forbs exceed, the nutritional requirements of ungulates. So while they are a preferred group, their distribution on the landscape and the standing crop available in various communities, along with the bite size they afford, affect animal foraging tactics from one landscape to another.

Perennial forbs tend to allocate more resources to structural components, thereby creating greater differentials in quality between plant parts than is the case with annual forbs. Moreover, they do not generally accumulate previous years' growth as does shrubby browse. Because they are present through much of the grazing season, they are particularly vulnerable to over-use by forb-preferring ungulates. This over-use reduces the relative acceptability between plant parts, causing the plant to become more attractive to the animal. That is, handling time is reduced and bite size/quality is increased.

Browse presents itself in many forms: deciduous or evergreen, spineless or spiney, single leaves or compound leaves, short or tall, single stemmed or multi-stemmed, etc. Selective pressures on this food group are again dependent on the

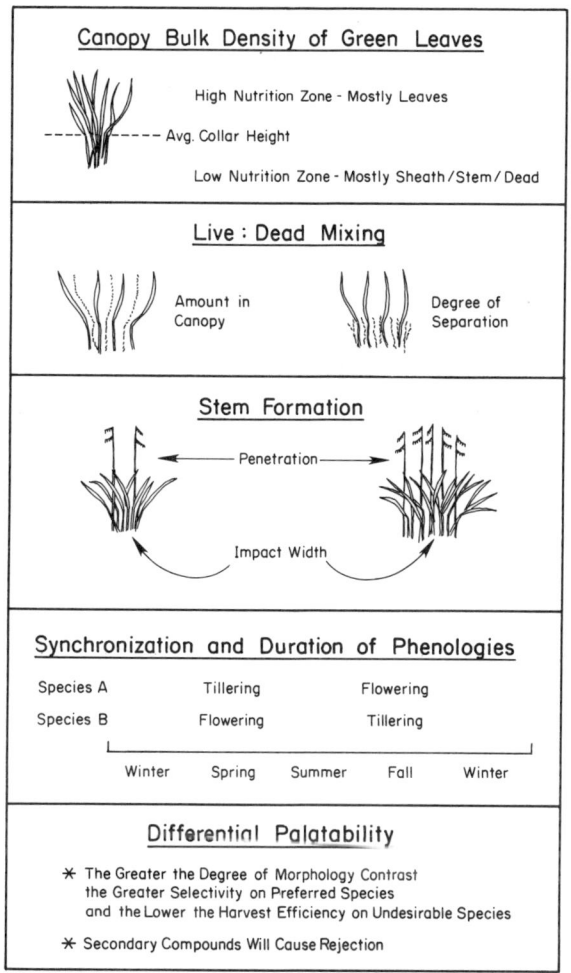

Figure 3.11. Structural characteristics of grasses influencing grazing decision.

associated animal species community. Browse-preferring ungulates (concentrate selectors and intermediate feeders, Van Soest 1982) have adapted prehensile and digestive organs to a level where height, spineyness, and secondary compounds are the principle plant characteristics affecting selective pressure on browse species (Cooper and Owen-Smith 1986). Generally secondary compounds play a major role in suppressing selective pressures on evergreen browse species. Spineyness, leaf size, and, to a lesser degree, secondary compounds, are important morphological/physiological attributes of deciduous browse species which influence the selection response. Again, the relative importance depends on the attributes of the particular animal species as discussed in Chapter 2.

Whatever the situation, animals are continually making choices among plants at the feeding station level. Choices are influenced by plants in the animal's view, the animal's short-term memory of plants in previous feeding stations, and the frequency with which positive reinforcement of that choice has been made while actively grazing. The kind of plant chosen is largely related to the kind of animal, the relative abundance of alternative food sources, and the complexity of the landscape relative to the water and thermal needs of the animal.

DIETARY PLASTICITY AND ITS IMPLICATIONS FOR INTERSPECIFIC COMPETITION

Most of the previous discussion focused on the plant/animal relationship with little regard for animal/animal relationships. There is an abundance of site-specific information on dietary overlap between animal species. However, the intent of this section is to focus on food group plasticity of the major groups of domestic grazing livestock.

Cattle have often been referred to as generalists, eating a wide variety of foods and plant parts. However, graze-out studies have indicated that they prefer the grass food group unless the availability of grasses is so marginal as to severely limit daily intake requirements (Launchbaugh et al. 1990). Analysis has shown that cattle maintain a grass-dominated diet over a wide array of herbage standing crops, switching to browse only after severe restriction in dry matter intake (Fig. 3.12) or to forbs when temporal flushes of desirable species occur.

Goats, on the other hand, show a high preference for browse regardless of availability and a negative-to-proportional response to grasses, i.e., they increase the amount of grass in their diets relative to its availability as the composition of browse and forbs decline (Fig 3.13). Like cattle, goats consume low quantities of forbs unless large flushes of highly desirable species emerge. Recent studies of Cashmere goats grazing cool-season, grass-legume pastures indicate that they consume mostly grasses, with legumes comprising less than 10% of their diet. No alternative browse was available to the animals in this study.

Although sheep have a rumen:body volume ratio similar to cattle, their principal dietary preference is the forb food group and, to a lesser degree, grasses (Hanley 1982; Demment 1982; Demment and Van Soest 1981). Browse is utilized more readily by sheep than cattle but generally does not comprise a major portion of their diet unless grass and forbs are in limited supply.

Increasing grazing pressure by one animal species can force another species into their less preferred food group resulting in reduced performance, decreased harvest efficiencies, or both. Cattle, because of their high demand for grass, regulate the level

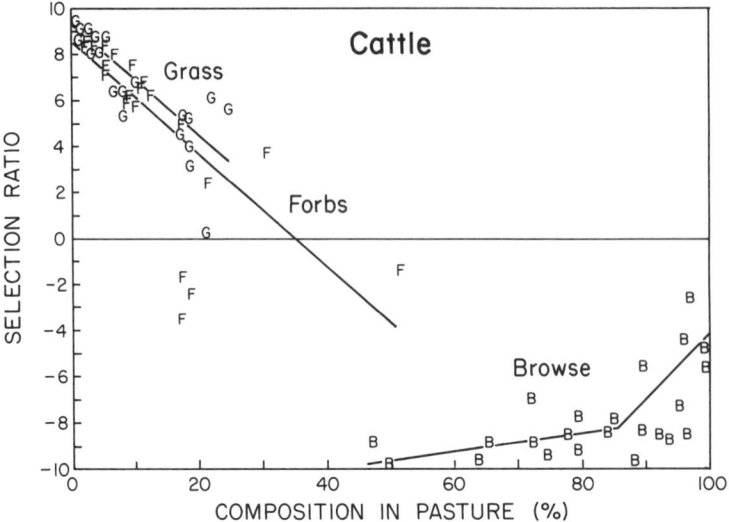

Figure 3.12. Preference status of the primary food groups of cattle as a function of their selection ratios in a grazed landscape.

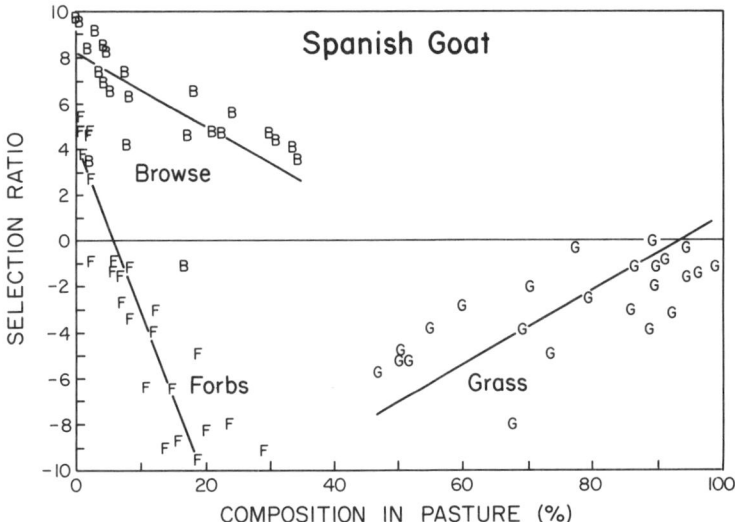

Figure 3.13. Preference status of the primary food group of goats as a function of their selection ratios in a grazed landscape.

of grasses consumed by the smaller goat or sheep (Rector and Huston 1982). If desirable browse species are maintained and stocking ratios properly balanced, little competition occurs between cattle and goats, resulting in an increased potential carrying capacity of a landscape (see Chapter 7). The same can be said for sheep and cattle or sheep and goats if an abundant and stable source of forbs is provided to the animals. However, since forbs comprise a low but constant portion of cattle and goats diets and are actively consumed by sheep, it would appear that the degree of competition for food between these species is determined by the availability and desirability of the forb component of a landscape. As forb supply declines under combination grazing by cattle, sheep, and goats, the resulting impact depends on the time pastures are jointly shared by all species, the amount of alternative browse available, and the level of utilization of the grasses.

HIERARCHICAL FEEDBACK

The concepts covered in this chapter have focused on the grazing hierarchy as seen by the animal from the landscape level to the selection of individual plants. But the direct impact of grazing animals on physical properties of soils and growth of plants—together with the indirect impacts on soil aggregate stability, plant food reserves, and effective precipitation—also combine to markedly affect not only plant competition and community composition, but also subsequent spatial configuration of forage resources. These small shifts and changes in plant community composition and the spatial relationships of these plants lead to long-term changes in the landscape. These changes in turn alter the grazing animal's grazing behavior, a classic feedback loop as explained by systems analysis developers. It must be noted that while the relationships discussed in this section are short-term in nature with only gradually shifting foraging tactics from one year to the next, a long-term reconfiguration of landscapes occurs which alters grazing behaviors over the same time frame.

Patch grazing (i.e., small areas of intense defoliation) is often the most obvious early sign of landscape reconfiguration. The rapid growth of new green foliage in a patch previously grazed clean of competing vegetation offers the animal a highly desirable food source which in turn facilitates redefoliation in a relatively large, distinct area. This kind of intense patch defoliation and redefoliation is in marked contrast to the more typical selective grazing of individual plants in a mixed plant community. Such frequent defoliation in turn alters the hydrologic condition of the patch/community, ultimately to the point where plant species composition shifts and forage production declines. Furthermore, repetitive defoliation results in the expansion of the size of the patch and in some cases leads to the loss of most plant material and consequent development of eroded areas.

Once soil loss accelerates to the point of erosion, a permanent reconfiguration of the grazing area occurs. If, despite these landscape and terrain reconfigurations, the essential physiological needs of the animal can in some measure be met, the outcome is typically a reduction in animal population levels. However, in many cases of landscape terrain degradation the needs of the animal cannot be met, resulting in population relocation or extinction.

Landscapes are altered not only by grazing animals but often by other agents, including man as well. Such alterations quite commonly lead to change in animal use not only in terms of intensity of use but that of occupancy patterns. For example, logging of transitional areas between summer and winter range alter season of use by elk (Sheehy 1988). Such temporal shifts in landscapes typically result in either intensification of interspecific competition for resources or greater niche separation depending on the behavioral flexibility of the animal species involved.

Landscape configurations are also altered by natural events. Annual variation in precipitation patterns regularly results in major changes in landscape configurations and subsequent alterations in animal grazing patterns. For example, sporadic rainfall events in regions marked by soils with varying moisture-holding capacities or slope position create areas of lush green growth adjacent to communities with senesced forage. Depending upon intensity and location of precipitation events, these highly contrasting landscapes can shift throughout the year. Drought can accentuate this dynamic pattern and causes animals to range further from normal grazing areas, thereby exposing forage in distant sites to grazing pressures during periods stressful to the impacted plants. If drought persists, areas near water can be denuded, resulting in an ever-expanding ring of degraded forage resources around essential watering points.

In all of these various situations, hierarchical feedback shapes available plant communities and subsequent animal behavior. The processes are complex and interdependent. If management chooses to alter vegetation or facilities, landscapes are inevitably altered so the animals must readjust their grazing tactics. Care and consideration must be given to ascertain potentially destabilizing consequences in forage composition and relationship before changes in habitat and grazing management programs are implemented.

CONCLUSIONS

The grazing animal possesses a unique prehensile morphology to gather and processes food in a digestive system adapted to the primary food groups ingested (see Chapter 2). The grazing process used to gather food can best be described as a hierarchical system of diet selection interacting with the animal's physiological needs (water, thermal, balance, food, etc.), resulting in a unique pattern of use across a given landscape. The configuration of forage resources, water locations, thermal foci, and terrain constraints interact with the animal's hierarchy of needs to determine the overall impact of animal populations. The reaction of forage relative to animal grazing pressure provides both a short-term feedback suggesting the need to alter grazing tactics, and a long-term feedback in terms of the successional trends of the plant community. This interactive hierarchical system of plant-animal-soil interactions reflects management inputs relative to manipulation of animal populations and the vegetation matrix (Fig. 3.14). As pointed out in this chapter, foraging behavior involves tactics animals employ within this manipulated hierarchical system. The reaction of plants to these grazing tactics as both individuals and in aggregate is the subject of chapters 4 and 5.

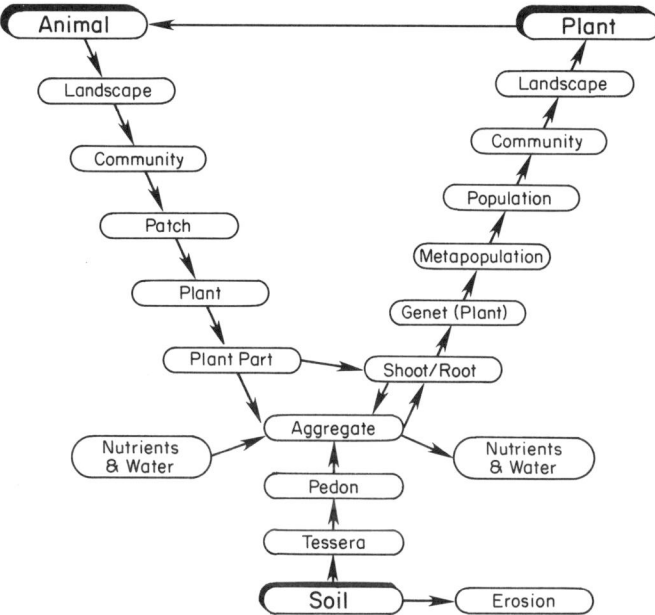

Figure 3.14. Hierarchical presentation of the components of the plant-animal-soil interface (adapted from Coleman et al. 1989).

CHAPTER 4

DEVELOPMENTAL MORPHOLOGY AND PHYSIOLOGY OF GRASSES

D. D. Briske

INTRODUCTION

The vegetation managed in livestock and wildlife production systems is produced by a series of developmental and physiological processes within individual plants. The series of structural changes displayed by organisms from inception to maturity—including cellular division, differentiation, and growth—are collectively referenced as **developmental morphology** (Esau 1960). The developmental morphology of plants defines their architectural organization, influences their palatability and accessibility to herbivores, and affects their ability to grow following defoliation. Physiological processes establish the capacity for solar energy capture and product synthesis necessary to sustain structural development.

The predominant impact of grazing on plant growth is a reduction in photosynthetic capacity associated with a decrease in leaf area. Species cope with grazing by minimizing the probability of being grazed and/or rapidly replacing leaf area removed by herbivores. Morphological attributes and biochemical compounds influence the probability and severity of grazing by affecting tissue accessibility and palatability. The capacity for rapid leaf replacement is conferred by physiological processes and meristem availability.

The inherent morphological and physiological attributes of individual species influence the structure and function of populations and communities by determining the extent of competitive interactions among plant species. Grazing alters competitive interactions among species by removing various amounts of leaf area and establishing the potential for differential growth rates following similar defoliation severities. Species composition is altered when a particular intensity, frequency, and/or seasonality of grazing shifts the competitive advantage from one group of species to another. Species composition changes subsequently influence livestock production and managerial strategies by affecting the quantity, quality, and seasonality of plant production. Consequently, the design and evaluation of grazing management strategies must be based in part on the developmental morphology and physiological function of the dominant plant species to conserve rangeland resources and maintain production stability.

MORPHOLOGY

Architectural Organization

The developmental morphology of grasses is remarkably similar among species with only minor morphological variations separating growth forms. Individual

phytomers, which consist of a blade, sheath, node, internode, and axillary bud, form the basic unit of growth (Hyder 1972; Briske 1986; Fig. 4.1). The size, number, and spatial arrangement of phytomers determines the architectural organization of individual tillers. A tiller consists of a series of phytomers successively differentiated from an apical meristem with the initial phytomer located nearest the soil surface. Individual grass plants are composed of an assemblage of tillers originating from the axillary buds of previous tiller generations.

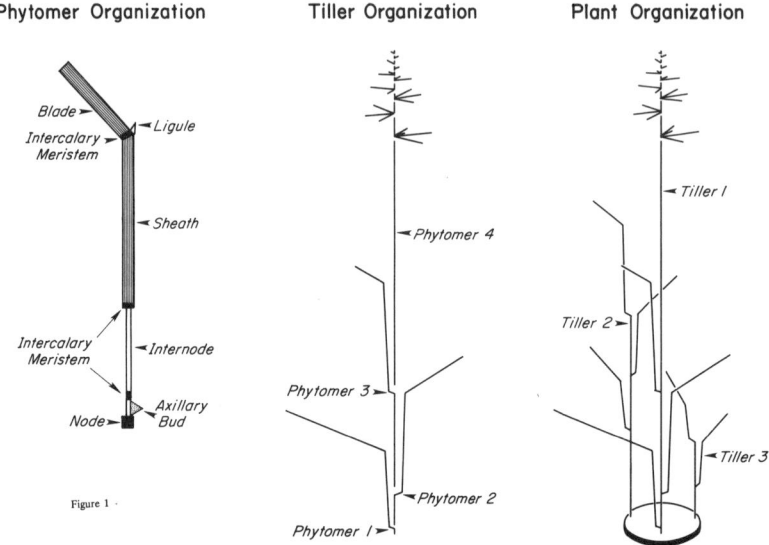

Figure 4.1. The developmental morphology of grasses originates from the successive differentiation of phytomers from the apical meristem of individual tillers. Individual plants are composed of an assemblage of tillers originating from axillary buds of previous tiller generations (adapted from Etter 1951).

Phytomers. Phytomers are differentiated from the **apical meristem** (growing point or shoot apex) by rapid cell division in the two outer layers of the apical meristem, the dermatogen, and hypodermis. The rudimentary phytomer expands into a crescent-shaped structure and eventually extends beyond the height of the apical meristem (Sharman 1945; Etter 1951; Fig. 4.2). Soon after the leaf primordium (i.e., blade and sheath of phytomer) has encircled the apical meristem, cells of the third innermost layer of the apical meristem, the subhypodermis, begin to divide. Cellular division in this layer forms an axillary bud within the axil of the subtending leaf primordium on the opposite side of the apical meristem. Phytomer differentiation continues as long as the apical meristem remains in a vegetative state, giving rise to a series of leaf primordia at progressive stages of development.

Initially, the entire leaf primordium is meristematic, but cellular division is quickly restricted to **intercalary meristems** (meristematic tissue separated from the apical meristem by a region of non-meristematic tissue) (Dahl and Hyder 1977). Intercalary meristems are located in narrow zones at the base of the blade, sheath, and internode (Sharman 1945; Etter 1951; Langer 1972). Intercalary meristem activity ceases within the blade when the ligule is formed and within the sheath when the ligule becomes exposed. Consequently, the blade ceases elongation prior to the sheath, while internode elongation is dependent upon species and phenology. The basal location of the intercalary meristem within the blade and sheath explain why leaf elongation can occur following defoliation without replacement of the leaf tip (Hyder 1974).

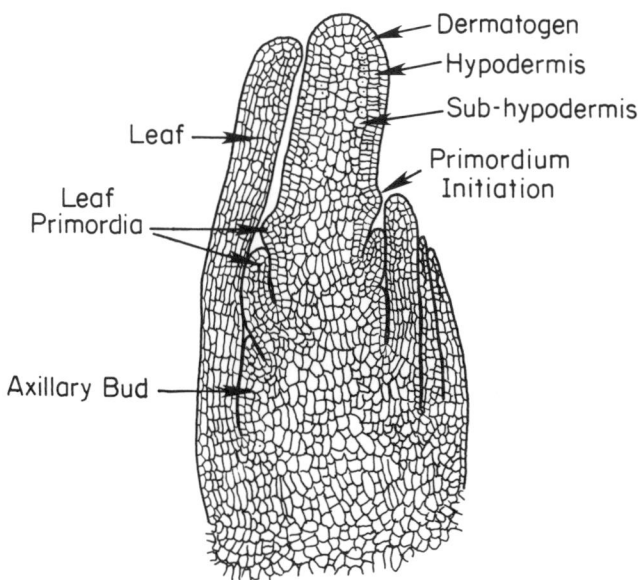

Figure 4.2. Leaf primordia (blade and sheath of individual phytomers) are differentiated from the upper portion of the apical meristem to form a series of primordia at successive stages of development (from Langer 1972).

Floral induction marks the transition of the apical meristem from a vegetative to a reproductive status (Sharman 1945; Etter 1951; Langer 1972). Floral induction occurs in response to a photoperiodic stimulus (i.e., day length) following a sufficient juvenile growth period. At the time of floral induction, both leaf primordia and axillary buds are rapidly differentiated, producing a double ridge appearance on the apical meristem. Spiklet primordia differentiate from the axillary buds while further development of leaf primordia is suppressed halting additional vegetative development. Vegetative growth can only occur from immature intercalary meristems of existing phytomers or from previously differentiated axillary buds in reproductive tillers.

Tillers. The accumulation of successive phytomers differentiated from a single apical meristem defines the **tiller** (Etter 1951; Hyder 1972; Briske 1986). Tillers are initiated from the axillary buds of ontogenetically, older parental tillers (Fig. 4.3). Following a juvenile period of development, tillers are potentially capable of initiating additional tillers from **axillary buds** differentiated with each phytomer. The largest, but ontogenetically youngest axillary buds develop to form tillers in crested and bluebunch wheatgrass (Mueller and Richards 1986). These observations support the contention that axillary buds may possess a relatively brief longevity following their development (Hyder 1974; Dahl and Hyder 1977). However, no evidence of bud senescence was observed in either of the wheatgrass species even though bud growth was arrested at about the time the associated leaf within the phytomer matured (Mueller and Richards 1986).

Morphological variation of individual tillers is largely a consequence of the number and length of phytomers comprising the tiller. Variation in tiller architecture among tall, mid- and short grasses does not originate from a major deviation in the pattern of developmental morphology, but rather results from a variable number and/or size of phytomers determining cumulative tiller height. Internode elongation increases phytomer size and is most frequently associated with reproductive tiller

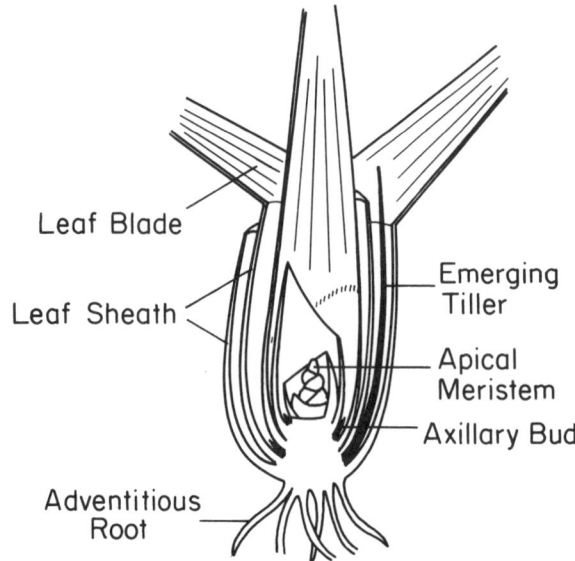

Figure 4.3. Tiller initiation from axillary buds in the crown of a grass plant. Axillary buds contain single rudimentary apical meristems capable of differentiating a complete tiller (from Jewiss 1972).

development, but may also occur in nonreproductive tillers in a small number of species. In **culmed vegetative tillers**, the apical meristem is elevated above the soil surface by internode elongation while in a vegetative condition. **Culm elongation** originates from the activity of intercalary meristems located at the base of the several uppermost internodes. In reproductive tillers, the inflorescence and several uppermost leaves are elevated above the soil surface presumably to facilitate wind pollination. The developmental morphology of reproductive tillers is similar to vegetative tillers prior to floral induction (Hyder 1974).

Plants. The spatial arrangement of tillers within the grass plant, in addition to morphological variation within individual tillers, is a major determinant of architectural variation within the grass growth form (e.g., bunchgrasses versus sodgrasses). Spatial arrangement of tillers within the plant is dependent upon the pattern of tiller development. **Intravaginal tiller development** within the subtending leaf sheath results in a compact spatial arrangement of tillers defining the **bunchgrass** (caespitose or tussock) growth form (White 1979; Briske 1986; Fig. 4.4). Contrastingly, **extravaginal tiller development** proceeds laterally through the subtending leaf sheath, contributing to greater inter-tiller distance and tiller angles within the plant. Extravaginal tiller development is a prerequisite to the formation of the **sodgrass** (creeping or spreading) growth form, which may be further accentuated by the development of rhizomes and stolons. These modified, horizontal tillers further increase inter-tiller distance within the plant depending on whether they are determinate or indeterminate in growth. The apical meristems of **determinate rhizomes** eventually emerge from the soil to form a tiller while the apical meristems of **indeterminate rhizomes** continue to grow parallel to the soil surface with individual tillers potentially developing from axillary buds located at the nodes. Stolon growth generally displays the indeterminate growth pattern (Hyder 1974).

The mechanisms determining whether a tiller, rhizome, or stolon develops from an axillary bud are not clearly understood. These shoot types are initiated from a finite

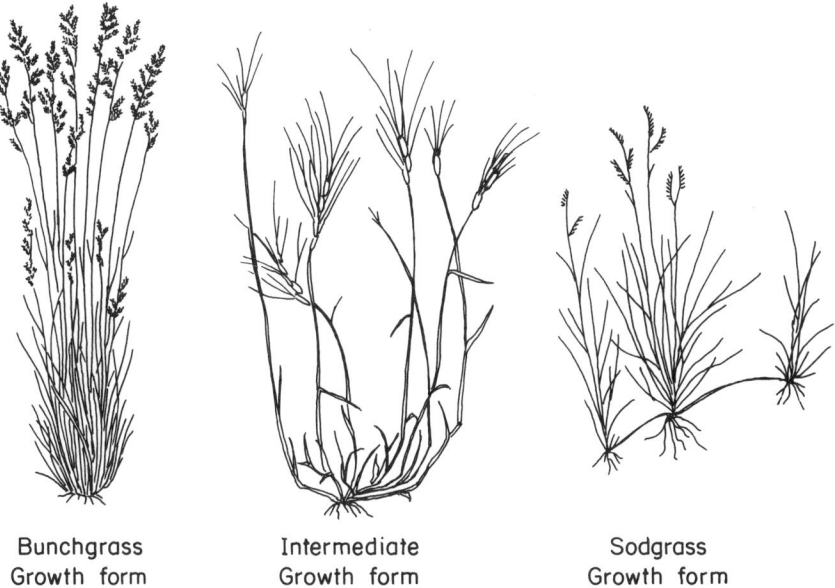

Figure 4.4. Variation within the grass growth form originates from the pattern of tiller emergence expressed by various species groups. The bunchgrass growth form originates from intravaginal tiller development, while extravaginal tiller development contributes to a more diffuse tiller arrangement and serves as a prerequisite to the sodgrass growth form. Stolons or rhizomes further increase inter-tiller distances within plants.

number of axillary buds which may partially explain the observed seasonality of tiller and rhizome recruitment (i.e., rapid tiller recruitment may potentially limit rhizome recruitment). High nitrogen availability, high temperatures, and short photoperiods promote tiller development to a greater extent than rhizome development in quackgrass (McIntyre 1967). Bermudagrass stolons exposed to a high ratio of red:far-red radiation display a more upward curvature, increased leaf and internode elongation, and lower carbohydrate concentrations than stolons grown in darkness or in radiation with a low red:far-red ratio (Willemoes et al. 1987). These data suggest that **phytochrome** (a proteinaceous pigment sensitive to specific wavelengths) may regulate the differentiation of stolons, rhizomes, or tillers by affecting the distribution of photosynthetic products within the plant. Radiation quality has also been implicated in the regulation of tiller recruitment in several grasses (Casal et al. 1986; Kasperbauer and Karlen 1986).

Root Systems. Grasses produce two distinct root systems during their developmental history. The initial root system, referred to as the **seminal root system**, develops rapidly from the embryo upon seed germination (Langer 1972; Hyder 1974). Although the seminal root system is essential to the initial development of grass seedlings, it is relatively short-lived, generally surviving no longer than the growing season in which it originated. The **adventitious** or **nodal root system** consists of whorls of roots originating from nodes along the base of the tiller, forming the permanent root system of the grass plant. Adventitious root longevities vary from 1 to 4 years among species (Weaver and Zink 1946; Troughton 1981).

Adventitious roots differ from seminal roots in both diameter and mass per unit length. The large diameter of adventitious roots is associated with a greater cross-sectional area of xylem to enhance water transport to the shoot system (Wilson et al.

1976). The large root mass per unit surface area may explain why adventitious roots are not initiated until seedlings produce sufficient leaf area and photosynthetic capacity to support their development 1 to 3 weeks following seedling emergence (Wilson and Briske 1979). In the case of vegetative tiller development a seminal root system is not initiated. Juvenile tillers are supported by parental tillers until the adventitious root system develops (Welker et al. 1987). Adventitious root development is initiated at approximately the third to fourth leaf stage in little bluestem (Carman and Briske 1982).

Tiller Demography

Tillering. The perenniality and sustained productivity of grasses are conferred by the successive production of relatively short-lived tillers (Langer 1963; Tomlinson 1974). Successive tiller recruitment produces a series of connected tiller generations referred to as **tiller hierarchies** or families (Langer 1963). The number of generations within a hierarchy is determined by the rate of tiller recruitment and tiller longevity as influenced by genetic and environmental constraints. The number of tillers per hierarchy and number of hierarchies per plant define the size and architectural configuration of the plant. With increasing plant size and age, these tiller hierarchies become separated as the initial tiller generations die and decompose (Gatsuk et al. 1980). Each plant fragment is capable of survival and may continue tiller development as previously described. The **hollow crown phenomenon** characteristic of many perennial grasses is very likely a natural consequence of the developmental morphology of grasses and not a symptom of plant stress (Gatsuk et al. 1980). The disproportionate initiation of tillers on the plant periphery eventually reduces the density of tillers and axillary buds necessary for continuation of tiller recruitment within the plant interior (Butler and Briske 1988, Olson and Richards 1988a).

The number of live tillers per plant or per unit area is determined by the rate and seasonality of tiller recruitment in relation to tiller longevity. Changes in tiller density occur when recruitment lags behind or exceeds mortality (Fig. 4.5). The density of live tillers defines the potential for biomass production, within the constraints of resource availability, by determining the number of intercalary meristems, apical meristems, and axillary buds available for growth (Olson and Richards 1988b). The continued existence of perennial grasses is also dependent upon the successive development of axillary buds capable of initiating subsequent tiller generations and perpetuating the plant. If tiller recruitment was suspended for an interval equivalent to the maximum longevity of the most recently developed tillers, the plant would lose all meristematic potential and cease to exist. Growth could potentially resume from axillary buds, but their longevity in relation to that of parental tiller is not known.

Tiller Recruitment. Tiller recruitment in temperate, perennial grasses is most prevalent in the spring and fall, yielding two tiller generations annually (Langer 1956; Butler and Briske 1988). However, only a single tiller generation is initiated in the fall in ungrazed crested and bluebunch wheatgrass populations (Mueller and Richards 1986). The relatively short growing season and limited summer precipitation in the Intermountain West may account for this discrepancy. The seasonality of tiller mortality is also highly correlated with periods of maximum tiller recruitment and reproductive tiller development.

Reproductive tiller development terminates the development of leaf primordia and is followed by tiller death as the existing phytomers senesce (Noble et al. 1979; Fig. 4.6). Vegetative tiller mortality, coincident with flowering of parental tillers within the plant, is assumed to be a consequence of the shading of smaller vegetative tillers

Developmental Morphology and Physiology of Grasses 91

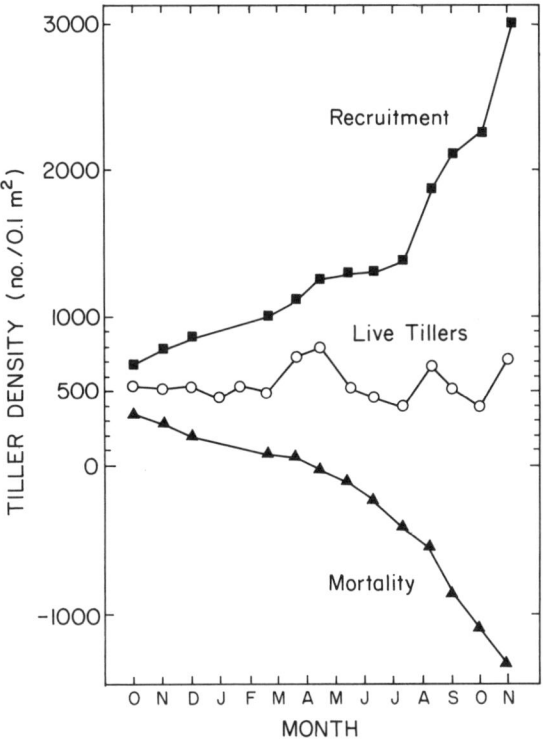

Figure 4.5. Live tiller density as a consequence of tiller recruitment and mortality within a population. Tiller density increases when recruitment exceeds mortality and decreases when recruitment lags behind mortality (from White 1980).

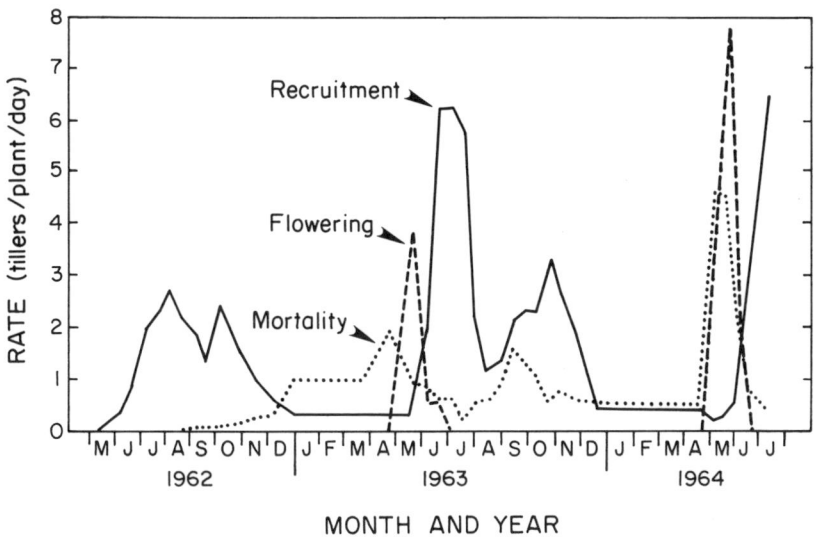

Figure 4.6. Seasonality of tiller recruitment, flowering, and mortality within a population of tall fescue. Synchronization among all three processes is indicative of a cause-effect relationship (adapted from Robson 1968).

and a reduction in resource import as culm and inflorescence development increase resource demand within reproductive tillers (Ong 1978). It is the smallest, but not necessarily the youngest, tillers which experience mortality when the plant is stressed. The relationship between recruitment and mortality is generally described as a density-dependent process with either mortality increasing resource availability for the recruitment of new tillers, or recruitment contributing to the mortality of existing tillers by resource depletion (Robson 1968).

Tiller Longevity. Tiller longevity in temperate perennial grasses is approximately 1 year and does not exceed 2 years (Langer 1956; Robson 1968; Butler and Briske 1988; Briske and Butler 1989). Longevity is directly influenced by season of tiller recruitment and phenological development. Tillers recruited early in the growing season will have the greatest probability of becoming reproductive and terminating growth at the end of the season in which they were recruited. Tillers initiated later in the season apparently do not surpass the juvenile growth requirements necessary to respond to the long-day photoperiodic stimulus. Consequently, these tillers may overwinter in a vegetative stage and resume growth the subsequent spring. Tiller growth, including dry weight, leaf number, and seed yield, is greatest in tillers initiated in the latter portion of the previous season or early in the season of reproductive development because they experience a longer period for growth and development (Langer 1956).

Leaf Demography

Leaf longevity also displays seasonal patterns of recruitment and mortality (Vine 1983; Chapman et al. 1983, 1984). Leaves initiated when growing conditions are most favorable, spring and early summer in temperate environments, have shorter longevities than those initiated during periods of less favorable environmental conditions. Leaves of perennial ryegrass and browntop exhibited mean longevities of 60–70 days when initiated in the spring and summer compared to 70–105 days when initiated in the fall and winter (Chapman et al. 1984). Leaf longevities of less than 90 days during favorable growing conditions indicate that grazing must closely follow leaf initiation to optimize the harvest of live leaves. The environmental conditions experienced by emerging leaves may program their subsequent development to a larger extent than the conditions during growth or maturity.

Synchronous leaf initiation and senescence maintains a relatively constant leaf number per tiller throughout much of their developmental history. Generally, a tiller possesses an emerging leaf, immature leaf, mature leaf, and senescing leaf (Anslow 1966; Chapman et al. 1984). The net difference between leaf initiation and senescence represents the number of live leaves per tiller. Leaf demography, by determining the amount of live leaf area per tiller, influences both the potential photosynthetic capacity of the tiller and the amount of leaf biomass available for consumption by herbivores. Leaf and tiller demography collectively determine the rate of biomass turnover (i.e., production versus senescence) within the community.

Benefits of Vegetative Growth

Grasses exhibit **vegetative growth** or reproduction by the successive recruitment of tillers from previous generations. Each tiller establishes a shoot and root system (adventitious) to acquire resources from the environment. Vegetative growth may confer several ecological advantages originating from resource allocation between and among tillers within individual plants (Pitelka and Ashman 1985). The capacity for continuous tiller replacement and site occupation based upon parental support of juvenile tillers is perhaps the most significant ecological benefit. Resource allocation from parental to juvenile tillers confers a greater likelihood of establishment and survival in comparison with seedlings which must become established from energy and nutrient reserves available within the endosperm (Tripathi and Harper 1973).

Survival and growth of stressed tillers are also enhanced by resource import from associated non-stressed tillers within a plant (Gifford and Marshall 1973; Ong and Marshall 1979; Welker et al. 1987). Root growth, as estimated by both total length and numbers of roots, of recently recruited juvenile tillers progressively decreased as the juvenile tiller, the parental tiller, or the parental tiller and all remaining tillers within little bluestem plants were defoliated (Carman and Briske 1982). Greater growth rates were observed in tall fescue plants following defoliation as the percentage of undefoliated tillers within the plant increased (Matches 1966). Both observations support the conclusion that defoliated tillers were deriving support from associated, nondefoliated tillers.

Vegetative growth theoretically confers plants with potential immortality. Individual plants of hard fescue have been estimated to be greater than 1000 years old (Harberd 1962). However, the few age estimates available for North American perennial grasses indicate that their life spans are relatively short. Estimates of individual plant longevities on the Jornada Experimental Range in New Mexico—including tobosa grass, black grama, and red threeawn—indicate that maximum plant longevity does not exceed 30 years (Wright and Van Dyne 1976). These estimates of plant longevity are corroborated by the work of West et al. (1979) on the U.S. Sheep Station in Idaho, and Canfield (1957) on the Santa Rita Experimental Range in Arizona.

GRAZING RESISTANCE

Grazing resistance is an ambiguous term used to describe the relative ability of plants to survive grazing. However, strategies to cope with grazing vary greatly in form and expression among plant species. Additional insight can be gained by organizing grazing resistance into a tolerance and avoidance component (Stuart-Hill and Mentis 1982; Briske 1986; Fig. 4.7). **Avoidance mechanisms** reduce the

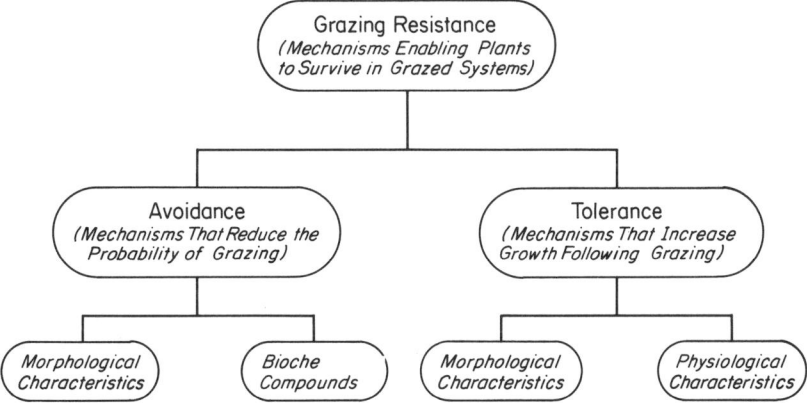

Figure 4.7. Organization of grazing resistance into avoidance and tolerance components. Avoidance mechanisms decrease the probability of grazing while tolerance mechanisms increase growth following grazing (adapted from Briske 1986).

probability and severity of plant defoliation (i.e., escape mechanisms), while **tolerance mechanisms** facilitate growth following defoliation (i.e., mechanisms of rapid leaf replacement). The ability of a species to survive grazing undoubtedly results from a combination of these two components, but in certain species and under specific environmental conditions, one component may predominate over another.

Grazing resistance within ecological plant groups may be generally ranked as follows: herbaceous monocots > herbaceous dicots > deciduous shrubs and trees > evergreen shrubs and trees (Archer and Tieszen 1986). This ranking is based upon both morphological and physiological considerations. Apical and intercalary meristems within monocots are less vulnerable to large herbivores because of their basal location within the plant. Meristems are located at terminal and lateral positions of shoots in dicots, increasing their susceptibility to large herbivores. Many woody plants, evergreen species most notably, possess slow growth rates and low rates of resource acquisition which require that individual leaves be retained for long periods (Chapin 1980). Consequently, these species are less efficient in replacing photosynthetic surfaces removed by herbivores and frequently rely on avoidance mechanisms rather than tolerance mechanisms to cope with grazing.

Grazing Avoidance

Mechanical Mechanisms. Avoidance mechanisms primarily influence plant accessibility and palatability to specific herbivores. At the individual tiller level of organization, the probability and severity of defoliation may be reduced by a number of avoidance mechanisms originating from a variety of morphological parameters (Fig. 4.8). Tissue accessibility is primarily a function of tissue proximity to the soil surface as influenced by the length and angle of leaves and tillers. Species possessing culmed vegetative shoots are especially susceptible to defoliation because apical meristems are elevated above the soil surface and readily accessible to herbivores (Branson 1953; Rechenthin 1956; Booysen et al. 1963). Mechanical deterrents including spines, awns, and epidermal characteristics (e.g., silica bodies, pubescence, and cuticular waxes) directly influence palatability (McNaughton et al. 1985; Cooper and Owen-Smith 1986; Young 1987b). Leaf anatomy, primarily the presence of vascu-

Level of Organization	Avoidance Mechanism	Morphological Characteristics
Tiller	Tissue Accessibility	Number and Length of Elongated Internodes; Tiller Angle; Leaf Length and Angle; Leaf/Culm Ratio
	Mechanical Deterrents	Awns or Spines of Inflorescence
	Epidermal Characteristics	Pubescence; Waxes; Silicification
	Leaf Anatomy	Leaf Tensile Strength; Presence of Vascular Bundles
Plant	Tissue Accessibility	Accumulation of Culms; Vegetative/Reproductive Tiller Ratio
	Interspecific Association	Differential Expression of Avoidance Mechanisms Among Species

Figure 4.8. Categories of avoidance mechanisms and associated morphological characteristics at the tiller and plant levels of organization that may potentially reduce the probability of being grazed (from Briske 1986).

lar bundles associated with the C_4 photosynthetic pathway (i.e., Kranz leaf anatomy), has been demonstrated to influence species selection in herbivorous insects (Caswell et al. 1973). Greater cell wall thickening in the bundle sheath cells (specific mesophyll cells) of warm-season grasses limits digestibility by impeding microbial access to cellular contents within the digestive tract (Akin and Burdick 1977; see Chapter 2).

Avoidance mechanisms may also originate at the plant level of organization from reduced tissue accessibility or by interspecific association. At the level of the individual plant, tissue accessibility is primarily a function of tiller height, number of culmed tillers and the amount of dead material which has accumulated within the plant. Grazing intensity is inversely related to the basal area of individual crested wheatgrass plants (Norton and Johnson 1983). Grazing intensity in large plants (basal area > 200 cm^2) is reduced, relative to their canopy volume, by the accumulation of dead culms and litter. Grazing intensity becomes proportional to canopy volume when this material is removed. Shrub species may develop a "hedged" appearance with frequent grazing through the initiation of numerous shoots which form a mechanical barrier protecting leaves and meristems within the canopy.

The association of palatable species with less palatable species may also influence the frequency and intensity of plant defoliation (McNaughton 1978). The protection afforded to grasses growing within the canopy of low-growing shrubs serves as a frequently observed example (Davis and Bonham 1979). Although not well documented, there is no reason to suspect that this phenomena does not occur among herbaceous plant species as well. This implies that not only the relative amount, but also the spatial distribution of herbaceous species may be an important parameter regulating plant utilization by herbivores.

Biochemical Mechanisms. A diverse array of biochemical compounds referred to as **secondary compounds** or metabolites may also contribute to grazing avoidance. Secondary compounds which deter herbivores in low concentrations (< 2% dry weight) by interfering with herbivore metabolism are referred to as **qualitative compounds** (Rhodes 1979, 1985). This category of compounds, including alkaloids, glucosinolates, and cyanogenic compounds, are produced at a relatively low cost to the plant, and concentrations may increase rapidly in response to grazing. Conversely, the second category of metabolites known as **quantitative compounds** are present in relatively large concentrations (5–20% dry weight). This group of compounds, including tannins, lignin, and resins, are more costly to produce and increase in concentration only slowly, if at all, in response to grazing. Secondary metabolites are known to be effective in deterring specific herbivores from grazing grasses, forbs, and shrubs (Simons and Marten 1971; Provenza and Malechek 1983).

Optimal defense theory indicates that the most apparent plants within a community rely on quantitative defenses because they are easily located by herbivores (Rhodes 1979; Provenza and Malechek 1983). Less apparent plants frequently possess the less costly qualitative compounds since they have a lower probability of being grazed. The observation that approximately 80% of woody perennials contain tannins in comparison with only 15% of the herbaceous dicots serves to substantiate this point (Rhodes and Cates 1976).

Grazing avoidance mechanisms do not necessarily remain constant, but may increase with an increase in grazing intensity. The qualitative biochemical compounds which increase in response to grazing, referred to as **inducible defenses**, have previously been referenced (Rhodes 1985; Young 1987b). Similarly, grazing can modify morphological parameters influencing avoidance mechanisms. Long-term grazing has been observed to function as a selection pressure against the tall, upright growth form in several perennial grasses (Detling and Painter 1983; Carman and Briske 1985). The remaining growth forms, characterized by a large number of smaller statured tillers with reduced leaf numbers and blade areas, are better able to avoid grazing because less biomass is removed by herbivores and a greater number of meristems may escape grazing to facilitate growth following defoliation. Contrastingly, plants characterized by a small number of large tillers with large leaf areas are more competitive in environments with dense canopies. It is currently uncertain whether this herbivore-induced shift in growth form is a result of genotypic selection or phenotypic plasticity.

Grazing Tolerance

Morphological Mechanisms. **Leaf replacement potential**, defined as the rate at which leaf area is reestablished following defoliation, is largely a function of the number, source, and location of meristems within a plant following defoliation (Fig. 4.9). Growth will occur most rapidly from intercalary meristems, followed by newly developed leaf primordia, and least rapidly from newly initiated axillary buds (Cook and Stoddart 1953; Hyder 1972; Briske 1986). Growth from intercalary meristems results from the expansion of previously differentiated cells, whereas growth from axillary buds is delayed by the time required for differentiation and growth of leaf primordia. However, axillary buds ensure perenniality by providing a meristematic source for the production of subsequent tillers in contrast to the limited activity of intercalary meristems (i.e., phytomer growth only). Consequently, species possessing a high ratio of reproductive or culmed vegetative tillers are best suited to intermittent defoliation rather than continuous grazing (Hyder 1974). When the apical meristem assumes reproductive status or is removed by grazing, leaf replacement must originate from axillary buds which require the greatest time interval following defoliation.

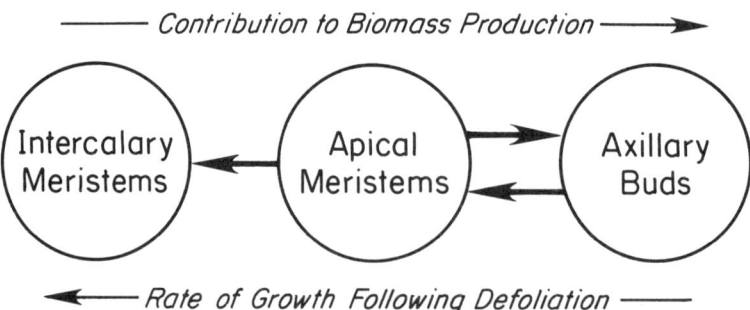

Figure 4.9. Sources of meristematic activity in a grass plant. The relative growth rate from each source following defoliation is established by the extent to which tissue differentiation has previously occurred. Axillary buds confer perenniality to the plant while intercalary meristems are relatively short-lived (from Briske 1986).

Physiological Mechanisms

Compensatory photosynthesis. Grazing alters the age structure of leaves within plant canopies in addition to reducing total leaf area. This has direct consequences for the photosynthetic capacity of plants because leaves generally exhibit maximum photosynthetic rates at about the time of full expansion and decline thereafter (Caldwell 1984). Consequently, leaves of defoliated plants may display greater rates of photosynthesis than nondefoliated plants because many of the leaves are chronologically younger and more efficient photosynthetically. However, the carbon gain capacity of plants is a function of both total leaf area and photosynthetic rate.

Increased rates of photosynthesis following partial defoliation relative to similar aged leaves of nondefoliated plants is referred to as **compensatory photosynthesis** (Nowak and Caldwell 1984). Compensatory photosynthesis has been observed in a number of species with leaf photosynthetic rates increasing within the range of 15–50% in comparison with nondefoliated leaves (Gifford and Marshall 1973; Dyer et al. 1982; Wallace et al. 1984; Nowak and Caldwell 1984). Maximum photosynthetic rates occur several days following defoliation, a portion of which is attributable to decreasing photosynthetic rates of nondefoliated leaves as they age and senesce. Photosynthetic response is influenced by leaf position relative to the location of plant defoliation. Photosynthetic rates of the remaining portions of defoliated leaves generally decrease (Detling et al. 1979; Dyer et al. 1982), while photosynthetic rates of nondefoliated leaves on tillers from which associated leaves have been removed, or nondefoliated tillers within plants from which associated tillers have been removed, generally increase (Gifford and Marshall 1973; Dyer et al. 1982). Potential mechanisms contributing to compensatory photosynthesis include increased carboxylase activity, increased leaf conductance to carbon dioxide, and a decrease in feedback inhibition resulting from a greater demand for carbon following defoliation (i.e., greater sink strength) (Hodgkinson et al. 1972; Gifford and Marshall 1973; Wallace et al. 1984).

Although compensatory photosynthesis does occur, its significance to grazing tolerance appears limited. Compensatory photosynthesis occurs in both crested and bluebunch wheatgrass, but is most evident in the oldest leaves which comprise a small percentage of the total photosynthetic area (Nowak and Caldwell 1984). The total amount of compensatory photosynthesis resulting from this portion of the canopy is insufficient to explain the difference in grazing tolerance between these two species.

Carbon allocation. The allocation of photosynthetic products within grasses is consistently altered by defoliation. The proportion remaining above-ground to reestablish photosynthetic tissues increases relative to the proportion allocated below-ground (Ryle and Powell 1975; Detling et al. 1979). Flexible patterns of carbon allocation may increase grazing tolerance by increasing the rate of leaf replacement and reestablishing the photosynthetic capacity of plants. For example, crested wheatgrass, known to be more grazing tolerant than bluebunch wheatgrass, exhibits a greater capacity to reallocate carbon to reestablishment of photosynthetic tissues while temporarily decreasing allocation below-ground (Caldwell et al. 1981; Richards 1984). The reallocated carbon is apparently converted into leaf area following the activation and development of axillary buds (Mueller and Richards 1986). These data also indicate that a short-term reduction in root growth is not necessarily detrimental to the leaf replacement potential or competitive ability of grazed plants.

Carbohydrate reserves. The significance of carbohydrate reserves to the grazing tolerance of grasses has been investigated for half a century. The major premise for monitoring carbohydrate reserves has been to provide an index of leaf replacement potential (i.e., vigor) based upon the assumption that the depletion of carbohydrate reserves by excessive defoliation reduces growth, and in extreme cases, causes plant mortality (Weinmann 1948). **Carbohydrate reserves**, referred to as total nonstructural carbohydrates or total available carbohydrates, are a product of photosynthesis in excess of growth and maintenance requirements (White 1973). Carbohydrates are composed of fructosans, a starchlike fructose polymer, and sucrose in grasses of temperate origin (C_3 photosynthetic pathway), while starch and sucrose are the primary storage carbohydrates in grasses of tropical origin (C_4 photosynthetic pathway). However, temperature is known to influence the proportion of starch, sucrose, and fructan accumulation within a species (Chatterton et al. 1987). Carbohydrates are stored in living parenchyma cells in organs both above- and below-ground.

Carbohydrate reserves are utilized for plant growth and maintenance when photosynthetic capacity is limited, as evidenced by the reduction in reserves following defoliation (Deregibus et al. 1982). However, a substantial amount of information has developed suggesting that the contribution of carbohydrate reserves to the leaf replacement potential of perennial grasses may not be as large as previously assumed (May 1960; Ryle and Powell 1975; Atkinson and Farrar 1983; Caldwell 1984). Investigations utilizing labelled carbon indicate that carbohydrates allocated to the root system are not capable of being remobilized for subsequent use above-ground following defoliation (Davidson and Milthorpe 1966).

The amount of reserve carbon is not directly related to leaf replacement potential in crested and bluebunch wheatgrass and does not account for the wide variation in grazing tolerance between them (Richards and Caldwell 1985). Carbohydrate pools (tissue mass × carbohydrate concentration) within the crowns of the two wheatgrass species only contain an amount of carbohydrate equivalent to that produced in approximately 3 days of photosynthesis. Consequently, plant growth is more dependent upon current photosynthesis than stored carbon within 3 days of defoliation. In addition, rates of leaf elongation or tillering are not dependent upon carbohydrate concentrations in leaves, crowns, or roots of tall fescue, orchardgrass, or canarygrass (Sambo 1983; Zarrough et al. 1984; Volenec and Nelson 1984). These findings support the view of May (1960), who indicated that use of the term "reserve" evokes a false conception of their contribution to growth following defoliation (Deregibus et al. 1982).

The magnitude of carbohydrate reserves necessary to ensure plant survival and maintain maximum leaf replacement potential has not been established for individual species or species groups. Carbohydrate concentrations of 1–6% have been suggested as minimum reserve levels in grasses, but these estimates are far from conclusive (Caldwell 1984). In addition, total **carbohydrate pools** must be quantified by determining both carbohydrate concentrations and weight of the associated plant organ(s) (Richards and Caldwell 1985). Defoliation could potentially reduce total carbohydrate availability within a plant by reducing the weight of crown tissue without necessarily altering carbohydrate concentration.

It has been suggested that plant growth may be limited to a greater extent by the availability or activation of axillary buds than the amount of reserve carbon (Watson and Casper 1984; Richards and Caldwell 1985). A **meristematic growth limitation** would promote the accumulation of carbon reserves because photosynthesis would exceed growth and respiration requirements (White 1973; Caldwell 1984). Although insufficient carbon undoubtedly limits plant growth, a direct relationship between the amount of carbohydrate reserves and plant growth has not been established.

Inconsistencies associated with analytical techniques for carbohydrate extraction have also curtailed research progress. Estimates of nonstructural carbohydrates derived by acid extraction may be two to three times greater than for boiling water extraction because a greater amount of structural carbohydrates are apparently digested (Richards and Caldwell 1985). Compounds other than nonstructural carbohydrates (e.g., proteins, hemicellulose, and organic acids), which are not currently evaluated, may also contribute energy and organic intermediates to plant growth following defoliation. Additional investigation of the analytical techniques and reserve compounds within the context of whole plant carbon balance is required to define the relationship of carbohydrate reserves to grazing tolerance.

Plant water status. A reduction in transpirational area following grazing has prompted the suggestion that grazing conserves soil water, potentially prolonging the growing season (McNaughton 1983b). It is difficult to assess the validity of this assumption based upon the limited number of investigations conducted. It appears that soil water may be conserved by the reduction in transpirational area associated with grazing, but the increase in available water is not expressed as improved plant water status (Archer and Detling 1986; Svejcar and Christiansen 1987; Wraith et al. 1987). Plants apparently posses the ability to partially compensate for differences in soil water availability by extracting water from various portions of the soil profile or adjusting transpiration rates per unit leaf area without altering their water status. Defoliation may also conserve water by decreasing root depth or root density. However, soil water conservation does not necessarily convey an ecological advantage. The postponement of soil water extraction may potentially decrease the efficiency of water use as temperatures increase during the growing season (Caldwell et al. 1983). In addition, selective grazing may increase the proportion of water utilized by the less preferred species within the community.

Root growth and function. Root growth and function is dependent upon energy provided by photosynthesis. Consequently, the suppression of root growth is generally proportional to the intensity and frequency of above-ground defoliation (Crider 1955; Cook et al. 1958; Youngner 1972). A single defoliation removing 50% or more of the shoot volume retarded root growth for 6–18 days in seven of eight perennial grasses investigated (Crider 1955). A single defoliation removing 80% and 90% of the shoot volume stopped root growth for 12 and 17 days, respectively. Multiple defoliations detrimentally influenced root growth to a greater extent than single defoliations. The initial removal of 70% of the shoot volume followed by three subsequent clippings per week stopped root growth for the entire 33-day investigation in all three species subjected to multiple defoliations. Cessation of root growth has been observed to occur within hours of defoliation (Davidson and Milthorpe 1966; Hodgkinson and Baas Becking 1977).

Root growth cessation affects both lateral and vertical development of root systems (Schuster 1964; Smoliak et al. 1972) as well as detrimentally influencing root initiation, diameter, branching, and total production (Biswell and Weaver 1933; Jameson 1963; Evans 1973; Carman and Briske 1982; Richards 1984). Root mortality has also been observed following defoliation (Weaver and Zink 1946; Hodgkinson and Baas Becking 1977; Troughton 1981). These detrimental responses collectively serve to reduce the total absorptive surfaces and soil volume explored for water and nutrients.

The capacity for nutrient absorption per unit length in temperate, perennial grasses parallels the growth responses following defoliation. Phosphorus absorption, root elongation rate and respiration rate remained suppressed for an 8-day observation period following defoliation of orchardgrass to a height of 2.5 cm (Davidson and Milthorpe 1966). Similar responses of root growth and function to defoliation originate from the dependence of both processes on energy produced in plant respiration (Caldwell et al. 1987). Respiration, in turn, functions upon substrate produced in photosynthesis.

Apical dominance. **Apical dominance** within the annual grasses, teosinte and barley, was initially described by Leopold (1949) as the production of auxin within the apical meristem which suppressed axillary bud expansion. This line of experimentation was apparently based on the work of Thimann and Skoog (1933) who had established that auxin controlled branching in dicots. This single investigation (Leopold 1949) has largely shaped our perception of how the tillering process is regulated in grasses, but has been criticized as being less than definitive from an experimental perspective (Williams and Langer 1975). Aspinall (1961) later forwarded the **nutritive theory** which suggested that inter-organ competition for nutrients inhibited axillary bud development. This theory was deemed untenable based on the relatively small metabolic demand presented by both the apical meristem and axillary buds. These two theories were eventually combined and extended into the **nutrient diversion theory** of apical dominance (Jewiss 1972; Hillman 1984). This theory proposed that growth regulators control both the supply of photosynthetic products received by axillary buds and the rate of cellular division and expansion within the buds. The discovery that cytokinins and potentially other growth regulators are involved in the regulation of bud growth marked a significant advance in the understanding of apical dominance (Phillips 1975). The principle role of auxin produced in the leaf primordia of apical meristems is to limit the availability or utilization of cytokinin within axillary buds thereby inhibiting growth. Although the complete mechanism of apical dominance is not thoroughly understood, the direct suppression of axillary bud growth by auxin is no longer an accurate interpretation of the phenomena.

It would appear unduly restrictive to presume that only the removal of apical meristems by grazing could serve as an environmental cue to induce tillering. How would tillering occur in plant populations subjected to limited grazing? Tiller recruitment has been observed to occur in response to grazing even though apical meristems were insufficiently elevated to be removed by livestock (Butler and Briske 1988). Conversely, removal of apical meristems from tillers of crested and bluebunch wheatgrass did not always result in accelerated tiller recruitment (Olson and Richards 1988b; Richards et al. 1988). In contrast to the temperate species, lateral bud growth was stimulated by both apical meristem and canopy removal in three tropical grasses. Expanding leaves and either the inflorescence or elongating culm were observed to be the source of apical dominance in vegetative and reproductive ryegrass tillers, respectively (Laidlaw and Berrie 1974). These conflicting observations attest to the complexity associated with the regulation of tiller recruitment in perennial grasses (Youngner 1972).

Light quality has been implicated in the control of axillary bud expansion in several grass species (Casal et al. 1985, 1986; Deregibus et al. 1985; Kasperbauer and Karlen 1986). This photomorphogenetic response is presumably mediated by phytochrome, as is flowering and branching in many dicotyledonous species. A decrease in the ratio of red:far-red radiation associated with increasing canopy development may signal the diminishing availability of resources and suppress additional tiller recruitment (Deregibus et al. 1985; Simon and Lemaire 1987). Partial removal of the plant canopy by grazing may increase the ratio of red:far-red radiation and promote tillering without disturbing the apical meristem. The versatility of phytochrome as a mechanism regulating tiller recruitment is yet to be proven, but the direct inhibition of axillary buds as presented by Leopold (1949) is overly restrictive.

Grazing does influence both the seasonality and total number of tillers recruited by affecting axillary bud development. Tiller recruitment in ungrazed populations of little bluestem in central Texas is restricted to spring and fall coincident with the bimodal precipitation pattern of the region (Butler and Briske 1988; Briske and Butler 1989). An intermediate severity of grazing extended the period of tiller recruitment throughout the spring and summer. Grazing did not significantly increase total tiller recruitment, however, when the number of recruited tillers were summed over the entire growing season. Similarly, grazing promoted tiller recruitment in the spring in addition to the normal pattern of fall recruitment in crested wheatgrass (Olson and

Richards 1988b). Grazing at the time of culm elongation or thereafter reduced subsequent tiller recruitment, while grazing prior to culm elongation produced little affect in comparison with ungrazed plants (Olson and Richards 1988c; Busso et al. 1989). The absence of a positive tillering response in a species with the demonstrated grazing tolerance of crested wheatgrass confirms the conclusion of Ellison (1960) that grazing generally inhibits tillering over the long term. Removal of a large portion of the photosynthetic surfaces from a plant apparently reduces the amount of resources available for tiller growth irrespective of the mechanisms regulating the tillering process (e.g., apical dominance, light quality, etc.; Youngner 1972).

Compensatory growth. Most available information fails to support the contention that grazing-induced modifications of plant function (McNaughton 1979) increase growth of grazed plants over that of ungrazed plants (Ellison 1960; Jameson 1963; Belsky 1986; see Chapter 1). Belsky (1986) indicates that of the reports in the literature referencing above-ground production in response to grazing, 34 reported a decrease in production, 5 reported no change, and 9 reported an increase in production. However, growth of a warm-season African sedge was increased 3-fold when grown in specific environmental conditions and subject to specific defoliation treatments (McNaughton et al. 1983; Wallace et al. 1985). Similarly, scarlet gilia, a biennial forb, displayed a 3-fold increase in flower and seed production when grazed at the time of stem elongation (Paige and Whitham 1987).

Ill-defined terminology has undoubtedly contributed to the conflicting viewpoints on this topic. **Compensatory growth** may be generally defined as any positive plant response to injury (Belsky 1986). **Overcompensation,** more specifically, describes an increase in the cumulative total dry weight of grazed plants, including the biomass removed by defoliation, in excess of that produced by ungrazed plants (Belsky 1986). In other words, an increase in the rate of plant growth following defoliation does not in itself constitute greater production. It must also be established that the growth increase is maintained for a sufficient period to offset the reduction in biomass resulting from defoliation.

An additional source of confusion is associated with the potential mechanisms contributing to compensatory growth. Although evidence for the involvement of inherent physiological mechanisms does exist (e.g., compensatory photosynthesis), the modification of competitive interactions among plants following defoliation is probably of far greater consequence. In greenhouse experiments, Belsky (1986) noted that in the few cases where compensatory growth was observed, it occurred when short grasses were growing in competition with tallgrass species. Defoliation to a uniform height may have placed the tall grasses at a competitive disadvantage rather than directly inducing a physiological response which enhanced growth of the shortgrass species. The benefit a plant derives from defoliation of its neighbors coincident with its own defoliation has been termed **competitive fitness** (Belsky 1986).

Cost of Grazing Resistance

The "costs" associated with grazing resistance are most clearly defined in relation to the production of secondary compounds which deter herbivores (i.e., avoidance mechanisms) (Coley 1986). Tree seedlings with high concentrations of tannin were grazed to a lesser extent, but displayed lower rates of leaf production than seedlings with low tannin concentrations. The growth reduction can be interpreted as the cost associated with the production of tannins. A trade-off may also exist between competitive ability and grazing resistance for similar reasons. Plants of white clover possessing cyanogenic compounds were less effective competitors than acyanogenic plants in the absence of grazing because of the energy diverted to grazing avoidance (i.e., cyanogenic compounds) (Dirzo and Harper 1982). In the presence of grazing, however, the cyanogenic plants would presumably be better competitors because cyanogenic compounds would reduce the intensity of grazing.

Morphological and physiological resistance mechanisms do not represent such clear costs to the plant. What costs are incurred by the decumbent growth-form of western wheatgrass resulting from long-term grazing (Detling and Painter 1983) or the short-term increase in resource allocation to the shoots of crested wheatgrass following defoliation (Richards 1984)? These mechanisms may not represent a cost in terms of the diversion of previously assimilated energy, but rather a reduction in the potential for subsequent production and resource acquisition. Hyder (1972) has cautioned that placing excessive emphasis on grazing-resistant species may decrease productivity. Wise strategies of grazing management must be coupled with inherent grazing-resistance mechanisms of existing species to optimize plant and animal productivity in grazed systems.

SPECIES REPLACEMENT

Grazing-induced modifications in species composition have been documented in numerous grasslands throughout the world (Voigt and Weaver 1951; Branson and Weaver 1953; Ellison 1960; Williams 1969; Noy-Meir et al. 1989). Compositional changes frequently involve the replacement of higher successional species by lower successional species (Canfield 1957). The lower successional species are frequently mid- or shortgrass species held in a subordinate position by competitive interaction with species possessing greater stature (Arnold 1955, Belsky 1986). Grazing reduces the competitive ability of the mid- and tallgrasses thereby increasing the relative abundance of lower successional grasses and forbs and establishing the potential for shrub invasion. This scenario of species replacement in response to grazing has frequently been inferred from field observation, but has not been experimentally verified. A scenario incorporating elements of developmental morphology, grazing resistance, competitive interactions, and population structure is presented as a mechanistic explanation for species replacement in grasslands. Herbaceous retrogression is emphasized in this section while its implications to woody plant encroachment are discussed in Chapter 5.

Competitive Interactions

Plant species do not grow or respond to grazing as isolated individuals, but rather as members of a population and community. It has previously been demonstrated that individual grass plants consist of an assemblage of phytomers and tillers (Fig. 4.1). Similarly, grass populations reflect the number of plants per unit area and the number of tillers per plant (Fig. 4.10). Grassland communities are further composed of an aggregation of populations variously arranged in terms of abundance and space.

The first, and most direct, mechanism by which grazing alters competitive interactions involves the differential utilization of populations within the community in response to the relative display of avoidance mechanisms. Species grazed severely are placed at a disadvantage when competing with associated species grazed less severely. Production of bluebunch wheatgrass plants subjected to 50% canopy removal just prior to culm elongation was equivalent to the production of nondefoliated plants growing with full competition when associated vegetation

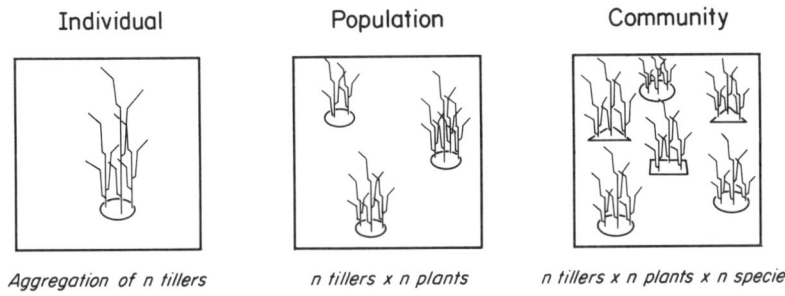

Figure 4.10. Hierarchical organization of grassland communities is dependent upon the density and spatial arrangement of tillers in individual plants and within and among species populations.

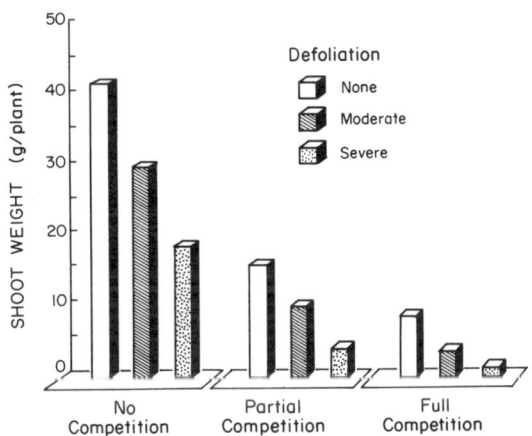

Figure 4.11. Response of bluebunch wheatgrass to three defoliation intensities in the presence of full, partial, or no competition from associated vegetation. Competition from associated species exerts a greater influence on growth following defoliation than defoliation intensity (adapted from Mueggler 1972).

within a 90 cm radius of the defoliated plants was clipped at ground level (Fig. 4.11). When competition from associated vegetation was removed by tilling within a 90 cm radius, defoliated plants produced three times the biomass of nondefoliated plants growing with full competition. These data clearly demonstrate that the ability of plants to respond to defoliation is not only determined by an inherent suite of morphological and physiological characteristics, but also by competitive pressure from associated species (Caldwell 1984).

The second mechanism by which competitive interactions among plant species may be altered in response to grazing involves the differential ability of species to grow following a similar intensity of defoliation based upon the possession of various tolerance mechanisms. Species rapidly replacing photosynthetic surfaces gain a competitive advantage over associated species that grow more slowly following defoliation. Crested wheatgrass exhibits greater leaf replacement potential than bluebunch wheatgrass following an equivalent intensity of defoliation. This response is due in part to the ability of crested wheatgrass to rapidly initiate a greater number of tillers and to allocate carbon to reestablish photosynthetic surfaces while temporarily decreasing allocation below-ground (Caldwell et al. 1981, Richards 1984). Inequitable responses to defoliation between these two species can be attributed to the differential expression of tolerance mechanisms because similarity in plant architecture minimizes the influence of avoidance mechanisms (Caldwell 1984).

Species grazed less severely (i.e., avoidance mechanisms), capable of growing more rapidly following defoliation (i.e., tolerance mechanisms) or possessing a combination of these two resistance components, realize a competitive advantage within the community. These species, through the possession of a greater canopy area, are able to intercept greater amounts of solar energy and assimilate greater amounts of carbon, further enhancing their competitive ability. By allocating a greater amount of carbon below-ground, grazing-resistant species may more effectively explore the soil profile for water and nutrients and preempt resources that may have been utilized by associated species prior to defoliation (Mueggler 1972; Eissenstat and Caldwell 1988).

Grazing management partially governs the intensity of competitive interactions by regulating the relative frequency and intensity of defoliation among plant species within grassland communities. Lenient grazing may not alter species composition appreciably, even though species may be grazed non-uniformly or respond differentially, because the intensity is insufficient to alter plant growth and subsequent competitive interactions. However, as stocking rate and defoliation intensity increase, differential utilization and growth among species becomes intensified, altering competitive interactions and ultimately contributing to species replacement. In addition, the species or combination of herbivore species affect the relative frequency, intensity, and seasonality of grazing within communities based on preference and behavioral differences (see Chapters 2 and 3). Sheep grazing sagebrush steppe, for example, can shift community composition toward grass dominance more rapidly than the total exclusion of grazing (Laycock 1967).

Season of grazing in relation to the progression of phenological development among species plays a major role in determining the outcome of competitive interactions. Species grazed throughout their entire growth period are placed at a competitive disadvantage in the presence of species possessing growth periods which do not coincide entirely with the grazing season. This is very likely the reason Texas Wintergrass, a cool-season species, increases in relative abundance in grasslands of central Texas which are grazed most intensively during the spring and summer (Launchbaugh 1955). Similarly, fall grazing has been observed to favor dominance of warm-season grasses in relation to shrub species in the cold desert of the western U.S. (West et al. 1979).

Given these considerations, it is to be expected that competitive interactions among species are modified on a relative rather than an absolute basis. Consequently, it is possible for an individual species to decrease in relative abundance in one community, but increase within another in response to grazing. The inherent mechanisms of grazing resistance probably remain constant over the distributional range of a species, but the expression of relative resistance mechanisms change in relation to the resistance mechanisms of associated species. The response of individual species has been observed to vary depending upon the intensity of grazing and topographic position within a mixed-grass prairie community. Relative abundance of western wheatgrass decreased regardless of grazing intensity or topographic position, suggesting that it possessed limited grazing resistance relative to the associated species (see Chapter 5, Fig. 5.4). By contrast, the relative grazing resistance of buffalograss varied in relation to topographic position. Relative abundance decreased in the swale, but increased on the ridge. Livestock behavior and environmental variables may further influence competitive interactions among plant species to produce an array of plant responses (see Chapter 3).

Population Structure

Grazing-induced modifications of competitive interactions are eventually expressed at the population level through the modification of basal area and tiller demography of individual plants. A reduction in basal area of individual plants may be the initial and predominate response contributing to the decline of bunchgrass populations in response to grazing (Butler and Briske 1988). Grazed populations of several perennial grasses have been observed to consist of individuals with smaller basal areas in comparison with ungrazed populations (Pond 1960; Hickey 1961). This decrease in individual plant basal area is very likely a consequence of the fragmentation of individual large plants into smaller units (Fig. 4.12). Consequently, plant density may remain constant or even increase while basal area per plant decreases. Further, an increase in tiller number per unit of remaining basal area may initially offset the decrease in total basal area thereby maintaining a constant tiller density. However, with continued severe grazing the decrease in individual plant basal area may become so great that tiller density declines within the population.

This decline in population structure is very likely paralleled by a decrease in resource acquisition within the community. Consequently, the more grazing-resistant species within the community utilize a greater proportion of the available resources (Caldwell et al. 1987). The continued existence of populations composed entirely of plants with reduced basal areas may be jeopardized by the inability to effectively com-

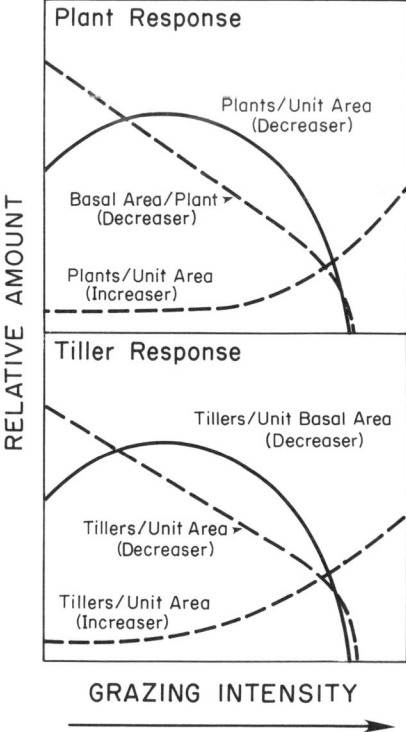

Figure 4.12. Relative response of plants and tillers of increaser and decreaser species to increasing grazing intensity. Plant density and tiller number per unit of remaining basal area initially increase with grazing intensity, but eventually decrease as increasing grazing intensities reduces total basal area of decreaser species.

pete with populations of less severely grazed species and increased susceptibility to extreme abiotic conditions (e.g., drought or temperature extremes). Although bunchgrass populations composed of numerous, small plants appear very persistent, a large reduction in basal area may predispose the population to elimination from the community.

Species replacement influences the quantity, quality, and variability of biomass production by altering the initial harvest and subsequent flow of energy through the ecosystem (see Chapter 1). In many rangeland systems, the ratio of unpalatable to palatable species increases with increasing grazing severity (Noy-Meir and Walker 1986). Although this may not decrease total productivity of the system, it reduces the proportion of energy transferred through the grazing food chain (i.e., plants to herbivores). Consequently, as the proportion of unpalatable species increases a greater proportion of the solar energy captured in photosynthesis is transferred into the decomposer compartment following plant senescence or is incorporated into woody biomass.

Compositional changes involving an increase in the abundance of annual and short-lived perennial species limits the amount of solar energy captured by reducing the proportion of the growing season during which a canopy is present. Additional difficulty is encountered in efficiently harvesting biomass in annual systems based on the large variability and limited duration of the growth period. Finally, it is possible for severe grazing to modify the production potential of a site through soil erosion and alteration of hydrological properties (see Chapter 6). In this case, fewer resources are available to sustain biomass production.

SUMMARY

Plant growth originates from the incorporation of photosynthetic products into cells and tissues differentiated from meristems. Tillers represent the sum of all tissues differentiated from a single apical meristem to form the basic unit of growth. The grass plant is composed of an assemblage of tillers initiated from axillary buds of previous tiller generations. Variation in the size and number of phytomers comprising the tiller and the pattern of tiller emergence contribute to the architectural distinction among grass growth forms (e.g., bunchgrass versus sodgrass).

Grazing resistance can be organized into avoidance and tolerance components. Avoidance mechanisms consist of morphological characteristics or biochemical compounds which reduce the probability and severity of plant defoliation at several levels of vegetation organization. Tolerance mechanisms are conferred by physiological processes at the tiller and plant levels to enhance growth following grazing. Both components contribute to grazing resistance, but the relative magnitude and associated cost of each component are poorly understood.

Grazing management modifies competitive interactions by influencing the frequency, intensity, and seasonality of plant defoliation (Fig. 4.13). Plant species grazed less frequently and intensively, or with a greater capacity to grow following defoliation, display greater leaf areas for photosynthesis and attain a competitive advantage. Grazing-induced modifications in competitive interactions are eventually expressed as modifications in plant and population structure. A decrease in total basal area, plant density, or tiller density of a given species is ultimately manifested in a relative reduction in resource acquisition within the community. Shifts in species composition subsequently alter the quantity, quality, and variability of plant production by modifying the amount and pattern of energy flow through the ecosystem (see Chapter 1).

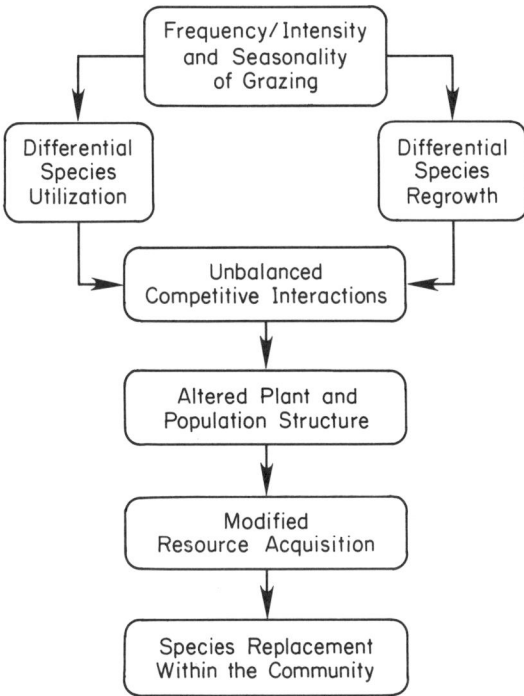

Figure 4.13. Proposed cause-effect relationship between grazing-induced competitive interactions and modified plant and population structure as it may regulate species composition at the community level.

Vegetation response to grazing may be investigated at one or more levels within the **hierarchical organization** of grasslands (e.g., tiller, plant, population, or community). For example, plant productivity may be reduced by a decrease in individual tiller weight, tiller number per plant, or plant density in response to grazing. Consequently, insight into mechanisms occurring at higher hierarchical levels (e.g., community) requires that processes at lower hierarchical levels (e.g., population, plant, and tiller) be investigated (Archer and Tieszen 1986; Brown and Allen 1989; see Chapter 5). Hierarchical levels of vegetation organization may respond in a comparable manner to affect vegetation dynamics, but frequently additional complexity is encountered by the occurrence of opposing responses between or among levels. Grazing has been observed to increase tiller density, but concomitantly decrease individual tiller weight (Jones et al. 1982) or increase plant density while reducing basal area per plant (Butler and Briske 1988). It is essential that several hierarchical levels be considered when evaluating vegetation responses to grazing to avoid incomplete or erroneous conclusions.

Research oriented at the population level of vegetation organization possesses the potential for integrating the divergent sources of information available from individual plant and community studies. These two research perspectives have not been effectively unified into an information base for vegetation management in grazed systems. Community level investigations describe species composition shifts and biomass dynamics, but do not yield insights into mechanistic cause-effect relationships. Conversely, reductionist investigations at the individual plant level yield mechanistic insights, but are frequently too narrow in scope to identify interactions and properties of systems at levels of organization suitable for vegetation management. A major limitation to the extrapolation of plant level studies is the minimal amount of information concerning competitive interactions and population ecology of dominant plant species.

CHAPTER 5

ECOSYSTEM-LEVEL PROCESSES

Steve Archer and Fred E. Smeins

INTRODUCTION

Plant species composition and productivity within a region largely reflect the prevailing climate, whereas seasonal and annual variability in rainfall and temperature play a central role in dictating the dynamics of populations over time. However, substantial spatial variability occurs across landscapes, and broad-scale climatic variables cannot account for the spatial patterns which shape vegetation form and function on a local scale. Soils and topography exert a strong influence on patterns of plant distribution, growth, and abundance over the landscape through regulation of the availability of moisture from precipitation, which also affects nutrient availability. Grazing influences are superimposed on this background of topo-edaphic heterogeneity and climatic variability to further influence community-level processes. As a result, species whose adaptations to the prevailing climate and soils would make them the competitive dominants of the community under conditions of light grazing may assume subordinate roles or even face local extinction as grazing pressure increases.

Grazing animals affect plants directly and indirectly. Direct effects of grazing are those associated with alterations in plant physiology and morphology resulting from defoliation and trampling (see Chapter 4). Grazing also influences plant performance indirectly by altering microclimate, soil properties (see Chapter 6), and plant competitive interactions. These indirect affects accentuate plant response to defoliation in ways not readily simulated by clipping experiments.

Over time, the combined direct and indirect effects of grazing on plant growth and reproduction are manifested in plant population dynamics. Herbivores affect the productivity, composition, and stability of plant assemblages through mediation of plant natality, recruitment, and mortality and may cause directional changes in community structure and function. A community may be relatively stable and resistant to changes produced by grazing up to a certain threshold point(s). Beyond these thresholds, changes are rapid and often augmented by climatic events. The pathway of succession following relaxation or removal of grazers may differ substantially from the pathway of retrogression, depending on the mobility and availability of propagules, soil conditions, and climatic variables. In addition, the probability of ecosystem recovery to previous states may be greatly reduced beyond certain critical threshold levels of disturbance or change. The goal of grazing management for sustained yield is to identify these critical thresholds and manage landscapes so as not to exceed them.

Management and manipulation of plant communities for sustained livestock and wildlife production (see Chapters 7 and 8) requires the seasonal integration of information on plant species composition and production across expansive, often heterogeneous areas (landscapes) and over extended planning horizons (decades). If the structural and functional aspects of grazed ecosystems are to be understood at spatial and temporal scales appropriate for long-term sustainability, key plant and population processes must be identified and linked across time to community and landscape levels of integration. Chapter 4 summarizes direct effects of grazing (defoliation) on plant and population level processes developed from short-term, small-scale controlled experiments. In this chapter we review long-term, large-scale changes in plant communities on grazed landscapes. At this level of organization, herbivore effects on microclimate, hydrology, energy flow, nutrient transformation and translocation, and soil physical/chemical properties operate against a backdrop of climatic variability to influence plant species interactions and cause fluctuation, retrogression and succession in communities.

ORGANIZATION OF ECOLOGICAL SYSTEMS

Ecosystems are dynamic, complex, and difficult to define or delimit in space and time. Hierarchical ordering has been applied to multilevel ecological systems to provide a conceptual framework for practical definition. **Systems**, including ecological systems, are groups of interacting, interdependent parts operating together for some purpose. Systems have unique characteristics or **emergent (nonreducible) properties** which are manifest only when sub-components interact to produce larger functional wholes. Hence the axioms "the whole is greater than the sum of the parts" and "a forest is more than just a collection of trees." A system's principal attribute is that we can only understand it fully if we view it as a whole. Water, for example, might be considered a system comprised of hydrogen and oxygen subcomponents. However, the physical/chemical attributes of hydrogen and oxygen by themselves could not be used to predict the unique physical characteristics of H_2O. In ecological systems, the study of individual organisms does not reveal the unique properties of higher levels of organization that emerge from interactions of organisms with each other and their environment.

Hierarchy Theory

Because ecological systems are complex and composed of many interacting parts, it is useful to view their organization as a **hierarchy**, or a graded series with several levels of organization, for example, organisms, populations, communities, ecosystems, and landscapes (Rowe 1961; MacMahon et al. 1978; Allen and Starr 1982). Any level of organization in the hierarchy can be represented as a system, and interactions with the physical environment at each level produces a characteristic, functional system. The components of ecosystems (plants, animals, microbes, geologic substrates, soils, climate) interact and are dependent upon one another for the flow of energy and cycling of nutrients. Each level of organization has characteristic processes that operate at prescribed spatial and temporal scales (Woodmansee and Adamsen 1983; Woodmansee 1988) (Fig. 5.1).

Conceptual schemes such as those depicted in Figure 5.1 are important in that they explicitly identify levels of organization and hence give concrete meaning to the abstract concept of "ecosystem". To minimize misleading, confusing, or confounding comparisons, the same level of organization and the same processes, inputs, and outputs should be compared in ecological studies. At different levels of organization, other processes, inputs, and outputs should be evaluated. The goal of research is to

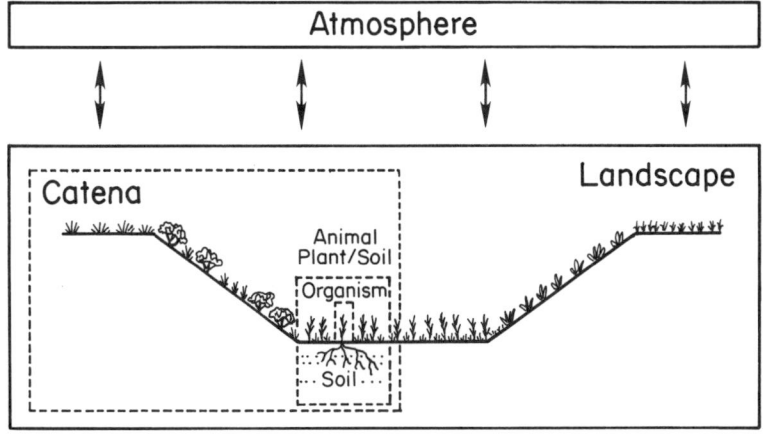

Figure 5.1. Conceptual integration of concepts from biological and pedological hierarchies. Each level of organization in the hierarchy is characterized by key processes, inputs and outputs. Energy and matter are exchanged between components within a level and between levels. In this scheme, an "ecosystem" is a biotic community (plants, microbes, herbivores, carnivores) in association with a given soil body. A "catena" is comprised of linked ecosystems. The landscape is a hierarchical level comprised of catenas.

understand the behavior of ecological systems at various levels of organization and ascertain the properties emerging at each level. Management can then focus on the processes and inputs that regulate these key properties.

In contrast to reductionism, hierarchy theory permits evaluation of a complex system without reducing it to a series of simple, disconnected subsystems. No single level in the hierarchy of an ecological system should be considered fundamental. Understanding a system at one level of organization requires knowledge of the levels both above and below the targeted level (Webster 1979; Allen et al. 1984). Interpreting the behavior of a system at one level of organization without consideration of adjacent levels may generate misleading results. Plant response to grazing illustrates this point (Archer and Tieszen 1986). Controlled defoliation studies have increased our understanding of adaptations that confer tolerance to leaf removal at the organism level (see Chapter 4). However, such studies do not account for other factors which come into play at higher levels of organization. For example, at the organism level, species tolerant to leaf removal in a controlled environment may disappear from the community if they are grazed more frequently or intensely than neighboring plants which are perhaps less tolerant of defoliation. Thus, the infrequently grazed or ungrazed plants may come to dominate the site by virtue of their competitive advantage over the species used more frequently. Studies at the individual plant level often do not take into account key processes operating at the community level of organization, for example, differential grazing of competing species under conditions of limiting resources.

Systems (holistic) and reductionist approaches to analyzing and managing ecological systems should not be viewed as mutually exclusive. Each provides a unique perspective. The reductionist approach dissects lower levels of organization, and provides mechanistic explanations and insights as to how systems work. The holistic approach, on the other hand, views a system in the context of the higher levels in which it is embedded, and provides insight into the significance of phenomena at lower levels. The search for mechanisms should therefore be balanced by concern for significance (Passioura 1979, Allen et al. 1984, Lidicker 1988).

Spatial Scale of Disturbance

Grazing influences interact with climatic variability and other variables to cause changes in plant communities at various spatial and temporal scales. On a large scale, precipitation and temperature regulate vegetation dynamics in arid and semiarid systems (MacMahon 1980; Austin et al. 1981; Sala et al. 1988). However, most plant communities and landscapes are extremely patchy (Belsky 1983), and broad-scale climatic factors cannot account for the existence of these small-scale patterns. Frequent, small-scale perturbations such as ant mounds, small animal activities, patch grazing, and dung and urine deposition (Coffin and Lauenroth 1988) occur within the context of larger-scale, less frequent disturbances such as fire and drought to produce a complex disturbance regime (Collins 1987). Thus, as spatial and temporal frames of observation are diminished, and resolution increased, edaphic heterogeneity and biotic processes (such as grazing) assume greater importance in determining community structure and function.

Herbivore contributions to patchiness include localized defoliation, urine and dung deposition, altered competitive interactions resulting from the differential utilization of plants variously tolerant of defoliation, trampling and the transformation and redistribution of nutrients. These frequent, small-scale perturbations associated with grazing contribute to the development of fine-grained mosaics of varying successional age-states across landscapes. Interpreting community composition and productivity is thus contingent upon our ability to understand the interactive role of concurrent, multiple-scale disturbances (Collins and Barber 1985; Loucks et al. 1985; Collins 1987). To fully appreciate the response of various levels of ecological organization to grazing impacts, it is essential to identify levels of landscape organization, the key processes that occur at each level, their interrelationships, and the influences of various disturbance factors and regimes.

THE ROLE OF CLIMATE

The type, magnitude, duration, frequency, and season of climatic change plays an important role in regulating the rate and direction of plant community changes. On grazed landscapes it is often difficult to assess the extent to which herbivory influences ecosystem processes relative to abiotic factors (McNaughton 1983a; Foran 1986). Despite numerous studies of secondary succession in grasslands, few generalizations have emerged, perhaps because vegetation dynamics in arid and semiarid systems is influenced so strongly by climate (MacMahon 1980). Infrequent but extreme climatic events may be especially important in masking or confounding patterns of vegetation change (Chew 1982).

The influence of grazing on species composition and productivity can be minor relative to the changes caused by variations in rainfall (Fig. 5.2). For example, had the data in Figure 5.2 been collected only from 1932–1937 (a long-term study by most standards!) and only from the heavily grazed site, one would have documented retrogression (Fig. 5.3) without knowing whether climate (drought) or grazing was

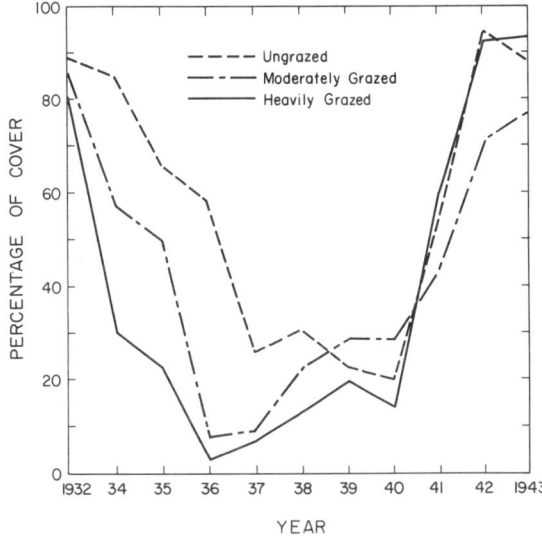

Figure 5.2. Changes in percentage basal cover in the short-grass steppe vegetation type during and after drought (after Branson 1985). The effects of grazing on plant basal cover are relatively minor compared to changes caused by variations in rainfall.

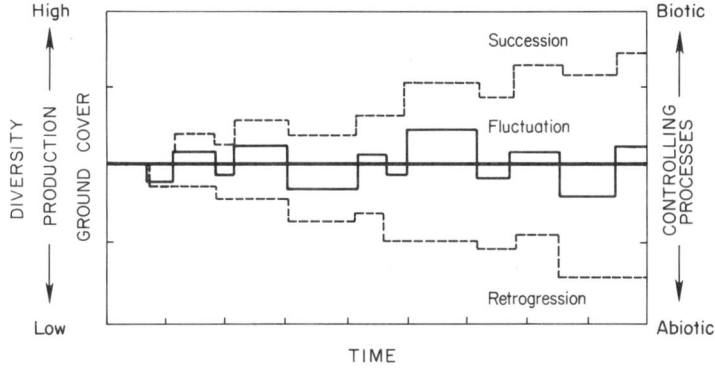

Figure 5.3. Hypothetical changes in community attributes over a number of climatic cycles. All lines illustrate the role of climatic variability in causing community composition to fluctuate through time. The upward path represents succession, whereas the downward path represents retrogression. Abrupt and potentially irreversible transitions may occur when succession or retrogression results in the development of new stable steady states.

the primary cause. Had data been collected only from 1940–1943, succession would have been characterized, but causes for the depleted state of the system at the beginning of the study (grazing versus drought) would not have been clear. When the data are viewed in their entirety (1932–1943 and three levels of grazing) we see that fluctuation in vegetation is a characteristic of this site, that the effects of climate substantially outweighed those of grazing, and that the system was quite resilient in the post-drought period.

Comparisons of vegetation change across grazing systems may therefore reflect differences in prevailing climatic conditions rather than intrinsic differences in management approaches (Herbel 1979). As a result, correlations between grazing and vegetation composition and production derived from short-term, small-scale studies may be spurious. For example, above-ground net primary production was increased under short-duration grazing relative to ungrazed pastures in one year, but reduced during the subsequent year (Heitschmidt et al. 1982). To properly assess the long-term effects of grazing on plant communities, treatments must be compared at the same point in each of a series of climatic cycles (Fig. 5.3).

In some cases, effects of climate have been erroneously ascribed to grazing (Hastings and Turner 1965; Western and van Praet 1973; Chew 1982; Branson 1985). In other instances, vegetation which established from seed under a previous climatic regime may survive under the present regime in a vegetative state. This phenomenon, whereby plants persist for periods of tens to hundreds of years under conditions very different from those under which they initially became established, is known as **biological inertia**. If climatic conditions today are such that the dominant plants cannot successfully reestablish from seed with sufficient frequency to maintain the population, community composition is destined to change regardless of grazing. However, rates of change may be accelerated with increasing grazing intensity. The apparent displacement of grasses by woody plants in desert grasslands of North America over the past 100 years may exemplify this phenomenon (Neilson 1986).

Where plants established under past climates have declined in abundance, their recovery or reestablishment following relaxation of grazing is not to be expected. The concepts of biological inertia and vegetation history thus have important implications with regard to realistically assessing site potential, range condition, and trend. Climate-based simulations predict present-day grasslands will become increasingly susceptible to desertification or woody plant encroachment if changes in temperature and precipitation associated with the greenhouse (warming) effect occur (Emmanuel et al. 1985a, 1985b).

The extent to which shifts in vegetation structure lag behind climatic changes which drive them, and the extent to which vegetation can ever be said to be in equilibrium with climate, are not easily identified (Davis 1982). This makes it difficult to form a baseline from which to judge the effects of grazing on community structure and function or range condition and trend.

GRAZING EFFECTS ON ECOSYSTEMS

Domestic Livestock

The impact of livestock grazing on ecosystems varies in relation to the evolutionary history of the site and the level of grazing pressure (Stebbins 1981; Milchunas et al. 1988). Intermountain grasslands of North America evolved with light grazing and have changed markedly since the introduction of livestock (Mack and Thompson

1982). In contrast, tall-, mixed- and shortgrass prairies of North America, which evolved with bison, pronghorn, and prairie dogs, have been relatively resistant to stresses associated with livestock grazing. Although plant species in ecosystems that evolved with grazing are well adapted to defoliation, domestic livestock can substantially impact their growth and persistence in numerous ways (Pieper and Heitschmidt 1988):

- In contrast to wild herbivores, whose numbers or patterns of grazing may vary substantially from year to year, concentrations of domestic livestock can be artificially maintained at consistently high levels.

- Fences prevent the emigration of livestock to new areas when the abundance of desired forages decreases, resulting in higher frequencies and intensities of defoliation than would occur otherwise.

- Supplemental feeding is used to
  - - Minimize animal mortality that would otherwise reduce grazing pressure when over-utilization of forage occurs; and
  - - Maintain grazing pressure over a greater portion of the year and over a higher frequency of years than would have occurred with native herbivores, both present and prehistoric.

- In grassland or savanna systems that occur in areas climatically and edaphically capable of supporting trees and shrubs, prolonged grazing may decrease the capacity of grasses to competitively exclude woody plants, while at the same time reducing fire frequency and (usually) intensity by preventing the accumulation of fine fuels.

Thus, while grazing has been an important selection pressure in many ecosystems, man has substantially changed its frequency, intensity, extent, and magnitude with the introduction of livestock. The result has been rapid and widespread changes in species composition and productivity of plant communities.

Other Herbivores

Herbivores other than livestock also influence plant community structure and productivity, and their effects on vegetation must be considered when setting livestock carrying capacity and interpreting the effects of livestock grazing on community structure. Activities of root-feeding nematodes, leaf-chewing grasshoppers, termites, herbivorous rodents, lagomorphs, large mammals, and granivores interact with livestock to affect rangeland vegetation. Seasonal and annual fluctuations in the abundance of wildlife populations are often marked, and accurate census data are difficult to obtain. As a result, estimates of numbers or biomass and the relative role of these organisms in regulating energy flow and nutrient cycling are seldom available.

Wild herbivores are often regarded as pests because they may compete with livestock for forage, consume seeds of desirable plants, impair restoration efforts, or disperse seed of undesirable weed and woody species. However, regarding these organisms as wholly detrimental over-simplifies their role. Beneficial activities include consumption of undesirable plants and their seeds, dispersal of desirable species, consumption of insect pests, loosening and aerating soils, and enhancing nutrient cycling (Huntly and Inouye 1988).

High densities of rodents and lagomorphs are perhaps indicators of degraded rangeland. Although the literature on the interactions among rodents, lagomorphs, livestock, and vegetation is sparse and diffuse, it generally appears that rodent and lagomorph densities are correlated with factors such as a high incidence of eroded ground, a high diversity of herbaceous or woody dicots, and low grass cover (Fogden 1978). These organisms may therefore represent symptoms of rangeland deterioration rather than its cause. Potential competition between rodents or lagomorphs and livestock for grasses may be more than offset if these wildlife species retard invasions of undesirable woody plants or weeds.

Although most research has focused on the effects of above-ground grazers on vegetation, below-ground herbivores may actually consume more plant material (Coleman et al. 1976). In addition, below-ground grazers may have a proportionally greater impact on total primary production than would be predicted on the basis of their consumption rates. For example, root-feeding nematodes, grubs, and scarabaeid larvae reduce above-ground productivity by adversely affecting plant growth, metabolism, and nutrient and water uptake (Ridsdill Smith 1977; Ueckert 1979; Detling et al. 1980). Some evidence also suggests aboveground herbivory may change plant-soil-microbial interactions in a manner that benefits soil herbivores (Seastedt et. al. 1988). The role of these important but unseen herbivores in causing vegetation change relative to conspicuous above-ground grazers is not well understood (Anderson 1987).

MEDIATION OF ECOSYSTEM PROCESSES AND CHARACTERISTICS

Direct effects of defoliation on plant growth and development are addressed in Chapter 4. Indirect effects associated with grazing activities further influence plant growth and community composition. These include:

- Alteration of energy flow through the detrital pathway.
- Redistribution and transformation of nutrients.
- Modification of microclimate.
- Alteration of edaphic physical and hydrologic properties.
- Destabilization of plant competitive interactions.

Acting in concert, these factors associated with the grazing process dictate the net impact of plant defoliation on community processes. When direct and indirect effects are considered together, it is easy to understand why activities of grazing animals may exaggerate variations within the community in ways not easily simulated by clipping experiments on individual plants.

Microenvironment and Hydrology

Modifications of site microclimatic and hydrologic properties by grazing constitute potentially important feedbacks that regulate plant responses to defoliation, community composition, and productivity. Grazing and related activities reduce litter accumulation and decrease plant cover which results in increased bare ground. The result may be a warmer, drier, and more extreme microenvironment facilitating an increase in short-lived perennials, annuals, or more xerophytic plants adapted to such conditions.

Microclimatic comparisons of grazed and ungrazed sites are few, but they generally indicate that air and soil temperatures are higher and ground-level wind speeds greater on grazed sites (Whitman 1971). Such changes affect primary productivity and species composition over time. For example, in arid and semiarid systems where water is commonly a limiting factor, herbivore alteration of microclimate may affect availability and utilization of water by primary producers. When the leaf area index is high, grazing may remove transpiring leaf tissue and reduce canopy interception losses of precipitation (see Chapter 6), thereby enhancing soil moisture and enabling plants to sustain growth over longer periods. Defoliation may also increase the water potential of remaining plant parts and contribute to increased rates of leaf expansion (Hodgkinson 1976; Wolf and Perry 1982).

On the other hand, higher rates of transpiration and evaporation on grazed sites resulting from higher radiant heat loads, soil temperatures, and wind speeds may contribute to depletion of soil moisture. Defoliation during periods of low soil water availability may also accentuate plant water stress by reducing root initiation, extension, and activity (see Chapter 4). However, removal of leaf tissue can increase the ratio of root:leaf area, improve the water relations of remaining tissues, and result in conservation of soil water (Archer and Detling 1986; Svejcar and Christiansen 1987).

The net result of grazing-induced modifications of microenvironment on community structure and function has been estimated with the ELM ecosystem-level grassland simulation model on tallgrass prairie. The model showed that cattle weight gain per head, above- and below-ground plant production, transpirational water loss, and standing dead biomass decreased, while soil temperature, water content, and soil water loss increased with increased grazing intensity (Parton and Risser 1980). Results of the simulation are consistent with field observations.

Where grazing has reduced plant and litter cover, sealing of soil surfaces via raindrop impact and hoof compaction may reduce infiltration and increase erosion and runoff (see Chapter 6). In addition, germination and survival of perennial grasses may be greatly reduced on such sites and recovery of surface soil properties following cessation of grazing may require decades (Braunack and Walker 1985; Salihi and Norton 1987). The rate and direction of plant succession following relaxation of grazing may therefore depend upon the degree to which soil properties have been altered in addition to the climatic factors discussed previously.

Nutrient Cycling

Grazers influence nutrient inputs, outputs, and transformations. Consumption of foliage diverts above-ground biomass from the litter component and modifies microclimate, both of which affect activity of soil microbes. In addition, defoliation affects the below-ground nutrient exchange pool by reducing root initiation and extension and increasing root mortality (see Chapter 4). Nutrients are also exported from the system when livestock are moved to other pastures or sold. Grazing-induced changes in microbial activity and the local distribution, form, and abundance of nutrients may then feed back and intensify plant response to defoliation and contribute to changes in species composition. Over the long term, changes in plant species composition or diversity additionally affects litter quality, mass, and seasonal dynamics of decomposition such that a positive feedback loop develops. Under conditions where erosion and runoff increase because of grazing (Chapter 6), nutrient losses from a site may be greatly accelerated.

Nutrient cycling via grazing animals can be important in enhancing or maintaining soil fertility (Floate 1981). Cycling of nutrients through grazers may help keep a pool of readily mineralizable organic nutrients near the soil surface where they are more accessible to plants and microbes (Botkin and Wu 1981). Consumption of vegetation and subsequent defecation could also increase the turnover and availability of various elements that would otherwise remain in recalcitrant organic forms. The fact that shoots of plants on grazed areas may have higher nutrient concentrations than plants from comparable ungrazed areas (Coppock et al. 1983; McNaughton 1984) may be a consequence of:

- Enhanced nutrient uptake by defoliated plants (Ruess 1984).

- Increased nutrient availability resulting from:
  - - Input from dung and urine;
  - - Enhanced microbial activity associated with higher soil temperature and moisture (Parton and Risser 1980);
  - - Decreases in root biomass which limits carbon availability to decomposers causing decreased microbial biomass and increased net mineralization (Holland and Detling 1990).
  - - Elevated levels of root exudation resulting from defoliation. Leakage of high quality organic compounds from roots stimulates microbial activity and speeds nutrient cycling (Ingham et al. 1985; Klein et al. 1988).

Plants require substantial amounts of nitrogen (N) for photosynthesis. When adequate water is available, nitrogen is typically the resource that most limits plant production in arid and semiarid systems. As a result, factors affecting its availability and retention in forms used by plants are important determinants of ecosystem processes. Cattle may return 80–85% of the N ingested with plant tissue to the soil system via urine and feces. Urine generally contains 50–80% of the N excreted by cattle, and vegetation on urine spots may have foliar concentrations of nitrogen 2–3% greater than that of plants on non-urine spots over a 40–60 day period (Stillwell and Woodmansee 1981, Thomas et al. 1988, Day and Detling 1990). Inorganic N levels in soils of urine spots may remain elevated for up to 90 days and above-ground standing crop increased 3- to 7-fold.

Cattle defecate as many as 14 times in a 24-hour period (Weeda 1967), with each defecation impacting an average of 225–600 cm^2 of ground surface (Welch 1985; Brown and Archer 1987). Urination patches of 0.28 m^2 can affect grass growth over an area of 1 m^2 (Wilkinson and Lowrey 1973). Thus, substantial portions of a landscape can be impacted, depending upon the number of animals and their temporal and spatial patterns of movement. Fences, water, shade, and topography dictate animal distribution patterns (see Chapter 3) and hence the pattern of nutrient translocation and redistribution (Senft et al. 1985). For example, Hilder (1964) found 33% of the total feces are concentrated in less than 5% of the area grazed by free-roaming sheep. Productivity may therefore be enhanced on some portions of the landscape by rapid cycling and nutrient import. Conversely, long-term net removal of nutrients may contribute to a reduction of productivity and shifts in species composition in other locales.

Nutrients ingested by grazing animals not returned via excreta and mortality may be lost from the system via harvesting of animal products, animal emigration, and transformation to labile or volatile forms. At stocking rates typical of tallgrass prairie, removal of N in animal biomass is small, amounting to about 0.08 g $N/m^2/y$ [assuming 60 kg weight gain per animal (Parton and Risser 1980, Svejcar 1989) and an N concentration in gained tissue of 0.014 g N/g wet body mass (Berg and Butterfield 1976)]. By contrast, net annual deposition inputs of N to tallgrass prairie have been estimated at 0.44 g/m^2 (wet fall only; NADP 1987) to 1.7 g N/m^2 (Seastedt 1985).

Losses of nutrients consumed in foliage is probably significant, but the magnitude is difficult to quantify. Approximately 12% of the N consumed by ruminants escape as gaseous products of digestion (Church 1969). Of the N excreted by ungulates, 50% may be lost via ammonia volatilization (Woodmansee 1978). However, subsequent analyses by Schimel et al. (1986) suggest such losses may be much lower and spatially variable. In addition, volatilization of N from urine and dung may not represent a net loss, because volatilization from unconsumed vegetation may exceed losses from animal excreta (Detling 1988). In modeling primary and secondary productivity of a tallgrass prairie, Parton and Risser (1980) assessed the various inputs and outputs of nitrogen and predicted that volatilized losses of nitrogen from cattle urine and feces would decrease the net nitrogen balance of a site as grazing pressure increased. Urinary N not taken up by plants or volatilized can be quickly converted to nitrates which are vulnerable to loss by leaching where precipitation is high (Stillwell and Woodmansee 1981).

Grazing may also effect a decrease in the input of "new" plant-available N by impacting lichens and blue-green algae. These nitrogen-fixing organisms constitute potentially important sources of plant-available N in arid and semiarid systems, although their input is highly variable and pulsed. Cryptogams also play a significant role in enhancing surface soil stability and water infiltration. Trampling associated with grazing reduced the number of moss and lichen species by 50% and reduced moss, lichen, and algal cover by 90%, 60% and 50%, respectively, in desert shrub communities (Anderson et al. 1982).

Over the long term, excessive levels of grazing can potentially reduce nitrogen fixation; increase ammonia volatilization, leaching and erosional losses; and cause a net directional transport of nutrients to localized portions of the landscape. The result can be a decrease in overall site fertility and increased heterogeneity of primary production. Reductions in site fertility constitute a positive feedback accentuating defoliation stresses, augmenting shifts in species composition, and determining the rate and direction of succession following relaxation of grazing.

Plant Community Dynamics

Competition. Wooton (1908), reporting on the status of grazing lands in New Mexico, observed that "Stock eat the valuable forage plants and leave the poor ones, thus giving the latter undue advantages in the struggle for existence." Thus, when plants are defoliated in a community rather than as isolated individuals, it becomes difficult to tell whether the plants' response is to loss of leaf tissue or to changes in interference from surrounding vegetation (Jameson 1963). In this regard, Mueggler (1972, 1975) and Archer and Detling (1984) found that regardless of severity of defoliation, plants under competition had greater reductions in biomass and flower production and were slower to recover from defoliation than were plants clipped under conditions of reduced competition.

At the community level of organization, differences in animal forage preferences (see Chapter 3) often result in differential frequency and intensity of defoliation of plants on a site and cause a shift in competitive interactions. Plants grazed less frequently gain an advantage over plants that are utilized at greater frequency or intensity. Thus, grazing-induced shifts in competitive interactions contribute to changes in plant community composition over time. On a local scale, herbivores may therefore play a key role in mediating species composition in the community through the differential utilization of plants variously tolerant to defoliation.

The effects of grazing on plant distribution and abundance vary. Figure 5.4 illustrates plant distributions in 1982 along a topo-sequence in pastures stocked at different rates with cattle since 1939. Plant responses to grazing varied with stocking rate

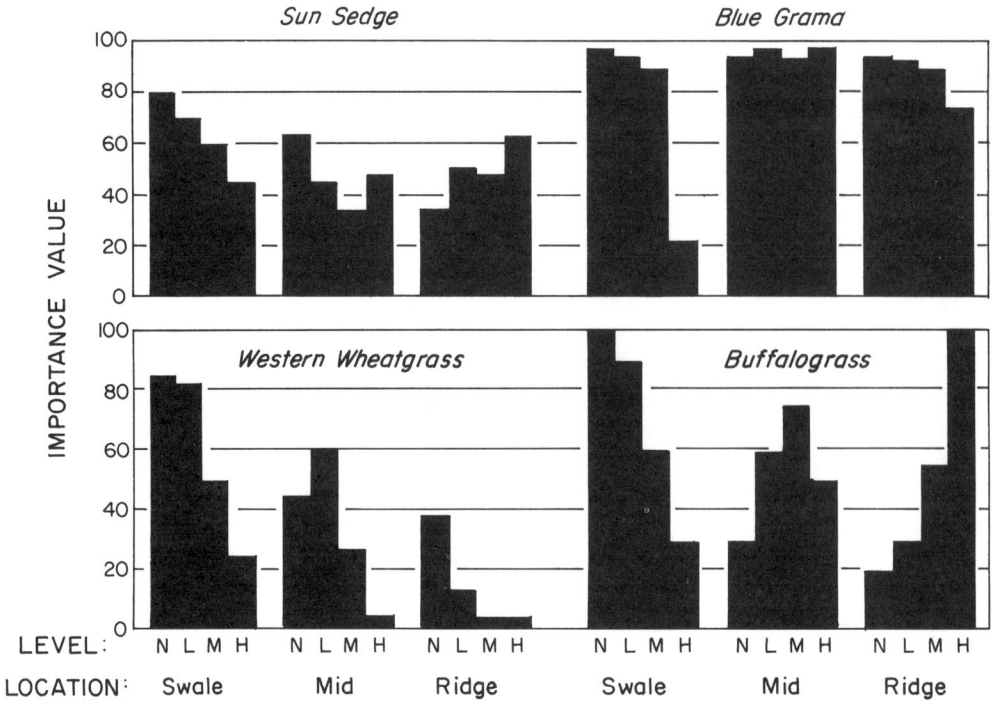

Figure 5.4. Mean importance values (relative cover plus relative frequency) of four graminoids along short-grass steppe topo-sequences in pastures subjected to different levels of cattle grazing since 1939 (N=none; L=light; M=moderate; H=heavy) (after Archer and Tieszen 1986).

and topographic position. Some species increased with increasing grazing pressure at some locations and decreased on others (e.g., buffalograss and sun sedge). Species tolerant of heavy grazing at some locations on the topo-sequence were less tolerant at other locations (e.g., buffalograss). Blue grama was able to maintain its status in the community regardless of grazing pressure, except in the lowlands under conditions of heavy grazing. Soil physical properties, nutrient content, and nitrogen mineralization rates differed significantly between swale, mid-slope, and ridge locations on this site (Schimel et al. 1985). Seasonal patterns of relative cattle grazing preference (ratio of percent of time grazing at a given topographic location to the percent of pasture area occupied by that topographic feature) ranged from 1.4 for swales to 0.4 for ridgetops on the site (Senft et al. 1985).

Differences in plant distribution and abundance along hill-slope gradients in these pastures subjected to different grazing regimes may be explained in several ways:

- Energy budgets, water relations and nutrient availability varied along the gradient and affected intrinsic plant tolerance to herbivory.

- The probability of being grazed at a given frequency and intensity by a given class of herbivore (e.g., nematodes, insects, rodents, ungulates) varied across the topo-sequence along with patterns of grazing (e.g., patchy versus uniform).

- The competitive neighborhood of a given plant species changed along the gradient.

Thus, at any given location, some species may be more or less tolerant to defoliation than others and they may be more or less likely to be grazed than others. As summarized in Chapter 4, plant adaptations to herbivory may fall into two general categories: **grazing avoidance** and **grazing tolerance**. Grazing tolerant plants have characteristics that facilitate the reestablishment of foliage following grazing, whereas plants that avoid grazing have characteristics that minimize the probability of defoliation (Archer and Tieszen 1980; Mooney and Gulmon 1982; Briske 1986). Although plants tolerant of defoliation would seem to have an advantage in grazed systems, such plants will eventually be disadvantaged when competing with avoidance-type species which, although less tolerant of grazing, are defoliated at a lower frequency or intensity. Where grazing intensity is high, the grazing avoidance-type plants will inevitably dominate the site because of the overriding influence of utilization. Evidence for this premise is most striking from southwestern hot deserts, the Great Basin and the Southern Plains, where extensive areas of grasslands characterized by grazing tolerant-type plants (perennial grasses) have given way to shrublands dominated by unpalatable grazing avoidance-type plants. In California, grazing-tolerant perennial grasses appear to have given way to annual grasses which could be viewed as employing an avoidance or escape-type strategy.

Diversity. In terrestrial plant communities, much of the theory of community organization has stressed the role of competitive interactions among plant species. Although plant species may occupy distinct niches in terms of their resource requirements, herbivores mediate species abundance and diversity through differentially utilizing plants variously susceptible to defoliation. Certain levels and combinations of grazing or disturbance increase overall plant species diversity by decreasing the capacity of competitive dominants to exclude other species and by creating gaps available for occupation by other species (Huston 1979; Archer et al. 1987; Collins 1987; Collins et al. 1987). Above certain frequencies or intensities, disturbance typically lowers diversity. This phenomenon of increased diversity at moderate levels of disturbance has been termed the **intermediate disturbance hypothesis** (Connell 1978).

Grazing can stimulate diversity by reducing the capacity of competitive dominants to exclude other species via defoliation, trampling, and dung deposition. The effects of grazing on plant diversity depends upon grazing intensity, the evolutionary history of the site, and climatic regimes (Milchunas et al. 1988). In semiarid grasslands with an evolutionary history of grazing, vertebrate herbivory appears to have a relatively small effect on community composition (Fig. 5.5). In these systems, defoliation enhances tillering and spread by rhizomes and stolons and seems not to affect competition for resources (McNaughton 1983a, 1984). In contrast, climatically similar grasslands with a short evolutionary history of grazing lose diversity at much lower grazing intensities. In regions with higher rainfall, a few tall species will be the competitive dominants of the community when grazing pressure is low. Moderate grazing increases diversity in these systems by creating mosaics of short grasses and forbs which occur on heavily grazed patches, mixtures of tall and short grasses on moderately grazed patches, and tall grasses on lightly grazed patches. Heavier grazing eventually causes diversity to decline as short grasses dominate an increasingly greater proportion of the community.

Community diversity has important implications for grazing management. Growth of each species in a community is limited by a different combination of environmental factors. Fluctuations in weather cause production of individual species to vary substantially from year to year. However, production of the whole

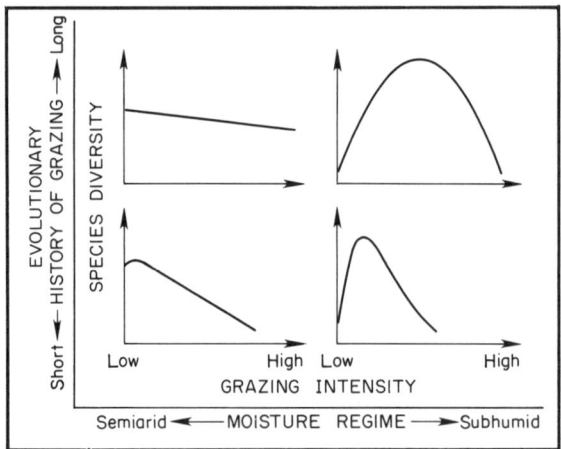

Figure 5.5. Hypothesized relationships between relative plant species diversity in grassland communities in relation to grazing intensity along gradients of moisture and evolutionary history of grazing (source: Milchunas et al. 1988).

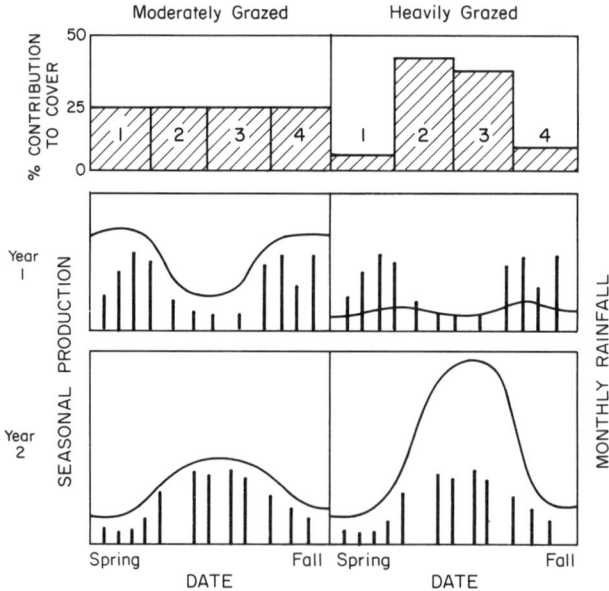

Figure 5.6. Hypothetical example of how changes in species abundance resulting from grazing (top panel) might affect the amount and variability of above-ground net primary production in years with contrasting patterns of rainfall (mid- and bottom panel). Vertical bars represent rainfall (Walker 1988).

community is more stable, because years favorable for growth of some species cause a compensatory decrease in growth of other species (McNaughton 1977; Chapin and Shaver 1985; Collins et al. 1987). Conversely, in stressful years, the loss of productivity of some species is compensated for by growth of others. As a result, changes in relative growth rates and abundances of co-occurring species tend to stabilize ecosystem processes such as primary production (Fig. 5.6). For example, C_3 grasses and shrubs may be physiologically active early in the spring, whereas C_4 species maintain growth for a greater proportion of the warmest, driest portions of the growing season. The result, at certain latitudes or elevations, is that a diverse mixture of C_3 and C_4 plants may give more stable and sustained productivity than a monoculture of either.

Certain levels of grazing may enhance above-ground net primary production (ANPP) (McNaughton 1979) if diversity increases. In other cases ANPP is relatively unaffected over a wide range of grazing intensities and durations (Whicker and Detling 1988). Above certain grazing intensities, the level of ANPP is reduced and its variability increased by either reducing productivity of dominant species or by increasing the proportion of short grasses, forbs, and annuals in the community. Species-rich plant communities are potentially more **resilient** (i.e., better able to regain functional characteristics after disturbance) to repeated defoliation than species-poor communities (Brown and Ewel 1988), perhaps reflecting the ability of species mixtures to use resources more fully than monocultures (Harper 1977).

Species Composition and Population Dynamics. Sustained productivity and long-term survival of plants in grazed systems depends upon successful reproduction in parental generations and the recruitment of new individuals. By mediating plant natality, recruitment, and mortality, herbivores affect the productivity, composition, and stability of plant communities. Changes in basal area, relative abundance, and species composition in grazed systems inevitably reflect differential recruitment, longevity, and survival of individuals comprising the community. Population parameters such as these provide a tangible, quantifiable link between individual plant and community-level processes and should forecast impending changes in community composition. Proper grazing management aids recruitment and persistence of desired species, whereas poor management hastens the demise of preferred species and leads to their replacement by other species (Jones and Mott 1980).

Traditionally, plants have been classified as decreasers, increasers, or invaders with respect to their response to grazing (Dyksterhuis 1949) (Fig. 5.7). However, the functional response of a given species to grazing can vary from site to site and across topographic and edaphic gradients (Fig. 5.4). Species that initially make up a large

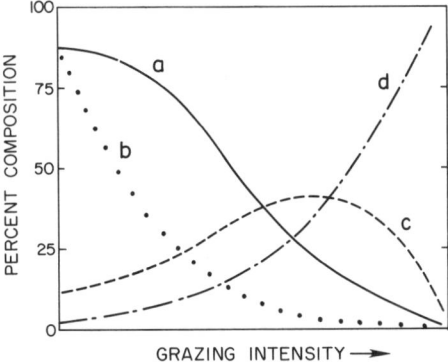

Figure 5.7. Plant species response to grazing at the community level of organization (after Dyksterhuis 1949). Species which are intolerant to defoliation or are more heavily grazed than others will decrease with increasing grazing, some more rapidly than others (a, b = decreasers). Plants less preferred or more tolerant to defoliation subsequently increase (c). If grazing intensity is maintained, canopy gaps created by the loss of perennials are occupied by unpalatable or ephemeral plants (including shrubs in many systems) previously absent or of limited abundance (d = invaders). It should not be assumed that the path of succession following relaxation of grazing pressure will necessarily result in the re-establishment of the earlier species composition (see Fig. 5.11 and related discussion).

proportion of the community, but which decline with increased grazing, are categorized as **decreasers**. Their decline with grazing is related both to defoliation tolerance and level of utilization relative to other species in the community. Species which **increase** in grazed communities do so because they are more tolerant of defoliation and/or they are less frequently grazed (less preferred, less palatable). Where annual rainfall supports tall- and midgrass species, there is often a decrease in the contribution of the competitively dominant taller species and an increase in more defoliation-tolerant, short-statured, or prostrate species or growth forms with grazing (Clarke et al. 1947; Herbel and Anderson 1959; Detling and Painter 1983; Archer et al. 1987; Smeins and Merrill 1988; Noy-Meir et al. 1989). Reductions in basal area, root biomass, mulch levels, and leaf area of dominants by grazing creates establishment gaps which benefit other growth and life-forms. As a result, an influx of species which were absent or of minor importance at lesser levels of grazing occurs. These **invaders**, typically annuals or unpalatable (and sometimes toxic) perennial herbaceous or woody plants, are often undesirable for livestock production because they displace more palatable species, are of lower nutritive value, or have low, erratic, or highly seasonal productivity.

Rates of change which accompany grazing are not well quantified. Shifts in species composition can be abrupt, non-linear (Fig. 5.3), and punctuated by fluctuations in rainfall (Fig. 5.2). Changes in physiognomy from tall to short grasses can occur within a few years (Archer et al. 1987; Smeins and Merrill 1988; Thurow et al. 1988b). Woody species pose special problems in many grasslands and savannas around the world. Significant physiognomic changes from grass to woody plant domination can occur within decades.

The Role of Woody Plants. Woody plants are a common component of most rangelands around the world. However, trees and shrubs have traditionally been viewed negatively, because they are presumed to reduce herbaceous production and because their presence increases the difficulty of livestock manipulation. For these reasons, shrubs and trees are frequently the targets of vegetation manipulation technologies aimed at improving livestock carrying capacity or handling. However, when assessing whether to invest in efforts to reduce woody plant cover or density, the following points should be considered:

- In many regions or landscapes within a region, the shrub or tree life-form is best adapted to the prevailing biotic and abiotic conditions, and woody plants play a key role in primary production and nutrient cycling while stabilizing soils and providing habitat for wildlife. On sites such as these, it may not be realistic to expect management or manipulation to enhance herbaceous production.

- Are woody plants reducing herbaceous production or is the increased abundance of woody species the result rather than the cause of decreased herbaceous production associated with improper grazing management? Effects of woody plants on herbaceous production varies from (a) stimulation beneath tree or shrub canopies relative to interstitial zones (Christie [1975] for poplar box; Barth and Klemmedson [1978] for algarrobo; Scifres et al. [1982] for huisache; Stuart-Hill et al. [1987] for *Acacia karroo*; Ludwig et al. [1988] for various hot desert shrubs; Belsky et al. [1989] and Weltzin and Coughenour [1990] for umbrella thorn and baobab); to (b) no effect (Harrington [1973] for *Acacia hockii*); to (c) decreased standing crop (Jameson [1967] for pines; Walker et al. [1986] for poplar box and narrow-leaved iron bark; Clary [1987] for pinyon-juniper woodlands; Scanlan [1988] for subtropical thorn shrubs).

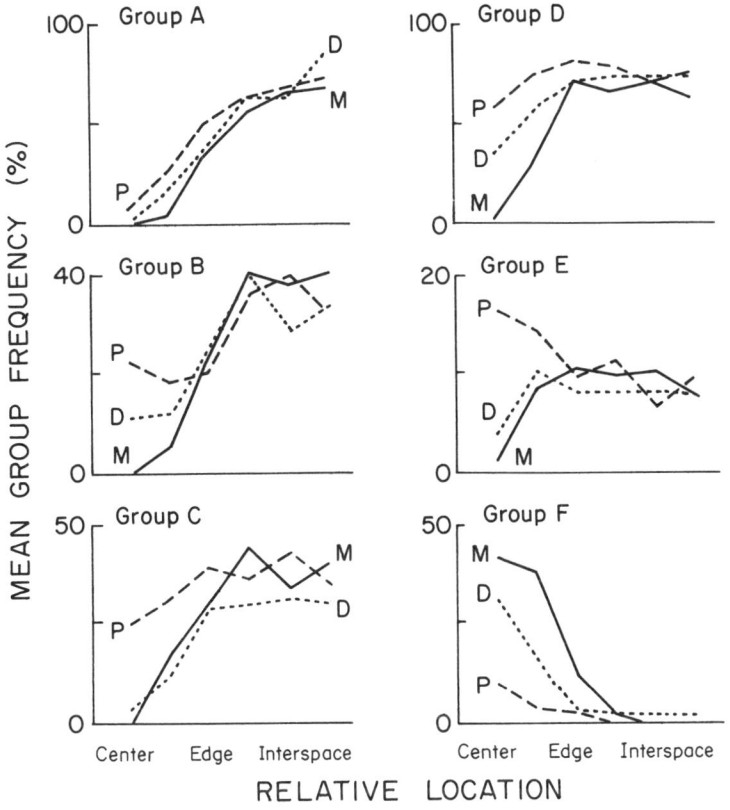

Figure 5.8. Changes in the frequency of occurrence of herbaceous plants accompanying the successional development of woody clusters in a savanna parkland in southern Texas. A classification analysis of 19 herbaceous species generated the six response groups shown here (Group A–Group F). Spatial gradients extend from the center and edge (dripline) of woody clusters into neighboring interstitial zones. Temporal gradients reflect increased density and diversity of woody plants in clusters (P = pioneer mesquite and 2 to 4 other woody species; D = developing woody cluster [8–10 woody species]; M = mature cluster [12–15 woody species]) (after Scanlan 1988).

- The effect of shrubs on herbaceous production depends upon the site (topoedaphic features), the climate (especially the temperature and rainfall regime), the species of woody plant and its growth-form (i.e., single- vs. multiple-stemmed), canopy architecture, size, and density (e.g., Frischknecht 1963; Scifres et al. 1982; Aucamp et al. 1983; Walker et al. 1986; Belsky et al. 1989). Herbaceous species respond differently to the presence of woody plants (Fig. 5.8), and manipulation of woody species may adversely affect key forbs and grasses. For example, mesquite savannas in the Rolling Plains of Texas were characterized by C_3 grasses beneath mesquite canopies versus C_4 grasses in zones between trees. Following herbicide application C_4 grasses displaced C_3 grasses beneath mesquite plants (Heitschmidt et al. 1986). As a result forage production in spring was delayed and greatly reduced, necessitating extended supplemental feeding.
- The presence of woody plants may:
  - - Reduce grazing pressure on grasses and provide protection for heavily utilized herbaceous species (Welsh and Beck 1976; Davis and Bonham

1979; Jaksic and Fuentes 1980);
- • • Reduce supplemental feed requirements during cold or dry periods (Cook 1971);
- • • Aid in meeting nutritional requirements of herbivores;
- • • Enhance soil nutrient status and water infiltration;
- • • Provide shade; and
- • • Improve habitat for game and non-game wildlife.

- • Will woody manipulation stimulate herbaceous production and increase livestock carrying capacity sufficiently to offset treatment costs? If so, how much time will be required before a follow-up treatment? Will treating one problem perhaps create another (i.e., single- vs. multiple-stemmed growth habit; replacement of non-sprouting species [e.g., big sagebrush] with sprouting species [e.g., rubber rabbitbrush])? Answers to these questions vary between species and sites (Scifres et al. 1988) and the type, timing, and sequencing of vegetation manipulation technologies (Scifres et al. 1983).

- • Desirable woody plants that provide cover and palatable forage for livestock and wildlife may have been reduced in abundance or eliminated by excessive grazing (Orodho et al. 1990) or by application of non-selective brush manipulation practices and replaced by other woody species (Fulbright and Beasom 1987). When woody plants are sprayed with herbicides, forbs utilized by wildlife (for example, deer and quail) are often adversely affected.

- • Aesthetic value.

Shifts in Grass and Woody Plant Abundance. Grazing animals in concert with human activities have caused the degradation of woodlands in Africa, Asia, and India (Tothill and Mott 1985). On the other hand, grazing has also been implicated in the spread of bush in Africa, desert and thorn scrub in North and South America, and acacia and eucalyptus woodlands in Australia, at the expense of grasslands and savannas. In the latter instances, increased grazing intensity may favor woody plants by decreasing herbaceous standing crop, reducing fire frequency and intensity, and enhancing the dispersal and germination of woody plant seeds. Alternative hypotheses regarding the balance between contrasting life-forms (grasses versus shrubs and trees) are centered on climatic (frequency, amount, and seasonality of rainfall) and edaphic factors (soil texture, nutrient status, moisture).

Quantitative and historical assessments indicate woody plant abundance has increased substantially in grasslands during the last century in Africa (Kelly and Walker 1976; van Vegten 1983), Australia (Harrington et al. 1984), India (Singh and Joshi 1979), North America (Smeins 1984), and South America (Schofield and Bucher 1986; Bucher 1987). Remaining grasslands and savannas may become increasingly susceptible to woody plant encroachment in response to anticipated global changes that may generate warmer, drier climates characterized by greater variability (Emmanuel et al. 1985a, 1985b). Although encroachment of woody plants into grasslands has been widely recognized, the rates, patterns, and dynamics of the process have seldom been quantified. However, in western and southwestern North America the transformation of grasslands and savannas to shrublands or woodlands appears to have proceeded exponentially within the last 200 years (Blackburn and Tueller 1970; Herbel et al. 1972; Young and Evans 1981; Madany and West 1983; Williams et al. 1987; Archer 1989).

Factors regulating the balance between graminoid and woody plant life-forms include climate, soils, disturbance (e.g., grazing, fire), and their interaction. Changes in one or more of these factors may enable woody plants to increase in abundance. A shift from grassland or savanna to shrub or woodland may result if:

- Climate or disturbance regimes change to enable native woody species to
  - • Extend their geographic range (e.g., creosote bush [Hunziker et al. 1977]);
  - • Increase in stature and density within their historic ranges (for example, honey mesquite [Johnston 1963]; one-seeded juniper [Johnsen 1962]; baccharis [Williams et al. 1987]).
- Introduced woody species successfully establish and reproduce (e.g., McCartney Rose [Scifres 1980]; mimosa [Lonsdale and Braithwaite 1988]).

Characteristics common to many woody species that increase in grazed environments include:

- High seed production
- Seeds that persist in soil for many years
- Ability to disperse over long distances
- Ability to sprout following top removal
- Tolerance to low levels of water and nutrients
- Low palatability.

Evidence suggests climate may have changed in recent history to favor woody plants over grasses in portions of North America (Hastings and Turner 1965). Grasslands and savannas which established under previous climatic regimes may have been only marginally supported by the recent climate and were perhaps prone to woody plant invasion (Neilson 1986, 1987). In addition, oscillations between different climatic regimes in recent history (Mitchell 1980) may have effected shifts in plant recruitment patterns to promote episodes of woody plant seed production and seedling establishment, but not necessarily their local extinction (Neilson and Wullstein 1985). However, changes or fluctuations in broad-scale climatic regimes

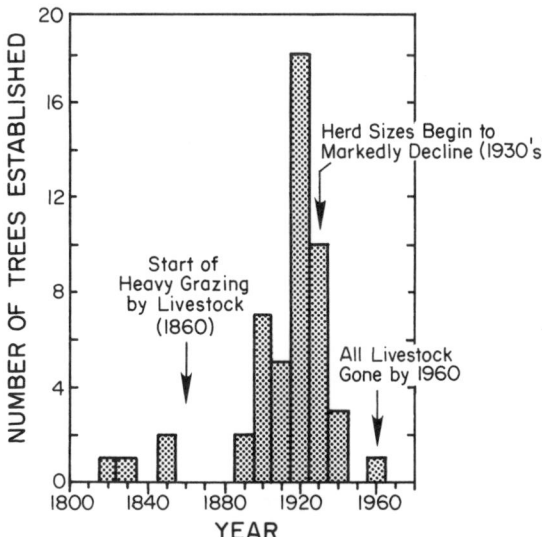

Figure 5.9. Establishment dates for ponderosa pine indicate the forest present on this Utah site developed soon after the introduction of cattle. Tree recruitment over the same period was low on nearby, edaphically similar pine savannas protected from grazing. The close geographic proximity of the sites ruled out climate change as a causal factor for vegetation change (after Madany and West 1983).

cannot explain how grasslands and savannas have persisted on some sites within a climatic zone but not others.

On a local scale, fire, grazing, and soil properties interact within a variable climate to determine the balance between grasses and woody plants. Madany and West (1983) have documented a case in which a savanna protected from cattle grazing was maintained despite low fire frequency (and possible climatic change), whereas nearby edaphically similar sites subjected to cattle grazing had substantially higher densities of woody plants which established after the introduction of livestock in the late 1800s (Fig. 5.9). Such data suggest climatic fluctuation in recent history may have been necessary, but was not sufficient, to have caused a shift from savanna to woodland.

Woody Plant Establishment. Numerous studies in North America indicate increased densities of woody plants in grasslands and savannas coincident with the introduction of domestic livestock (Chew and Chew 1965; Blackburn and Tueller 1970; Young and Evans 1981; McPherson et al. 1988; Archer 1989). Large numbers and high concentrations of livestock potentially favor establishment of woody plants in numerous ways:

- Increased dispersal of viable, germinable seed, either directly (Table 5.1) or indirectly, by creating changes leading to increased numbers of other seed-dispersing organisms such as rodents and birds.

- Compaction of surface soils may favor recruitment of woody plants over grasses (Braunack and Walker 1985).

- Defoliation of grasses can increase the probability of woody plant seedling establishment and subsequent rate of growth and development by:
  - • Adversely affecting the capacity of grasses to preempt resources above- and below-ground (Caldwell et al. 1987);
  - • Causing shifts in herbaceous composition to assemblages less effective in garnering site resources (Eissenstat and Caldwell 1988); and
  - • Reducing biomass and continuity of fine fuels and hence fire frequency.

- Time to reproductive maturity may be decreased (McPherson and Wright 1987) and the frequency and magnitude of seed production may be increased among woody plants on grazed sites.

Grazing management strategies for limiting woody plant encroachment should focus on reducing seed production and dispersal, enhancing the capacity of grasses to competitively exclude seedlings of woody plants and enabling periodic use of fire to induce mortality and minimize the number of woody plants that attain reproductive status.

Table 5.1. Survey results of mean ($\pm$SE) mesquite seed and seedling density in areas with and without cattle on an upland site in a Texas savanna parkland. *Canopy* refers to seeds and seedlings beneath adult mesquite plants; *open* refers to occurrences in herbaceous zones. The absence of seedlings on the area without cattle suggests native wildlife on the site were relatively poor agents of mesquite dispersal (after Brown and Archer 1987).

	With Cattle		Without Cattle	
	Canopy	Open	Canopy	Open
Area sampled (%)	8	92	9	91
Seedlings^{-2}	12 ± 2	15 ± 2	0 ± 0	0 ± 0
Seeds^{-2}	33 ± 7	11 ± 2	33 ± 8	0 ± 0

Although numerous studies have cited the importance of the relationship between grazing, graminoid biomass and woody plant invasion, these parameters have seldom been quantified. Herbaceous retrogression that accompanies grazing has been established in numerous studies. Yet few guidelines have been developed to predict how grazing at different stages of herbaceous retrogression and levels of stocking alters the susceptibility of sites to invasion by woody plants. It is often assumed that well-developed stands of grass can exclude woody seedlings and that sites with a history of heavy grazing are most likely to have woody invasion, other factors held constant.

To test these assumptions, honey mesquite (hereafter called mesquite) establishment was quantified on edaphically similar areas with different grazing histories (long-term heavily grazed vs. protected from livestock grazing for 40 years) and with various levels of herbaceous defoliation (none, moderate, and heavy). Mesquite seedling emergence and survival after two years was comparable and high (>75%) on plots subjected to moderate and heavy defoliation, regardless of site grazing history (Brown and Archer 1989). Above-ground growth, photosynthesis, conductance, and water potentials were comparable among one-year-old mesquite seedlings on grazed and protected sites, even though differences in herbaceous species composition and above- and below-ground biomass on the two sites were substantial. These data suggest competition for soil resources between grasses and woody plants such as mesquite may be minimal early in the life cycle of mesquite. The ability of woody plants such as mesquite to establish in grass-dominated stands may be related to the rapid development of a root system (Table 5.2) which enables the plant to access soil moisture at depths not effectively utilized by grasses (Table 5.3).

Table 5.2. Mean ($\pm$ SE) above-ground height and biomass, tap root length and biomass of mesquite seedlings. Seedlings were newly emerged (cotyledons only = NE) and recently established with at least one true leaf on 3 May (EST) and approximately 4 months old (4 mo) on 22 August. Excavations were terminated at 50 cm. Nearly 90% of the herbaceous root biomass was within 30 cm of the surface on this site (after Brown and Archer 1990).

	Shoot		Tap Root	
Age	Height (cm)	Biomass (g)	Length (cm)	Biomass (g)
NE	2.8 $\pm$ 0.2	0.11 $\pm$ 0.01	5.5 $\pm$ 0.0	0.04 $\pm$ 0.00
ESTA	7.4 $\pm$ 0.8	0.63 $\pm$ 0.11	20.7 $\pm$ 3.7	0.45 $\pm$ 0.12
4 mo	8.1 $\pm$ 1.1	0.68 $\pm$ 0.15	41.3 $\pm$ 5.1	0.73 $\pm$ 0.14

Table 5.3. Best two-factor regression models for variables regulating maximum daily conductance in one-year-old mesquite seedlings, mature mesquite plants, and tillers of the C_4 grass, hooded windmillgrass. Variables considered were % soil moisture (SM) in upper (0–30 cm), middle (30–60 cm), and lower (60–150 cm) horizons, air temperature, vapor pressure deficit (VPD), and light intensity. By the time mesquite seedlings were one year old, they were utilizing water at depths poorly exploited by grasses. Similar results were obtained for net photosynthesis (after Brown and Archer 1990).

Plant	Variable	Partial R^2	P
Hooded windmillgrass	SM-Upper	0.59	**
	SM-Middle	0.22	**
Mesquite seedlings	SM-Middle	0.47	*
	SM-Lower	0.24	*
Mature mesquite	SM-Lower	0.65	**
	VPD	0.15	*

*$0.01 < P < 0.05$ **$P < 0.01$

Establishment would be particularly enhanced when seeds germinate during periods when competition for soil moisture is minimal. These results, together with observations of Meyer and Bovey (1982) and Smith and Schmutz (1975), indicate grasses may not be effective at competitively excluding mesquite. The relatively unpalatable shrub big sagebrush also has great potential to increase in density regardless of ecological or management conditions short of periodic fires (Tisdale and Hironaka 1981; Johnson 1987).

Thus, in many instances, thresholds of herbaceous biomass production or composition required to limit establishment of woody seedlings may be exceeded even at low levels of grazing. Management focused on regulating levels of herbaceous utilization may slow the rate but may not curtail the invasion of aggressive woody species. Grazing systems which allow for the regular use of fire seem to be required to successfully regulate woody plant density, stature, and seed production on many sites. For woody legumes whose seeds are effectively distributed by livestock (e.g., mesquite and acacia), management directed at reducing seed dispersal might retard rates of encroachment. Pastures containing seed-bearing trees might be deferred during times of pod production. Livestock grazed on areas coincident with pod production should be detained long enough for seeds to pass through their digestive system (several days) before moving them to pastures where these plants are not a problem (Fig. 5.10).

Figure 5.10. The importance of livestock as agents of seed dispersal. (A) Fenceline contrast in the Mitchell grasslands of Queensland, Australia. Scattered prickly acacia plants were introduced from Africa in the 1940s for shade along bore drains in the pasture on the left. By the late 1980s the plant, spread primarily by cattle, dominated the property. The absence of prickly acacia in the pasture on the right suggests native fauna were not effectively disseminating seeds of this plant (photo by S. Archer). (B) Ground distribution of pods of honey mesquite beneath a tree whose canopy extended across a fence separating pastures with (right) and without (left) cattle. The accumulation of pods on the ground on the side without cattle illustrates the potential differences in the relative importance of livestock versus native fauna in seed dispersal on this southern Texas site (see Table 5.1) (photo by T. D. A. Forbes).

CONDITION AND TREND

Within a climatic area, differences in vegetation across a landscape reflect intrinsic variations in soils and topography. Resource managers have divided landscapes into **range sites**, which are comprised of one to several soil series potentially capable of producing the same kind, proportion, and abundance of late successional plant species (Shiflet 1973). The estimated climax community on a given range site has been used as a basis for determining **range condition**, which is defined as the composition of the current plant community relative to the known, presumed, or preferred climax. Additional details on rangeland condition in relation to plant succession can be found in Lauenroth and Laycock (1989).

Vegetation dynamics on range sites include fluctuation, retrogression, and succession (Fig. 5.3). **Fluctuation** represents reversible changes in dominance within a stable species composition, whereas succession and retrogression are directional changes in composition and dominance (Rabotnov 1974). As species composition and the contribution of various species to site production deviates from the idealized climax, range condition declines through the phenomenon of **retrogression** or **retrogressive succession** (Barbour et al. 1987). Grazing or changes in environmental conditions can cause retrogression which eventually leads to a loss of diversity, net primary production, and ground cover. As a result, site processes become increasingly coupled to and regulated by abiotic factors (Fig. 5.3). This, in turn, may accentuate fluctuation. Progressive directional changes, termed **succession**, occur in the opposite direction and represent the recovery of ecosystem structure following biotic or abiotic disturbance. As plant diversity, production, and ground cover increase through time, the plants themselves exert substantial control over microclimate, energy flow, nutrient cycling, and species interactions, thus dampening fluctuation associated with oscillation of weather and abiotic factors. Changes in range condition through time, termed **trend**, indicate whether succession or retrogression predominate on a site.

"Proper" management is assumed to be that which (1) minimizes the likelihood of a site retrogressing or being degraded to the point where primary and secondary productivity are adversely affected and soil resources are endangered; and (2) facilitates succession. As illustrated in Figures 5.2 and 5.3, determination of trend requires data over sufficient periods of time to distinguish fluctuation related to infrequent climatic events from directional change. Few such data exist. Collins et al. (1987) analyzed 39 years of vegetation data from two Oklahoma sand sagebrush-little bluestem-blue grama sites grazed by cattle. Each site contained an exclosure. The vegetation in the grazed portions of each pasture exhibited shifting patterns of abundance rather than sequential species replacement (i.e., fluctuation), as did one of the exclosures. The other exclosure exhibited a directional change from dominance by annuals to dominance by perennial grasses.

Terminology traditionally associated with condition classes (excellent, good, fair, poor) and species (desirable, undesirable) represent situation- or user-specific value systems. Terms such as "poor" and "undesirable" are relative and the reference point may vary substantially among different people. Species assemblages regarded as desirable for sheep may be undesirable for cattle, and vice-versa. Plants regarded as desirable for livestock may be undesirable for wildlife such as deer and quail, and vice-versa. A site in poor condition for livestock may be in excellent condition for wildlife. Because the connotation of these terms varies between user groups, they should be well qualified when used.

The idea underlying the range condition concept was to provide a base line from which to evaluate changes in ecosystem attributes through time (Dyksterhuis 1949). It does not necessarily follow that management for climax is necessary, desirable, or achievable for several reasons:

- The original state of vegetation and natural fire and grazing regimes prior to settlement is seldom known. In most cases it is not clear what constitutes climax vegetation.
- Sub-climax or introduced plant species assemblages may be desirable (i.e., palatable, productive, seasonally available) when managing for certain herbivores.
- Maximum plant productivity is not necessarily higher at later stages of succession.
- Productivity of many large herbivores (e.g., white-tailed deer, moose, cattle, sheep) may be greatest in vegetation states not considered "climax" (see Chapter 1).
- Late-successional or climax communities are not necessarily more stable than mid- or even early-successional communities (McGinty et al. 1978).
- The concept of single equilibrium communities that progress steadily toward or away from climax depending on grazing pressure does not apply in many arid and semiarid systems. Examples of the importance of stochastic events in shaping the path of succession, alternative steady states, and discontinuous and irreversible transitions are abundant (see Westoby et al. 1989).
- The rate and path of succession varies depending on the extent to which soil properties have been impacted. Return to the presumed climax or late successional communities may not occur over time frames relevant to management.
- Alternative steady states are possible on a given site. If disturbance thresholds are exceeded, positive feedbacks can cause systems to shift to a new steady state, regardless of grazing practices.

Retrogression

If climate becomes warmer and drier or if the frequency of drought increases, plant productivity, cover, and diversity decreases and species composition shifts toward an increasing proportion of annuals and xeromorphic perennials. This represents **retrogressive succession** (Barbour et al. 1987). Grazing can also induce retrogressive succession, as palatable grasses, forbs, and shrubs succumb to repeated defoliation (see Chapters 3 and 4) and are eventually replaced by other growth and life-forms. Grasslands are able to tolerate a moderate degree of grazing intensity before changing in composition, diversity, or productivity. However, as grazing intensity is increased or becomes continuous, tall and mid-grasses eventually give way to short-statured perennial grasses, which, in turn, give way to annuals and unpalatable perennials with a concomitant loss of primary and secondary productivity, diversity, cover, and soil. As discussed previously, the level of grazing required to cause a change from one of these states to another depends upon the type and numbers of herbivores, the plant species involved, and the evolutionary history of the site with respect to grazing. Retrogression associated with livestock grazing may be mitigated when growing season conditions are favorable or magnified in unfavorable years.

Retrogression implies degradation, and in range management it is typically used to describe the replacement of perennial species by annuals or the replacement of palatable species with unpalatable species. The latter case does not necessarily constitute ecological degradation, however. Instances in which grasslands and savannas become shrublands or woodlands may represent succession in that plant diversity, primary productivity, soil fertility, etc., are maintained or even enhanced on the site. This example is viewed as retrogression only in that forage plants valued for certain classes of livestock are lost. However, the change in species composition simply reflects the fact that the new assemblage of species on the site is better adapted to the prevailing environmental conditions which include livestock grazing.

The process of retrogression appears to be step-wise rather than linear. Thus, a community may be rather stable and resistant to change up to a certain threshold. Beyond certain threshold levels, changes can be rapid, dramatic, and potentially irreversible over reasonable time frames (Fig. 5.11). The goal of sustainable grazing management is to anticipate these critical thresholds and manipulate livestock so as not to exceed them. Once a threshold is exceeded, it may not be possible for the system to return to the previous condition, even with large inputs of energy and nutrients.

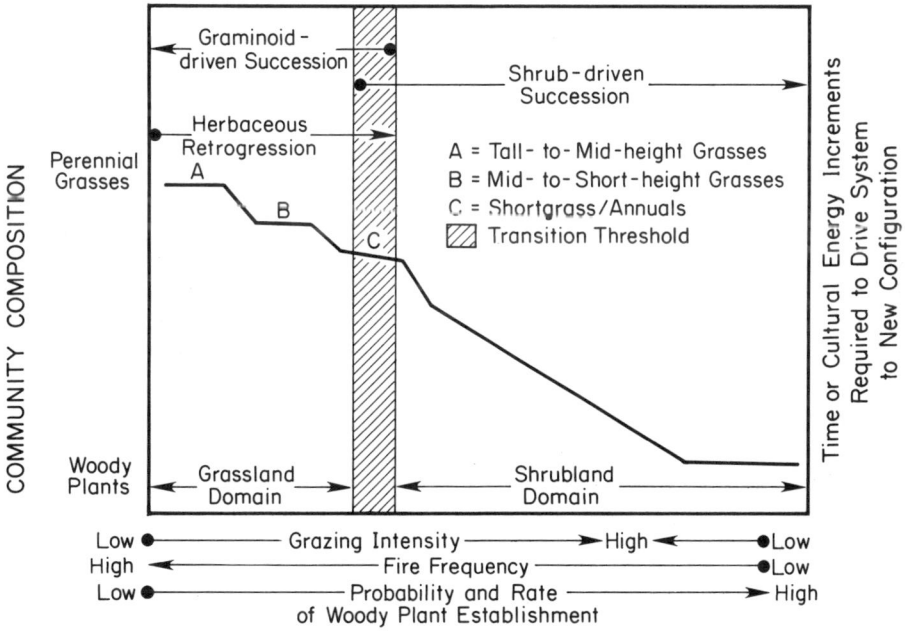

Figure 5.11. Conceptual model of changes in community structure as a function of grazing pressure (source: Archer 1989). Within the grassland domain, grazing alters herbaceous composition while decreasing fire frequency. If grazing pressure is relaxed prior to some critical threshold(s), succession toward higher condition grasslands can occur. In some cases such changes require decades. If sufficient numbers of woody plants become established, new successional processes and positive feedbacks may drive the system to a new steady state. Once in the shrub- or woodland domain, seed bank and vegetative regeneration potentials are altered and the site may not revert to grassland or savanna, even after terminating grazing. Herbicides and grazing management that allows for subsequent use of fire at regular intervals may be required to reduce woody cover and enhance herbaceous production (Scifres et al. 1983).

Succession and Resilience

Grazing management offers an opportunity to maintain or enhance species composition, diversity, and production, but hinges upon an understanding of processes regulating community succession, stability, and resilience. **Stability** is a measure of persistence in the face of disturbance. Two components of stability are resistance and resilience. **Resistance** describes the ability of the community to avoid displacement when a given type, frequency or magnitude of disturbance occurs (Begon et al. 1986). **Resilience** describes the speed with which a community returns to its former state after it has been displaced from that state. We know little of these attributes. How much stress can a particular community endure before significant changes in structure and function occur? Once disturbed, will the community return to its previous state? If so, what factors affect the rate of recovery? The concepts of resistance and resilience presume the existence of disturbance and transition **thresholds**. If thresholds exist, the magnitude of stress a system can absorb before changing to a less desirable configuration must be determined. Plant and population attributes which may forecast impending changes must also be quantified so that management can be adjusted to avert undesirable shifts.

Generally, as grazing pressure increases, a site assumes a new configuration or condition class (Figs. 5.7 and 5.11). In managing for livestock and certain classes of wildlife, changes in composition may be acceptable or even desirable, as long as site productivity and soil stability are maintained. When grazing pressure is reduced or removed, the rate of succession back to a previous configuration (higher condition class) depends on the extent to which soils, seedbank, and vegetative regeneration potential of the previous vegetation has been modified. It should not be assumed that the pathway of succession following reduction in grazing intensity will be a simple reversal of the pathway of retrogression. Reasons for this will be discussed later.

Rates of succession on sites released from grazing are highly variable, but generally proceed much more slowly than desired for management purposes. Some studies have reported significant, quantitative changes in vegetation (i.e., succession) to occur rapidly in the absence of grazing (Cooper 1953; Penfound 1964; Austin et al. 1981; Collins and Adams 1983; Potvin and Harrison 1984; Biondini et al. 1985). In other cases, the species present on retrogressed sites released from grazing may represent a new steady state and persist nearly unchanged for long periods (>30–50 y) following protection from grazing (Robertson 1971; Smith and Schmutz 1975; West et al. 1979; Glenn-Lewin 1980; Holechek and Stephenson 1983). For example, grazing within a live oak savannah caused mid-grasses (sideoats grama, Texas wintergrass, Texas cupgrass, and cane bluestem) to give way to short grasses, primarily the stoloniferous common curlymesquite (Smeins and Merrill 1988). Once established as a dominant, curlymesquite has persisted >25 years in the absence of grazing. Even when seed sources were present, midgrass species failed to re-establish in the curlymesquite community (Kinucan 1987). Thus, a relatively stable community has been maintained on a site previously supporting a higher successional class of herbaceous vegetation.

As discussed earlier, grazing, with the natural consequence of reduced fire frequency and intensity, can precipitate a physiognomic shift from grassland or savanna to shrubland or woodland. Once woody plants are established, new successional processes and positive feedbacks can cause a rapid conversion in physiognomy (Fig. 5.11). In many instances, cattle and sheep, the principal herbivores in most ranching enterprises, consume negligible amounts of woody vegetation and do little to inhibit its recruitment and growth. In some cases, livestock serve to increase the dissemination of seeds of woody plants (Table 5.1, Fig. 5.10). Over time, the capacity of defoliated grasses to carry fire or competitively exclude woody plants diminishes, after which a new, potentially stable community dominated by woody plants develops. Increases in the woody plant seedbank coupled with high vegetative regeneration potential and the deterioration of the herbaceous seedbank (Koniak and Everett 1982; Hobbs and Mooney 1986) and vegetative reserve make it highly unlikely this new plant community will revert to grassland, even if livestock are removed from the site (Niering and Goodwin 1974; Walker et al. 1981; Holechek and Stephenson 1983; West et al. 1984; Wester and Wright 1987). Once in the shrub- or woodland domain, these new communities may be highly resistant or resilient to fire or anthropogenic manipulation (e.g., herbicides, mechanical treatments). Apparent examples of regional shifts in North American vegetation to alternate steady states in recent history include the following:

- **The Great Basin**—from perennial bunchgrasses and open stands of sagebrush to dense sagebrush and annuals such as cheatgrass and medusahead.

- **Southwestern Desert Grasslands**—from tobosa and black grama grasslands to creosotebush, tarbush, or mesquite shrublands.

- **Southern Grasslands and Savannas**—from tallgrass prairies and savannas to oak, juniper, mesquite, or thorn scrub woodlands.

- **California Mediterranean Grasslands**—from perennial bunchgrasses to annual grasses.

As the preceeding examples illustrate, the pathway(s) of succession following removal of grazers are often not simply a reversal of the pathway of retrogression. As a result of this **hysteresis** phenomenon, large inputs of cultural energy (seeding, fertilization, mechanical and chemical manipulation, etc.) would be required to regain the desired composition. These management-imposed manipulations may increase the grass to woody plant ratio on a site, but the conversions may be short-lived and periodic follow-up treatments will be required (Scifres et al. 1983).

Transition Thresholds

The concept of community resistance and resilience portrayed in Figure 5.11 suggests the existence of transition thresholds. Output from the SPUR (Simulation of Production and Utilization on Rangelands) model (Wight and Skiles 1987) predicted the occurrence of a threshold in sagebrush-crested wheatgrass systems (Torell 1984). Simulated production of both crested wheatgrass and big sagebrush remained fairly constant over time when stocking rates varied from 1.12 to 0.56 ha/AUM. However, a slight increase in stocking rates from 0.56 to 0.50 ha/AUM resulted in a marked decrease in crested wheatgrass production accompanied by a sharp increase in sagebrush production (Fig. 5.12). The existence of such thresholds also account for the abrupt, non-linear development of woody plant communities in areas formerly dominated by grasses (Fig. 5.13) (Buffington and Herbel 1965; Herbel et al. 1972; Archer et al. 1988). Several studies suggest periods of drought followed by rainfall may trigger shrub encroachment on areas subjected to livestock grazing. If thresholds exist in the retrogressive process, we must determine the extent to which a system can be grazed before changing from one herbaceous state to another or from an herbaceous to a woody state.

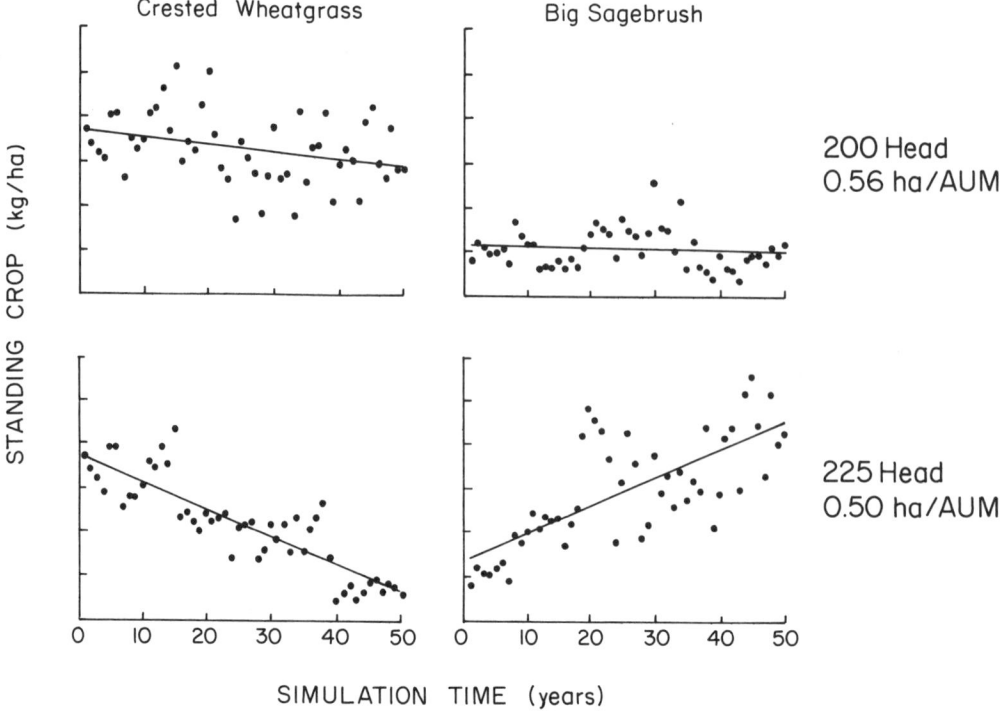

Figure 5.12. Predicted trends in crested wheatgrass and big sagebrush production under different stocking rates (SPUR model output, after Torell 1984).

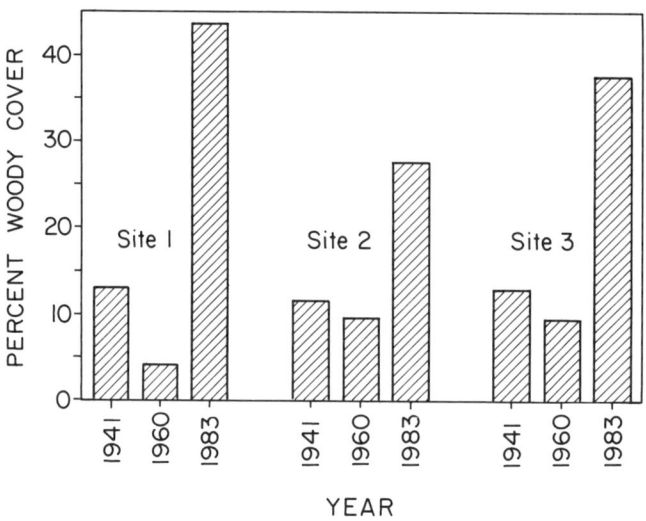

Figure 5.13. Changes in woody plant cover on three sites in a southern Texas savanna parkland between 1941 and 1983. The 1941–60 period included a major drought in the 1950s. Annual rainfall during the 1960–83 period was normal to above-normal. Cattle grazing was heavy and continuous through most of the 42-year period (after Archer et al. 1988).

RISK MANAGEMENT

The notion of resilience emphasizes the importance of maintaining community structure by maintaining diversity and variability. However, in agroecosystems, stable, sustained high yields are sought by employing intensive management schemes to minimize variability (see Chapter 10), but at the expense of resilience (Table 5.4). Over the short-term, yields are increased and planning is facilitated. However, problems may be generated over the long-term, for as a system loses its resilience, it may become increasingly sensitive to errors in management and to exogenous events. Reducing the biological variability of a system reduces its resilience and increases the probability that chance or rare events previously "absorbed" by the system will cause dramatic change. Relative to intensive management, extensive practices are less effective in reducing variability. As a result they typically generate fewer livestock products. However, over the long-term, integrity of the extensively managed system may be maintained in the face of perturbations in whatever unexpected form they might take. Given the long time spans required for natural recovery of many rangelands to desired states, and the high monetary and energy costs and often low probability of successful restoration following disturbance, this more conservative strategy is likely the most profitable and sustainable.

Table 5.4. Productivity, resilience, and variability in grazing systems (after Walker et al. 1981).

Grazing Pattern	Resilience	Herbivore Productivity	Variability in Time
Nomadic and native ungulate grazing	High	Intermediate	High
Settled peasant farmers, stock grazing	Intermediate	Low	Intermediate
Ranching, fixed number of stock, and range management	Low	High	Low

CONCLUSIONS

Most grazed systems in arid and semiarid regions are expansive and heterogeneous. Plant production is limited by climatic and edaphic factors. Because of the numerous, interactive, and to a large extent stochastic processes that regulate species composition and productivity in natural systems, it is difficult and misleading to propose standard prescriptions for vegetation management. Agronomic approaches, with their heavy emphasis on expensive cultural inputs and treatments, have a high rate of failure in arid and semiarid situations and are typically not economically feasible even when successfully implemented. It is therefore essential to acquire a functional understanding of the basic ecological processes that drive natural systems and develop flexible management strategies that work within constraints dictated by soils and a highly variable and unpredictable climate.

The traditional concept of single equilibrium communities that progress steadily toward or away from climax depending on grazing pressure seems not to apply in many arid and semiarid systems. Examples of alternative steady states, abrupt thresholds, and discontinuous and irreversible transitions are becoming increasingly abundant for both succession and retrogression. When one group of plants has been displaced by another as a result of altered climate-grazing-fire interactions, the new assemblage may be long-lived and persistent, despite progressive grazing management practices. The adage "an ounce of prevention is worth a pound of cure" thus has substantial application to vegetation management, particularly in situations where the probability of re-establishment of desired species composition and soil cover may be quite low once a change has occurred.

Abrupt transitions between various states of vegetation composition may be triggered by stochastic events related to the vagaries of climate, seed dispersal, and seedling establishment. Managers should seek to identify circumstances whereby desirable transitions can be augmented and facilitated and undesirable transitions mitigated or avoided. Westoby et al. (1989) liken grazing management to a continuous game where the object is to seize opportunities and avoid hazards. Such a philosophy is based on timing and flexibility rather than fixed policy. In systems where climatic variability is the rule rather than the exception, situations conducive to vegetation improvement or deterioration may arise infrequently and unexpectedly. Failure to recognize and respond to either situation constitutes missed opportunity. If the potential for transition to undesirable states is ignored, long-lasting, potentially irreversible impacts can result. Conversely, progressive and flexible management schemes which can capitalize on infrequent windows of opportunity for vegetation improvement or livestock production may realize long-term benefits in livestock and wildlife productivity.

PLANT SPECIES CITED

COMMON NAME	SCIENTIFIC NAME
Acacia	*Acacia* Mill. spp.
------	*Acacia karroo* Hayne
------	*Acacia hockii* DeWild
Algarrobo	*Prosopis juliflora* (= *glandulosa* Torr. var *Torreyana* [L. Benson, M. C. Johnst.])
Baccharis	*Baccharis pilularis* var. *consanguinea* (DC.) C. B. Wolf
Baobab	*Adansonia digitata* L.
Big Sagebrush	*Artemisia tridentata* Nutt.
Black Grama	*Bouteloua eriopoda* (Torr.) Torr.
Blue Grama	*Boutelous gracilis* (H. B. K.) Lag. ex Steud.
Buffalograss	*Buchloe dactyloides* (Nutt.) Engelm.
Cane Bluestem	*Bothriochloa barbinodis* (Lag.) Herter
Cheatgrass	*Bromus tectorum* L.
Common Curlymesquite	*Hilaria belangeri* (Steud.) Nash
Creosote Bush	*Larrea tridentata* (DC.) Cov.
Crested Wheatgrass	*Agropyron cristatum* (L.) Gaertn.
Eucalyptus	*Eucalyptus* L'Heritier spp.
Honey Mesquite; Mesquite	*Prosopis glandulosa* Torr. var. *glandulosa*
Hooded Windmillgrass	*Chloris cucullata* Bisch.
Huisache	*Acacia farnesiana* (L.) Willd.
Juniper	*Juniperus* L. *spp.*
Little Bluestem	*Schizachyrium scoparium* (Michx.) Nash
Live Oak	*Quercus virginiana* Mill.

COMMON NAME	SCIENTIFIC NAME
McCartney Rose	*Rosa bracteata* Wendl.
Medusahead	*Taeniatherum caput-medusa* (= asperum) Simonkai) Nerski
Mimosa	*Mimosa pigra* L.
Narrow-leaved Ironbark	*Eucalyptus crebra* F. Muell.
One-seeded Juniper	*Juniperus monosperma* (Engelm.) Sarg.
Pines	*Pinus* Lindl. spp.
Pinyon Pine	*Pinus monophylla* Torr. and Frem.
Ponderosa Pine	*Pinus ponderosa* Laws
Poplar Box	*Eucalyptus populnea* F. Muell.
Prickly Acacia	*Acacia* Mill. sp.
Rubber Rabbitbrush	*Chrysothamnus nauseosus* (Pall.) Brit.
Sagebrush	*Artemisia* L. spp.
Sand Sagebrush	*Artemisia filifolia* Torr.
Sideoats Grama	*Boutelous curtipendula* (Michx.) Torr.
Sun Sedge	*Carex eleocharis* Bailey
Tarbush	*Flourensia cernua* DC.
Texas Cupgrass	*Eriochloa sericea* (Scheele) Munro
Texas Wintergrass	*Stipa leucotricha* Trin and Rupr.
Tobosa	*Hilaria mutica* (Buckl.) Benth.
Umbrella Thorn	*Acacia tortilis* (Forsk.) Hayne subsp. *spirocarpa*
Western Wheatgrass	*Agropyron smithii* Rydb.

CHAPTER 6

HYDROLOGY AND EROSION

Thomas L. Thurow

INTRODUCTION

"Water is the driving force of nature." This fundamental observation by Leonardo da Vinci underscores the fact that water is the essential medium of biogeochemical cycles and of life itself (Smith 1974). Thus, the development of ecologically and economically sound grazing tactics requires a clear understanding of the interaction of grazing on hydrologic processes. The objective of this chapter is to review how grazing effects the hydrologic cycle.

PROPERTIES OF WATER

The importance of water to ecosystem function is largely a result of the water molecule's unique bonding structure. The strong dipolar nature of a water molecule results in an asymmetrical charge which strongly attracts one molecule to another. Van der Waals' forces (the attraction of a positively charged nucleus to the negatively charged electrons of neighboring molecules) and hydrogen bonds (the attraction of the two hydrogen atoms of one water molecule to the oxygen atoms of adjacent molecules) causes much stronger intermolecular bonding than is the case for most chemical bonds, such as covalent bonds which also bind oxygen and hydrogen atoms in water.

The combined effect of the water molecule's bond structure is that water is a substance possessing a unique set of physical and chemical properties. These properties are:

- Water has a very high **specific heat** (liquid ammonia is the only substance having a higher value). This means that a relatively large amount of energy must be applied to raise the temperature of one gram of water one degree centigrade; this amount of energy is defined as a calorie. This property gives water a great deal of temperature stability and facilitates maintenance of a steady temperature within an organism.

- Water has the highest **heat of vaporization** (540 cal/g at 100°C) known. The high heat of vaporization causes evaporating water to produce a cooling effect and condensing water to produce a warming effect. These effects are used by plants and animals to help regulate their temperature.

- Water has higher **surface tension** and **viscosity** than most other liquids. This is due to the high internal cohesive forces that exist between water molecules. This property is the key to the cohesion theory of sap movement in plants.

- Water is an important **solvent** of gases, minerals, and other solutes by virtue of the strong bonds it forms with other molecules. Consequently, water is a major reagent for many biochemical processes, such as photosynthesis and hydrolysis.

Because of these properties, water is an essential component of the protoplasm of plant and animal cells; thus, every life-sustaining process is directly or indirectly affected by water. This explains why an inadequate supply of water results in an immediate reduction of vegetative growth (see Chapter 4), and why hydrologic characteristics of an ecosystem are prime determinants of forage production and hence livestock yield (see Chapter 7).

In addition to water's vital role within an organism, water is also a major force influencing the topographic and functional attributes of landscapes. This is because:

- The kinetic energy of falling raindrops and overland flow has the potential for detaching and transporting soil. This erosive force makes water one of the most influential factors in the shaping of landscapes.
- Nutrient concentrations within an ecosystem are altered by water runoff transporting nutrients overland, and water percolation transporting nutrients through the soil profile.

Grazing by large herbivores can greatly affect the amount of water retained in or lost from a watershed. Thus, the maintenance or improvement of hydrologic and edaphic conditions of grazed ecosystems are important objectives in grazing management since they greatly affect both primary and secondary productivity. In addition to using the water and soil resources of a site to produce forage, grazing lands are an important source of surface water and groundwater recharge. The quantity and quality of water derived from grazing land watersheds are therefore key determinants of the fate of municipal, industrial, and agricultural sectors of society.

CONCEPTS OF THE HYDROLOGIC CYCLE

The hydrologic cycle is the process of water movement through the environment. On a worldwide basis, the amount of water is essentially constant, but the amount of water present at any given locale at any given instant varies depending upon how much water enters the system, the mode and rate of transit within the system, and how water exits the system. The fundamental avenues of input, transit, and output of water movement are outlined in Figure 6.1.

Inflow

Precipitation. Air heated at the earth's surface rises into the atmosphere and carries water vapor with it. As the air cools the water vapor begins to condense on small particles of matter suspended in the atmosphere. As the condensation process progresses clouds are formed. Coalescence of these droplets continues until the mass of the suspended droplets is sufficient for gravity to overcome the upward force of rising air. At this point, the droplets fall as rain, or if cold enough, snow or hail.

Underground and Overland Flow. Water originating from a source other than on-site precipitation may be an important component of an ecosystem's water balance. Water flowing in surface channels or in shallow groundwater reserves may be accessed by deep-rooted trees and shrubs. When roots reach a water table the plants

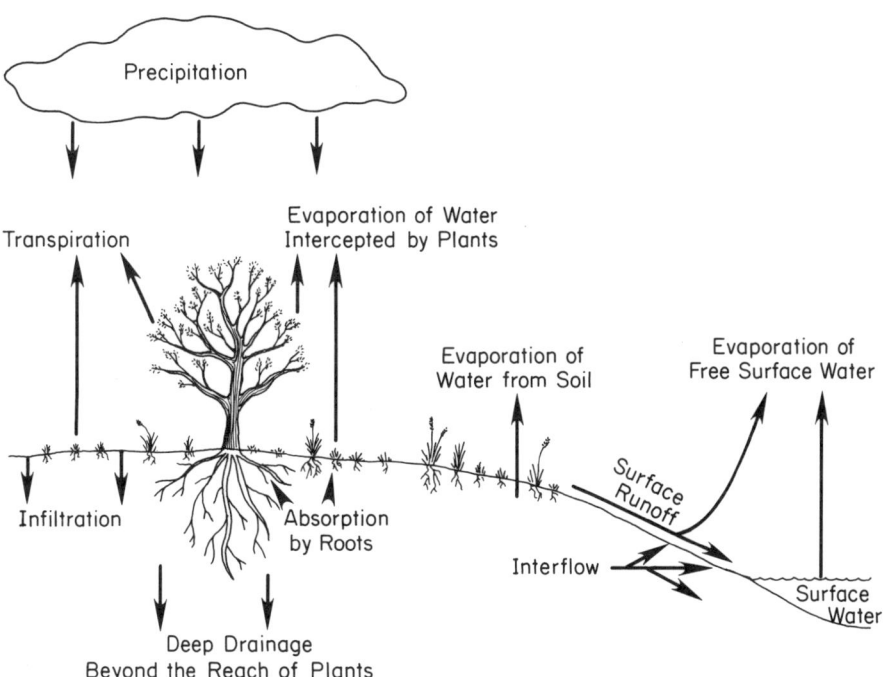

Figure 6.1. The water cycle showing major processes and pathways of water movement through a watershed. Water inflow = water outflow ± storage.

are referred to as **phreatophytes**. Phreatophytic communities (e.g., oases and riparian communities) are an important component of arid ecosystems.

Dew. Water vapor condensation forming dew on vegetation or soil is generally a negligible component of an ecosystem's water balance. However, dew may constitute a significant water source in localized coastal and montane regions such as in northern Kenya (Ingraham and Matthews 1988) and along the Pacific Ocean in North and South America (Azevedo and Morgan 1974).

Transit Through the Terrestrial System

Interception. As precipitation reaches the earth's surface it either strikes objects such as vegetation, litter, or rocks (i.e., **interception**) or falls unimpeded to the soil. Some of the intercepted water adheres to the objects and returns to the atmosphere via evaporation without ever reaching the soil. Some of the intercepted water runs down plants (**stemflow**) or drips off intercepting objects, resulting in a redistribution of water reaching the soil. All of the intercepted raindrops dissipate at least some of the kinetic energy associated with their fall.

The amount of water retained by an intercepting object is governed primarily by its water-holding capacity which varies primarily as a function of area (i.e., cover), morphological characteristics, and storm intensity. Interception loss, the amount of water intercepted and returned to the atmosphere without ever reaching the soil, is proportionately greater if the precipitation occurs as many small events since interception capacity may account for a large proportion of each such storm's total volume of precipitation. Interception loss is a relatively smaller proportion of the total volume of large storms since a large portion of the rain event occurs after the interception capacity of an object has been exceeded. The intensity with which raindrops strike an object is also important since a greater proportion of gentle rainfall is able to adhere to the cover surface compared to an intense driving storm.

The precipitation that reaches the soil may either run off, be stored in surface depressions, or enter the soil. Water stored in surface depressions or soil eventually exits the system via deep drainage, evapotranspiration, or seepage into surface or underground runoff streams. A small portion of water adheres to the soil particles (**hydroscopic water**) and is not available for plant use.

Surface Detention. Surface detention is a function of micro-relief, slope, soil texture, soil structure, and soil depth. Micro-relief formed by topography, vegetation growth, and accumulated litter influences the size, shape, and number of depressions and the amount of water that can be detained in them. Runoff generally increases as slope increases because of the associated decrease in size of detention storage sites. Water ponded on the surface is ultimately either lost via evaporation or enters the soil.

Infiltration. **Infiltration** is the process by which water moves into the soil. **Infiltration rate** is the quantity of water passing through the soil surface per unit of time. When the soil surface is saturated, infiltration stabilizes at a rate reflecting the interrelated hydrologic effects of the soil and vegetation characteristics of the site. This steady state rate is termed the **terminal infiltration rate**.

A low terminal infiltration rate is an indication of poor soil structure. **Soil structure** is the arrangement of soil particles and the intervening pore spaces. The structural characteristic of the soil is determined by the degree to which soil particles are held together in individual clusters which are termed **aggregates** (Fig. 6.2). Aggregation occurs when soil particles are mechanically bound by roots, fungal hyphae, and/or adhesive byproducts of organic matter decay and microbial syntheses. These mechanically bound particles are then cemented together by resistant humus components which form chemical bonds (Brady 1974). The **porosity** (pore volume) of the

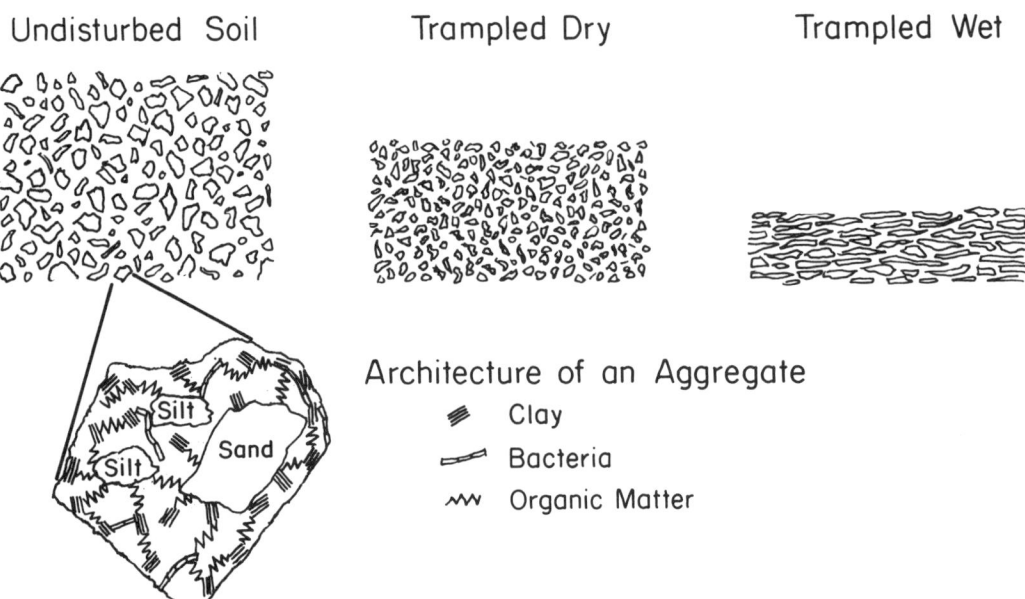

Figure 6.2. Conceptual architecture of a soil aggregate and the changes in soil aggregate structure caused by trampling under wet and dry conditions.

soil is a function of soil texture and the degree to which the soil is aggregated. Porosity and pore size determines the rate of movement of water into soil. Large macropores which aid high infiltration rates increase with improved aggregation (Allison 1973). The formation of soil aggregates is aided by any action that mixes the soil thus promoting contact between decomposing organic matter and inorganic soil particles. This action can be accomplished by wetting and drying, freezing and thawing, the physical activity of roots and burrowing animals, and soil churning by hooves or farm implements.

Aggregation alone is not a guarantee of high infiltration rate. The other key factor that must be considered is the stability of the aggregates. **Aggregate stability** is the collective measure of the degree to which soil particles are bound together and the stability of those bonds when wetted. Aggregate stability is used as an index of soil structure and as an empirical definition of aggregation (Kemper and Rosenau 1986; Boyle et al. 1989). The aggregates creating the soil pore structure must maintain their structural integrity when wet if infiltration through those pores is to occur. If the aggregate bonds are unstable when wetted, the clay particles disperse so the aggregate cluster begins to break into smaller pieces (slaking). These particles are then carried by the water and lodge in the remaining pores, making them smaller or sealing them completely (Lynch and Bragg 1985). This is one way in which soil crusts are formed. A "washed in" layer where clay particles have clogged soil pores to form a crust may reduce infiltration rate by as much as 90% (Boyle et al. 1989).

Aggregate bonds may be broken by the kinetic energy of raindrops striking the soil. Cover intercepts and dissipates raindrop energy before it strikes the soil, thereby protecting aggregate structure. Vegetation type affects the amount and structure of associated cover, therefore the infiltration rate differs among vegetation types (Figs. 6.3 and 6.4). The amount of cover, and hence the rate of infiltration, is usually greatest under trees and shrubs, followed in decreasing order by bunchgrass, shortgrass, and bare ground (Blackburn 1975; Thurow et al. 1986). Moreover, infiltration rates vary seasonally because of variation in growth dynamics (Thurow et al. 1988a). For example, cover on sites dominated by annual species is generally highly variable over time because the amount of cover rapidly increases during warm, moist periods that favor growth but rapidly declines during dormant seasons. The result is that the soil surface is poorly covered during some portions of the year and well covered at others. This is in contrast to the cover dynamics of perennial shrubs and grasses in that the amount of cover provided by these species fluctuates much less between seasons.

Another important difference between vegetation types relative to their affect on infiltration rate is related to the amount of litter. For example, bunchgrasses and shrubs tend to produce greater amounts of foliage than annuals and short grasses. The fallen foliage accumulates as litter which in turn leads to an increase in soil organic matter. Litter also creates a more consistent temperature and moisture microenvironment that favors microorganism activity. These factors enhance formation of stable soil aggregates which aids infiltration.

The hydrologic characteristics of various vegetation types can be expected to confer some competitive advantages to vegetation types with the greatest infiltration rates. Much of the water that flows overland does not reach a stream and leave the site. Rather, it flows for a short distance until it reaches an area of higher infiltration capacity that can accommodate both the falling precipitation and the overland flow. The net result of this process is that some areas receive more water than others. Since the infiltration capacity of some vegetation types is greater than others, the result is that in a vegetation mosaic the mineral soil near some species may receive more water from a storm producing a runoff event than soil near other species.

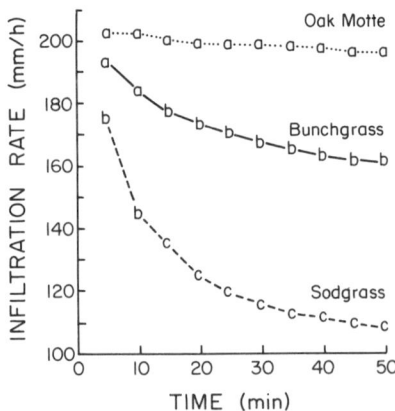

Figure 6.3. Mean infiltration rate for three vegetation types, Edwards Plateau, Texas. Means within a given time period with different letters are significantly different at P ≤ 0.05 (from Thurow et al. 1986).

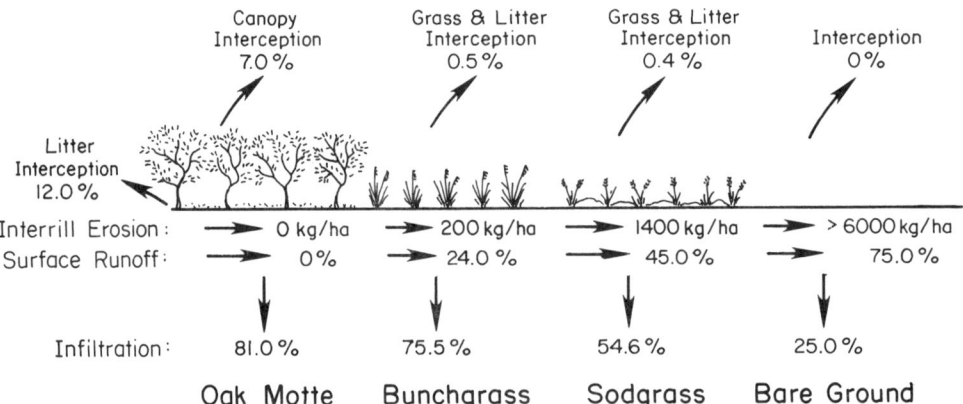

Figure 6.4. Water budgets and amount of interrill erosion, runoff, and interception from oak, bunchgrass, sodgrass, and bare ground dominated areas, Edwards Plateau, Texas. Based on 10 cm of rainfall in 30 minutes (from Blackburn et al. 1986).

Outflow

Runoff. Overland flow begins when the amount of water at the soil surface exceeds the amount of water entering the soil and when the storage capacity of surface depressions are filled. **Runoff** is the portion of precipitation that exits a watershed via overland flow. Water that exits the watershed without entering the soil is called **surface runoff**. Water that enters the soil before returning to a surface stream is called **interflow** or **seepage flow**.

Storm characteristics influence the amount of runoff. Storm intensity varies between different regions of the world and typically varies in different seasons within a region. For example, in the southwestern U.S. summer rainfall occurs predominately as thunderstorms, characterized as brief, intense rainfall events, whereas winter rainfall occurs predominately as frontal storms, characterized as prolonged, low-intensity rainfall events. The low-intensity characteristic of frontal storms results in precipitation reaching the soil at a slower rate than the rate at which it can enter the soil. Consequently, the precipitation from these storms is able to soak into the ground soon after it strikes the soil. Conversely, during the high-intensity rain associated with summer thunderstorms, the rainfall rate exceeds the rate at which water can enter the soil. Therefore, flash-flooding is most likely to occur in the summer. The direction of a storm, especially in mountainous regions, also affects the amount of runoff. For example, a storm moving down a drainage produces greater runoff and peak flow than a storm moving up a drainage since the rain input moves with the accumulating overland flow.

Evapotranspiration. **Transpiration** is the process whereby water vapor is released to the atmosphere by passing through permeable membranes or pores of living organisms. **Evaporation** is the process whereby water vapor enters the atmosphere from soil or from surface water. The term **evapotranspiration** encompasses the combined effect of both these processes.

The amount of water which returns to the atmosphere via evapotranspiration is affected primarily by soil and plant characteristics. Amount of transpiration is dependent upon climatic conditions (such as relative humidity, wind, temperature, etc.), soil moisture, morphological vegetation characteristics, stage of phenological development, and the amount of transpiring tissue (see Chapter 4). Generally, transpiration is lower from sites dominated by herbaceous vegetation than from sites dominated by trees or shrubs because of inherent physiological differences relative to water use efficiency and growth dynamics. For example, graminoides often enter a dormant state during periods of water- or temperature-induced stress so transpiration losses are minimal during such periods as compared to trees and shrubs which usually maintain a substantial number of green leaves. Also, trees and shrubs tend to have more extensive root systems than grasses thereby increasing access to soil water. Trees and shrubs intercept more water than grasses, therefore grass sites contain less intercepted water to lose via evaporation than trees or shrubs. For these reasons, conversion of shrub or forest cover to grass cover may increase the amount of runoff and deep drainage on a watershed (Hibbert 1983).

While evaporation from a moist soil surface proceeds at a rapid rate, a thin layer of dry soil at the surface can dramatically reduce the rate (Penman 1948; Veihmeyer 1964; Ripple et al. 1972). In the absence of soil cracks, soil water can be depleted by evaporation to a depth of about 10 cm in clay-textured soils and about 20 cm in sandy soils (Sosebee 1976). Other soil characteristics such as color (dark soils absorb more heat and lose more water through evaporation than light soils) may also have a significant effect on potential evaporation rates. But in general, evaporational water loss from unsaturated soils is a relatively small portion of the total outflow.

Deep Drainage. Water movement through a soil column after it has entered the soil is called **percolation**. Water that percolates beyond the reach of plant roots is termed **deep drainage**. The volume of deep drainage depends on the amount of infiltration, the evapotranspirational demand, and the substrate transmission characteristics which are primarily a function of regional geological features.

Amount and type of cover influence deep drainage to the degree that they effect infiltration rates and evapotranspiration loss. For example, from a 1-year study in south Texas, Weltz (1987) estimated that deep drainage accounted for about 10% of total precipitation on bare soil, 2% under herbaceous cover, and 0% under shrubs. The lower evapotranspiration loss on the bare plots (80% compared to 95% and 98% loss on herbaceous and shrub plots, respectively) offset the increased runoff on the bare plots (10% compared to 3% and 2% on herbaceous and shrub plots, respectively). Thus, the net effect was that more water was available in the soil column for drainage beneath the bare ground sites. On sites that have hard pans or caliche barriers, deep drainage is reduced to a small portion of precipitation regardless of cover type. In such situations the bare ground sites store more water because there is no transpiration loss via vegetation. The resultant higher antecedent soil moisture on the bare plots results in greater runoff (Carlson et al. 1990). These data from vegetated Texas shrublands bear a close similarity to the water balance of the environmentally similar *Burkea* savanna in South Africa (Whitmore 1971).

EFFECTS OF GRAZING ON HYDROLOGIC PROCESSES

The degree to which livestock grazing can influence water distribution over a landscape can be conceptually evaluated by examining grazing effects on the hydrologic cycle. The vegetation and soil characteristics of a site are the prime determinants influencing how water is partitioned to each of the potential pathways of water movement. Understanding how vegetation and soil factors affect the quantity and quality of water, and how grazing alters these factors, is essential to the development of grazing strategies aimed at sustained production by conserving the water and soil resource.

Inflow

Precipitation. Amount, intensity, duration, form, and temporal and spatial distribution of precipitation are intrinsic to climatic conditions, so they are viewed as being beyond the influence of grazing management. An exception may occur in some regions where prolonged, intensive grazing has caused serious land degradation. The extensive removal of vegetation cover and litter may cause an increase in local albedo (surface reflectivity) of the land surface (Otterman 1977), which reduces the amount of heat absorbed by the land surface and which in turn diminishes the convective activity in the lower atmosphere required for cumulus cloud formation. Furthermore, bare ground tends to cool more at night than vegetated soil surfaces which tends to reduce the opportunity for convective precipitation events (Charney et al. 1975). Extensive litter removal over a large region also reduces the population of bacteria such as *Pseudomonas syringae*, which are important sources of nuclei for the formation of raindrops or ice crystals in clouds. A reduction in the quantity of these raindrop nuclei may lead to reduced precipitation (Vali et al. 1976).

Transit Through the Terrestrial System

Interception. The amount and intensity of precipitation reaching a soil surface may be influenced by grazing to the degree that grazing alters the amount and type of vegetation cover on a site. The extent of rainfall interception differences among plant communities was documented for rangeland on the Edward's Plateau of Texas. The estimated annual interception loss for a site dominated by stoloniferous grass was 10.8% of annual precipitation. A site dominated by bunchgrass had an estimated annual interception loss of 18.1%, while oak trees and the litter beneath the trees intercepted about 46% of annual precipitation. However, due to the concentrating effects of stemflow, soil near the bases of trees received about 222% of annual precipitation whereas areas more than 100 mm from the trunk received only about 50.6% of annual rainfall (Thurow et al. 1987). Studies of eastern U.S. forests have shown that conifers intercept 3–8 times as much precipitation as deciduous trees (Douglass 1983). These data indicate that shifts in the kind or amount of vegetation associated with grazing and brush management affect interception, which in turn affects the amount and kinetic energy of the precipitation actually reaching the soil.

Surface Detention. The effect of large herbivores on surface detention of water are mixed and dependent on grazing intensity. For example, the hoofprints left by livestock increase micro-relief if the stocking intensity is moderate. If stocking intensity is heavy, micro-relief is reduced due to disaggregation of soil structure resulting in the soil surface becoming either loose dust if soil moisture is low, or flattened and compacted if soil moisture is great. Heavy grazing also often decreases micro-relief by reducing litter accumulation and bunchgrass growth forms.

Infiltration. The terminal infiltration rate is sensitive to the type of grazing management used on the site (Fig. 6.5). The key components affecting infiltration impacted by grazing are listed in Table 6.1. The extent to which livestock grazing effects these variables is largely dependent upon grazing intensity. The impacts of grazing and the magnitude of hydrologic response differ from region to region depending upon the interrelationships associated with the particular mix of these parameters. Livestock effects on the parameters that influence infiltration fall into two broad categories: vegetation impacts and trampling impacts.

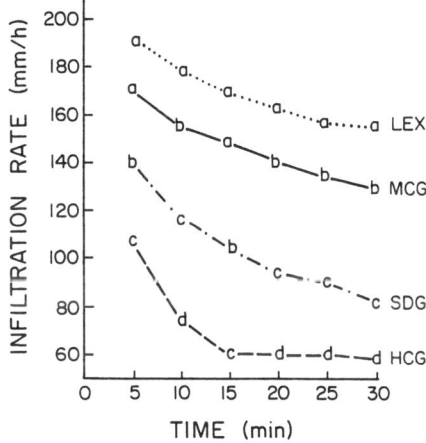

Figure 6.5. Mean infiltration rates for four grazing treatments six years after they were initiated on the Edwards Plateau, Texas. LEX = livestock exclosure; MCG = continuously grazed at moderate intensity; SDG = short duration rotation (14-pasture, 1-herd; 4 days on, 50 days rest) stocked 1.75 times the moderate intensity; HCG = continuously grazed, stocked at 1.75 times the moderate intensity. Means within a time period with different letters are significantly different at $P \leq 0.05$ (after Thurow et al. 1988a).

Table 6.1. Broad relationship between infiltration rate, grazing intensity, and various soil and vegetation attributes. Positive (+) and negative (−) symbols reflect relative relationships. For example, as aggregate stability increases, water infiltration rate increases (+), whereas as grazing intensity increases, aggregate stability decreases (−) thereby decreasing water infiltration rate.

	Infiltration Rate	Grazing Intensity
Soil Attributes		
Aggregate Stability	+	−
Bulk Density	−	+
Vegetation Attributes		
Standing Crop and/or Litter Biomass	+	−
Standing and/or Litter Cover	+	−

Vegetation impacts. The herbivorous nature of grazing animals clearly results in the removal of a portion of the vegetation. Removal of vegetation affects aggregate stability in several ways:

- A decrease in cover reduces interception. Consequently, less kinetic energy is dissipated prior to striking the soil with the consequence that greater force per storm is applied to the soil tending to break aggregate bonds.

- A decrease in above-ground biomass (standing crop and litter) results in less organic matter eventually being incorporated into the soil. As previously discussed, organic matter is an important factor in aggregate formation and stability.

- A decrease in above-ground biomass is eventually mirrored by a decrease in root biomass. Grass roots create a network physically binding soil particles together. Furthermore, grass roots induce aggregate formation by exuding biochemical byproducts which bind soil particles and distribute organic matter throughout the soil profile.

The degree to which grazing reduces cover, reduces litter, and changes the species composition of cover is dependent upon the frequency and intensity of grazing. This physical removal of vegetation by herbivores is superimposed on fluctuations of vegetation cover resulting from seasonal variation in growth dynamics as influenced by climatic factors. The combined effects of climate and livestock grazing intensity strongly influences seasonal fluctuations of cover that in turn contributes to seasonal fluctuations of infiltration rate (Thurow et al. 1988a).

The principle objectives of most livestock grazing systems (see Chapter 7) are to maintain or improve forage production and/or to improve forage harvesting efficiency. Maintenance or improvement of forage production is directly related to water infiltration rates. Indeed, drought stress caused by poor infiltration has been documented to be a major problem limiting production in the western U.S. (Boyle et al. 1989). Therefore, the long-term success of a grazing system depends on how well increased livestock harvest efficiency (which reduces cover and biomass) is balanced with the need to maintain aggregate stability (which is improved by increased cover and soil organic matter).

The extent to which grazing causes a change in species composition is a prime factor determining hydrologic condition of a site (Wood and Blackburn 1981; Gamougoun et al. 1984; Thurow et al. 1988a) (Figs. 6.3 and 6.4). Density of herbaceous perennials is an especially important indicator of hydrologic condition in many regions because their decline is usually associated with increased runoff. This insight provides an understanding of the basis for the range condition classification system outlined in Chapter 5, since infiltration rate in many respects is a synthesis of the measures used to evaluate range condition. For example, as grazing intensity in a mixed grassland increases, the vegetation composition shifts from midgrass to shortgrass dominance (Rhodes et al. 1964; Sharp et al. 1964; Thurow et al. 1988b), with the most severe changes being associated with heavy stocking (Ellison 1960). Heavy stocking tends to result in intense defoliation of palatable species resulting in their decline (Dyksterhuis 1949). Many bunchgrass species are palatable and nutritious, but if closely grazed the above-ground apical meristems are damaged (Sims et al. 1982). Therefore, heavy grazing intensity, regardless of grazing strategy, does not appear suited for long-term maintenance of hydrologically desirable bunchgrass species (Thurow et al. 1988b). Moderate or light grazing, regardless of grazing strategy, generally has little effect on bunchgrass cover (Ellison 1960; Rich and

Reynolds 1963) and thus has little affect on infiltration rate (Blackburn 1984).

Trampling impacts. Grazing animals also reduce infiltration by breaking the soil aggregate structure due to the force applied by hooves. A commonly asserted hypothesis is that intense trampling activity associated with high stock densities enhances infiltration (OTA 1982; Walter 1984). Intense trampling results from concentrating livestock on small areas for short periods of time, creating a "herd effect" (Savory 1978; 1979). The result of this "hoof action" is hypothesized by these authors to enhance infiltration rate and reduce erosion, even when conventional stocking rates are doubled or tripled (Goodloe 1969; Savory and Parsons 1980; Savory 1983).

Research conducted to date does not support the hypothesis that a hydrologic benefit accrues by increasing the magnitude of trampling. Length of rest, rather than intensity of livestock activity, appears to be the key to soil hydrologic stability (Warren et al. 1986a).

It is difficult to separate the disaggregating effects of the force associated with hoof impact from the force associated with raindrop impact. Studies aimed at determining the impacts of livestock trampling in the absence of concomitant removal of vegetation have generally shown that trampling increases soil compaction (i.e., bulk density) (Alderfer and Robinson 1947; Kako and Toyoda 1981; Willat and Pullar 1983); mechanically disrupts soil aggregates (Beckmann and Smith 1974); reduces aggregate stability (Knoll and Hopkins 1959); and destroys cryptogamic cover (i.e., cover provided by algae, moss, and lichens) (Loope and Gifford 1972; Brotherson and Rushforth 1983). Moreover, these studies have shown the magnitude of negative impacts from trampling increases as stocking intensity increases (Willatt and Pullar 1983). The degree of damage associated with trampling at a particular site depends on soil type (Van Haveren 1983), soil water content (Robinson and Alderfer 1952), seasonal climatic conditions (Warren et al. 1986a), and vegetation type (Wood and Blackburn 1984). For example, Warren et al. (1986b) show that repeated high-intensity trampling decreases aggregate stability and increases bulk density which in turn reduces infiltration rates and increases surface runoff and interrill erosion. Trampling dry soil did indeed churn the soil surface. However, this "hoof action" reduced the size of naturally occurring soil aggregates and increased the bulk density of the surface soil layer. Trampling moist soils destroyed existing soil aggregates by compacting them into a comparatively impermeable surface layer composed of dense, unstable clods (Fig. 6.2). Both of these outcomes were detrimental to infiltration rate and interrill erosion (Warren et al. 1986c) (Table 6.2).

Compacted trails may increase as the number of pastures is increased within an intensive rotational grazing system (Walker and Heitschmidt 1986; Andrew 1988). The low porosity of trails leads to a low infiltration rate resulting in concentrated runoff which eventually creates gullies. To prevent trailing on fenced or unfenced land, sufficient water points must be provided so that grazing pressure is evenly distributed. Livestock must regularly return to the same sites if water points are spaced too far apart, resulting in the creation of a series of radial paths leading to water. On sites of low productivity it is unlikely to be economical for a rancher to install the fences and water supplies necessary to avoid damage from livestock concentration. Yet it is these low-productivity sites which are often most fragile and susceptible to accelerated erosion. Therefore it may be in management's long-term interest to expend the capital necessary to disperse grazing if grazing use of these sites is to continue.

As previously discussed, one way in which crusts are formed is the clogging of the surface pores by disaggregated soil particles. Crusting is commonly associated

Table 6.2. Mean infiltration rate and interrill erosion in relation to trampling intensity and water content at the time of trampling on the Edwards Plateau, Texas. Based on 15 cm of rainfall in 45 minutes (from Warren et al. 1986c). 1× indicates moderate stocking intensity, 2× double the moderate intensity, 3× triple the moderate intensity. Values in the trampled dry and trampled wet columns followed by the same letter are not significantly different at the 95% level of significance.

Stocking Intensity	Trampled Dry	Trampled Wet
	Infiltration Rate (mm/hr)	
0	166a	160a
1×	140b	133b
2×	121c	99c
3×	117c	96c
	Interrill Erosion (kg/ha)	
0	976a	2007a
1×	2827b	2875a
2×	3438b	4274b
3×	4788c	5861c

with silty soils having a low organic matter content and low aggregate stability (Blackburn 1975). A stop-gap approach often used in an attempt to manage crusted soils is to concentrate grazing so that the soil surface is disrupted by hoof action (OTA 1982). Livestock trampling does indeed break the crust, incorporate mulch and seeds into the soil, and aid seedling emergence. However, this result is short-lived because the subsequent impact of falling raindrops re-seals the soil surface (i.e., the unstable soil pores will become plugged) after the first several minutes of an intense rainstorm. To effectively address a soil crusting problem, livestock grazing systems must concentrate on addressing poor aggregate stability which is the cause of the crusting. Livestock grazing systems that promote an increase in plant and litter cover and an increase in organic matter produces the only lasting effect in reducing soil crusts (Blackburn 1983). Not all types of crusts are bad. Some types of crusts formed by lichens, moss, and algae (i.e., **cryptogamic crusts**) play an important role in arid environments by stabilizing soils otherwise susceptible to wind erosion. The reduction of water infiltration associated with cryptogamic crusts may be offset by the benefits they provide in slowing runoff and evaporation, leading to a net soil water benefit (Johansen 1986). Cryptogamic crusts are prone to deterioration resulting from trampling or air pollution (Hawksworth 1971).

Outflow

Runoff. Management of grazing lands must be designed to cope with the storm characteristics of the region. For example, regions unlikely to experience runoff can afford to allow more protective cover to be grazed than regions where flood danger is extreme due to thunderstorms, slope, soil texture characteristics, etc. Areas of grazing lands susceptible to runoff and erosion may be best managed by minimizing grazing disturbance prior to and during the period when runoff and erosion danger is greatest. Protection of riparian areas is particularly important in regions where large seasonal runoff events are likely because these areas serve as buffer strips to filter sediment and slow the rate of overland flow. Riparian vegetation growth can also stabilize banks that would otherwise be susceptible to erosion by peak flows.

Evapotranspiration. Vegetation cover tends to reduce evaporation rates by shading the soil and reducing wind velocity. However, the greater interception and transpiration loss associated with the greater vegetative cover usually more than offsets the benefits of reduced evaporation (Lull 1964). Frequent, heavy grazing results in less transpiration loss per unit area due to removal of transpiring tissue. However, the water use efficiency of the remaining plant tissue is likely to be lower (Caldwell et al. 1983).

Water evaporates more rapidly from compacted soils than from well aggregated, friable soils. Consequently, soil compaction by grazing animals can increase evaporation from the soil (Sosebee 1976).

Deep Drainage. A decrease in percolation rate associated with increased bulk density may slow and reduce downward movement of water. Trampling normally does not alter bulk density much beyond a depth of 25 cm (Alderfer and Robinson 1947), although such effects have been recorded to a depth of 150 cm on wet soils (Lull 1959).

CONCEPTS OF THE EROSION PROCESS

Natural or **geologic erosion** results from climatic and topographic conditions and is independent of human activities. About 80% of the world's land surface considered susceptible to geologic erosion (Fig. 6.6) are classified as rangeland. **Accelerated erosion** is defined as an increase in soil erosion associated with human activities relative to changes in vegetation cover and/or the physical properties of the soil. The rate of erosion must be equal to or less than the rate of soil formation if sustained, long-term productivity is to be maintained.

The erosion process develops in three basic phases: detachment, transportation, and deposition. Different soils exhibit different responses to each of these phases; sand, for example, is more easily detached than clay, but clay particles are more easily transported. Erosion is a function of the **erosivity** (i.e., energy of the water or wind acting on the soil) of the detachment factor and the **erodibility** (i.e., a function of the soil's physical characteristics, topography, type of land use, and type of vegetation) of the soil (Smith and Wischmeir 1957).

Figure 6.6. Regions of the world most susceptible to geologic wind and water erosion (after Hudson 1981).

Water Erosion

Soil erosion by water is largely dictated by mean annual rainfall. The amount of rain in regions with annual rainfall under 350 mm is usually insufficient to cause serious erosion. At the other extreme, annual rainfall of over 1000 mm leads to such a complete vegetative cover that the soil is effectively protected from raindrop impact. Regions with annual precipitation between these two extremes are most susceptible to severe water erosion. Erosion is of course increased whenever the protective cover is removed from sites with higher precipitation (Fig. 6.7).

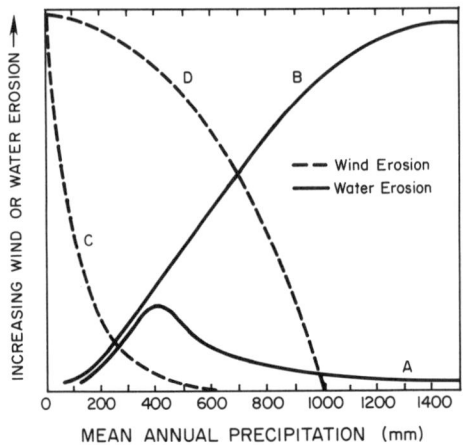

Figure 6.7. Relationship between water erosion (continuous lines) and wind erosion (broken lines) with mean annual precipitation. A and C represent natural vegetation cover: B and D represent bare ground conditions (from Marshall 1973).

Sediment production is closely related to runoff (Blackburn et al. 1986) since runoff is the principle agent of soil detachment and transport. Water erosion is typically characterized in stages corresponding to a progressive concentration of runoff. Evidence of erosion can be quickly determined by a variety of indicators (Fig. 6.8). The first stage, **interrill erosion**, combines the detachment of soil resulting from raindrop splash and its transport by a thin flow of water across the surface. This thin flow of water is highly turbulent as a result of raindrop impact and has a high erosive capacity. Extreme interrill erosion is evident when, for example, soil pedestals are formed by erosion around an area covered by a resistant material such as rock. The fact that the surrounding soil is eroded without undercutting the soil under the resistant cap illustrates that raindrop splash is the major transport mechanism, rather than surface flow. Clearly, therefore, although no running water is observed, water erosion may still be taking place. **Rill erosion** begins as the diffuse water movement causing interrill erosion concentrates into discrete flow paths. **Gully erosion** is generally defined as the point when rills increase in size to the point they can no longer be driven across by a truck. **Streambank erosion** is defined as soil displaced from the banks of rivers or streams.

The amount of interrill erosion varies depending upon graminoid growth form in that interrill erosion is less when equal cover is provided in bunchgrass vegetation types than sodgrass types. This is because the bunch growth form and the accumulated litter at the base of the bunch provide an effective obstruction to overland flow. By slowing or diverting the course of overland flow the kinetic energy of runoff is reduced, resulting in decreased sediment transport capacity. Bunchgrass clumps which are mounded above the level of surrounding soil indicate erosion and deposi-

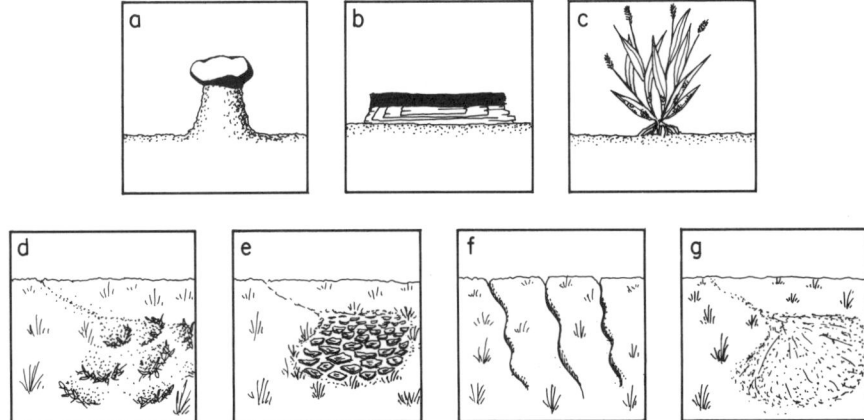

Figure 6.8. Visual evidence indicating accelerated erosion. a, Soil pedestals; b, differential charring of wood indicating how much soil eroded after the fire; c, base of plants discolored by soil contained in raindrop splash and exposed root crowns; d, miniature terraces and debris dams; e, puddled spots on soil surface resulting in fine clays forming a crust in every minor depression which crack as the surface dries and the clays shrinks; f, rill and gully formation; g, small alluvial cones formed by minor changes in slope.

tion of soil transported by splash or suspension from the exposed interspaces. Due to their diffuse basal characteristics, stoloniferous grasses or annuals generally do not have the capacity to catch and hold sediment. Consequently, while runoff is typically related to the amount of cover, interrill erosion is more strongly related to vegetation type.

Wind Erosion

Conditions necessary for soil erosion by wind are:

1. Soil must be dry to be blown by wind, so the site must be arid, and
2. Periods of strong prevailing winds, often associated with large land masses, must occur.

Wind erosion is a self-generating process that becomes increasingly more difficult to a halt as it develops. The process starts as fine soil particles become detached and strike other particles with enough energy to detach them too. This exposes the remaining larger particles, making it easier for them to also be detached. It should be noted that the suspended particles may contain over 3 times as much organic matter and nitrogen as the parent material left behind (Bennett 1939). Once larger particles have begun the process of **saltation** (i.e., movement of soil particles with a diameter of 0.05 mm to 0.5 mm by bouncing or being lifted off the soil for short distances) reestablishment of plant seedlings is difficult because of the abrasion associated with the soil movement.

EFFECTS OF GRAZING ON EROSION

The major pollutant from grazed watersheds is sediment. Livestock grazing generally does not significantly increase bacteria contamination of runoff as long as grazing intensity is light to moderate and animals are kept away from riparian zones (Doran and Linn 1979). If livestock grazing is heavy and not restricted from riparian zones, fecal coliform indicator bacteria counts may increase 10-fold above background counts (Tiedemann et al. 1987). Fecal coliform bacteria is not pathologically harmful itself, but it is a useful indicator of the likely presence and concentration of associated harmful bacteria which do constitute a threat to public health.

Because of the close association of runoff with water erosion, any practice that reduces runoff reduces erosion. Therefore practices that increase infiltration rate and surface detention reduce sediment loss. Changes in species composition associated with grazing therefore effect the amount of interrill erosion. Differences in interrill erosion within a vegetation type are related to the amount of cover and litter accumulation associated with contrasting grazing systems (Fig. 6.9). Increased trampling intensity reduces soil structure, making soil particles more susceptible to detachment (Table 6.2).

Reduction of cover and standing crop also exposes the soil more directly to the erosive force of wind. If the grazing intensity causes a reduction in cover, or if trampling disaggregates soil particles, wind erosion will increase.

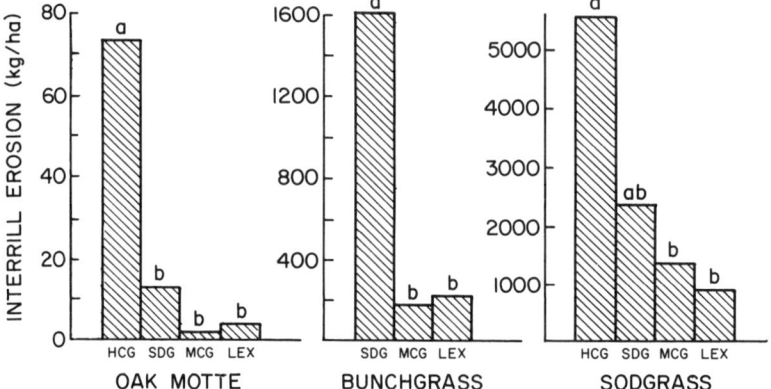

Figure 6.9. Mean interrill erosion for different grazing treatments on the Edwards Plateau, Texas. Bassed on 10 cm of rainfall in 30 minutes. LEX = livestock exclosure; MCG = continuous grazed at moderate intensity; SDG = short duration rotation (14-pasture, 1-herd; 4 days on 50 days rest) stocked 1.75 times the moderate intensity; HCG = continuous grazed stocked 1.75 times the moderate intensity. Means within a given vegetation type with different letters are significantly different at P − 0.05 (after Thurow et al. 1986).

DESERTIFICATION

Satterlund (1972) developed the critical point deterioration concept to explain what happens if erosion is not controlled (Fig. 6.10). Beyond the critical point, erosion continues at an accelerated rate which cannot be reversed by the natural processes of revegetation and soil stabilization, even if the initial cause of disturbance (e.g., intensive grazing) is corrected. As runoff and soil erosion increase, less water and fewer nutrients are retained to support the level of plant growth needed for surface soil protection. The plants that do grow are scarcer and hence receive greater focus of continued grazing pressure. The microclimate deteriorates leading to less microorganism activity needed for soil aggregate formation and a harsher environment for germination. These factors contribute to even more soil exposed to raindrop impact, further accelerating surface runoff and erosion. This spiraling pattern of deterioration (Fig. 6.11) eventually results in desertification. **Desertification** is the

Hydrology and Erosion 157

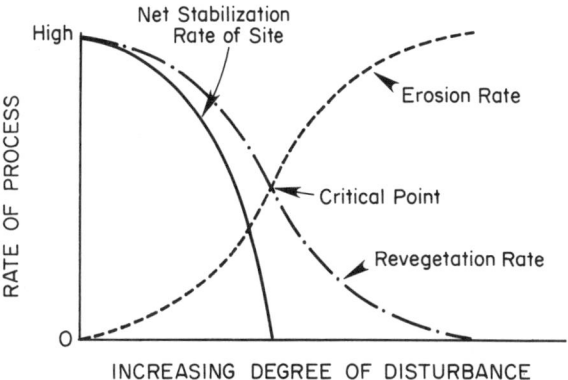

Figure 6.10. Relationship of deterioration of a site by erosion to the rate of revegetation (from Satterlund 1972©, copyrighted and reprinted by permission of John Wiley & Sons, Inc.).

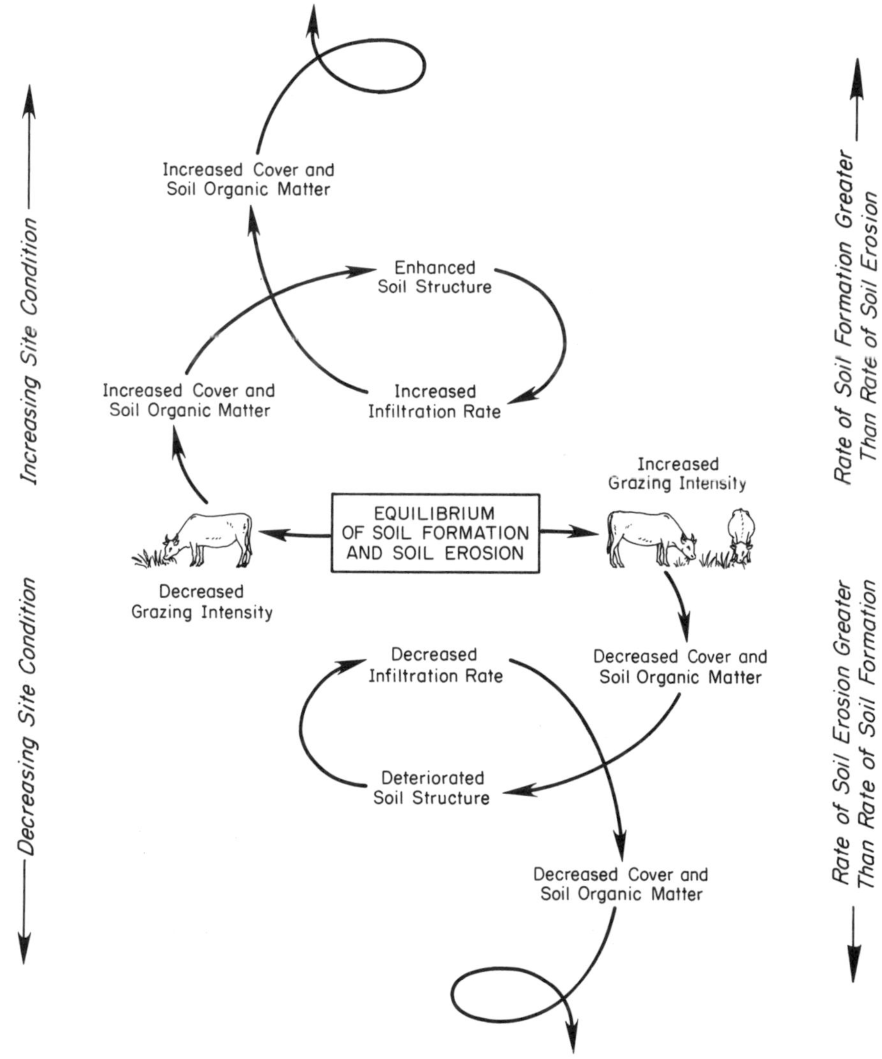

Figure 6.11. Conceptual pathway of changes in the hydrologic condition of a site.

diminution or destruction of the biological potential of land (Dregne 1987). A converse pattern occurs during recovery if the critical point has not been passed. Management intervention on most grazing lands is usually not economically viable once the critical point has been passed. Therefore, it is vital that management of the resource be sensitive to the hydrologic relationships of the site so that the desertification process is never initiated.

Land use practices and the concern for soil and water conservation vary greatly depending upon whether the management time horizon is focused on short-term economic gain or long-term sustained yield. Traditional economic theory asserts management and conservation of the soil resource will occur when the discount rate for future profit potential exceeds current returns. In overpopulated regions of the world the need for immediate survival outweighs consideration of future productivity. Consequently, people in these areas may indeed understand that the resource is deteriorating but are not in a position to change their use pattern since their immediate concern is to produce enough to stay alive. In developed countries the factors affecting discount rate decisions are usually not so dire; however, land use practices which cause accelerated erosion may be prompted by the need to generate high yield to remain financially solvent (see Chapter 9). Loss of soil fertility and water-holding capacity associated with erosion also reduces the magnitude of future benefits that could be gained through introduction of improved livestock breeds or better husbandry techniques.

The loss of the soil resource is potentially more serious than consumptive use of other commodities because once accelerated erosion is allowed to develop beyond the critical point, resource deterioration continues regardless of whether the destructive land use continues. Soil loss cannot be effectively restored through management since topsoil formation occurs at the rate of 1 in. formed every 300–1000 years. Soil accumulated in the past because conditions for its presence were in equilibrium with its environment. If grazing disturbance is not too great, stability will return and accelerated erosion gradually cease. However, if disturbance has been too disruptive the equilibrium is lost and will not return on its own, resulting in a "death spiral" towards desertification. Once this equilibrium point has been exceeded, the cost of restoring equilibrium is far greater than the initial energy which caused the original destabilization. For this reason, the first priority of grazing land management should be to maintain the soil resource and hydrologic condition of the site. These efforts form the foundation of sustained long-term productivity.

CONCLUSION

The hydrologic condition of rangelands is the result of complex interrelationships of soil, vegetation, topography, and climate. Maintenance or improvement of hydrologic condition and soil retention are critical determinants of long-term sustained production. It is possible to interpret and anticipate livestock affects on hydrology by understanding how livestock use of grazing lands impacts the soil and vegetation parameters.

The rangeland water balance is determined by the characteristics of incoming precipitation and by the outflow associated with runoff, evapotranspiration, and deep drainage. The amount of water available for forage production is dependent on the characteristics of the components affecting the water balance. The greatest potential impact that grazing has on rangeland hydrology is the effects on parameters which determine infiltration rate (broadly: soil structure, amount of cover, and type of cover). Runoff not only represents a loss of water, it is also an erosive force that transports topsoil and nutrients from the site. Water and soil leaving the site are resources unavailable for future forage production, which translates into reduced livestock and wildlife production. It is therefore evident that the long-term success of any grazing management strategy is dependent upon that strategy's ability to maintain or improve the hydrologic condition and soils of the site.

CHAPTER 7

LIVESTOCK PRODUCTION

R. K. Heitschmidt and C. A. Taylor, Jr.

INTRODUCTION

Level of livestock production on a given site is an integrated measure of energy capture, harvest, and conversion efficiencies (see Chapter 1). The principle factor affecting these efficiencies, and therefore livestock production, is grazing intensity which varies as a function of the:

1. temporal and
2. spatial distribution of various
3. kinds/classes and
4. number of livestock.

The objective of this chapter is to examine the relative effect of each of these four factors on livestock production.

CONCEPTS AND TERMINOLOGY

Livestock production is a dynamic process varying as a function of both plant and animal factors. The use of several closely related, often confusing terms is required to conceptualize and/or quantitatively describe the effects of these factors on livestock production (Booysen 1967; Hodgson 1979; Scarnecchia and Kothmann 1982; Soc. Range Manage. 1989). For the purposes of this chapter we use the following specific definitions:

- **Forage available** (FA) is any herbage and/or browse available for grazing.

- **Forage demand** (FD) is the amount of any specified forage required to meet the nutrient requirements of an animal over a specified period of time.

- **Grazing pressure** (GP) is the ratio of FD to FA for any specified forage at any instant (i.e., FD/FA).

- **Stocking density** (SD) is the number of specified animals/unit area of land at any instant.

- **Animal-unit** (AU) is any specified combination of animals with a total FD of 12 kg of dry matter/day.

- **Stocking rate** (SR) is the number of animals of a specified class or animal units/unit area of land over a specified period of time (i.e., SD integrated over time).

The fundamental inter-relationships between these variables in a simple 1-pasture, 1-herd grazing regime are presented in Figure 7.1. Section A reflects the effect of varying numbers and kinds of animals on SD and SR when SR is defined as number of a specific class of animal species. Section B demonstrates the effect of varying levels of FD on SR when expressed on an AU basis. For example, SR is equal (5 ha/AU/yr) for 1000 ewes or 200 cows if it is assumed the FD substitution ratio is 5. It is quite different, however, if the ratio is otherwise. Section C shows that at equal FD, SR is equal. Section D demonstrates the relative effects of changing levels of FA on GP. Section E demonstrates the effect of **preference**, which is a relative term

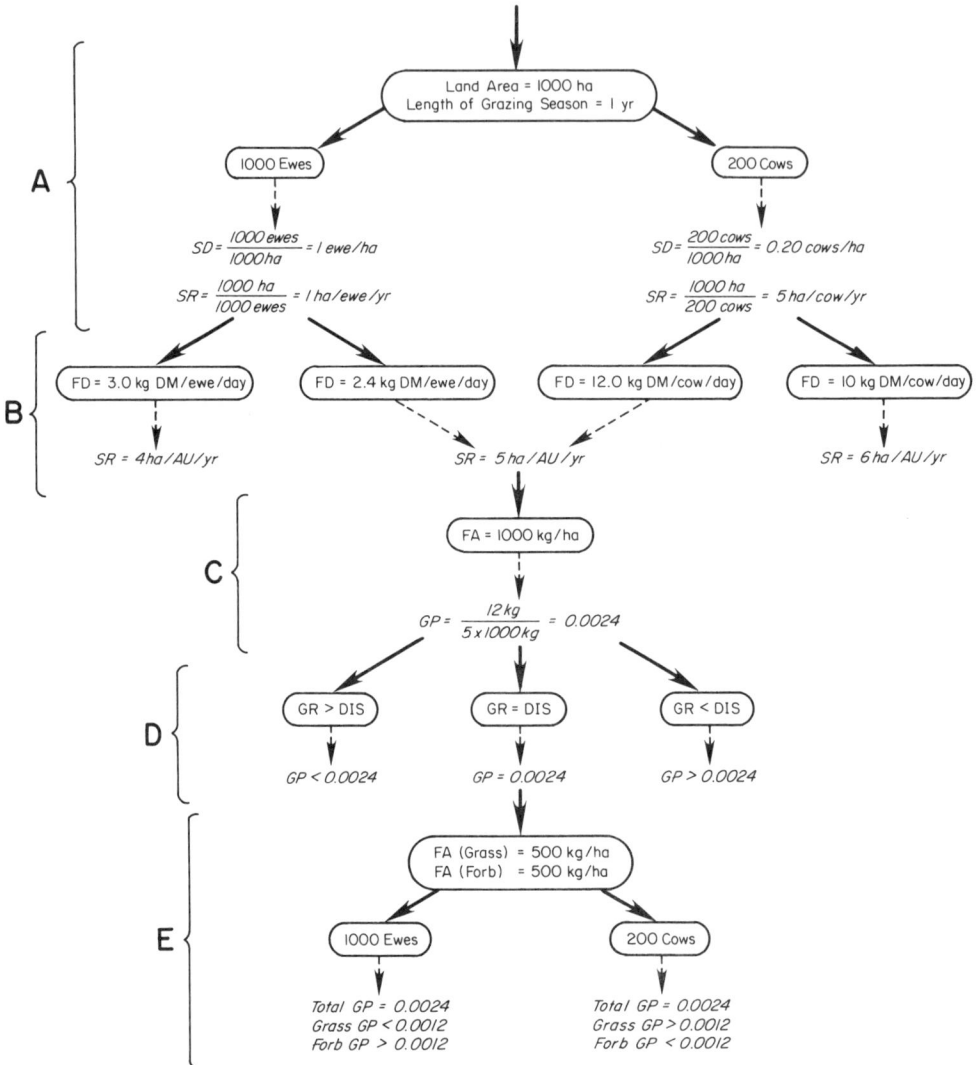

Figure 7.1. Flow chart depicting relationships among stocking density (SD), stocking rate (SR), forage demand (FD), forage available (FA), and grazing pressure (GP) as affected by varying numbers and types of animals and amounts of forage in a simple 1-pasture, 1-herd grazing regime. GR and DIS reflect relative rates of forage accumulation (growth) and disappearance, respectively. Broken lines are the equivalent of IF statements, and solid lines with arrows are the equivalent of THEN statements. For example, IF land area equals 1000 ha and length of grazing season equals 1 year, THEN stocking density (SD) equals 1 ewe/ha if stocked with 1000 ewes or 0.20 cows/ha if stocked with 200 cows.

describing the discretionary behavior of animals in the selection of various forages (Hodgson 1979), on GP if it is assumed sheep prefer forbs over grass and cattle prefer grass over forbs (see Chapters 2 and 3).

Although the definition of GP used herein differs slightly from traditional definitions (Hodgson 1979; Scarnecchia and Kothmann 1982; Soc. Range Manage. 1989), it is conceptually appropriate because it incorporates the concept of preference at the plant-animal interface. This concept is critical to meeting the objectives of this chapter because of the linkage between the basic principle of grazing management (i.e., control of the intensity of defoliation of individual plants), the four general principles of grazing management as outlined above, and traditional range condition concepts (see Chapter 5).

The concepts used to define and describe range condition have long been the subject of lively debate among range ecologists (Humphrey 1949; Costello 1964; Tainton et al. 1980; Smith 1988a; Wilson 1989; see Chapter 5), particularly with regard to the practical aspects of quantitative assessment and monitoring (Soc. Range Manage. 1983; Westoby et al. 1989). A major issue of concern is the concept that as range condition improves, livestock carrying capacity also improves (see definition of range improvement, Soc. Range Manage. 1989). Although this is an attractive concept because it seems intuitively reasonable and appropriate in certain instances (Danckwerts and Aucamp 1986), it has been shown to be inappropriate in many cases because it is not biologically possible to continually maximize energy capture (i.e., excellent range condition) and harvest and conversion efficiencies simultaneously (see Chapter 1).

In an attempt to avoid having to specifically address the absolute impact of range condition on livestock production throughout this chapter, we specifically define **range improvement** as any change in quantity and/or quality of available forage that facilitates sustained livestock production. Central to this definition is that optimum **range condition**, in terms of the forage complex, is that which maximizes livestock production/unit area of land on a sustained basis.

GENERAL PRINCIPLES

Number of Animals

Determining the proper number of animals to be placed on an area is the principal factor affecting the relative success of any grazing management strategy. This is so because number of animals affects not only individual animal performance but also production/unit area of land (see Chapter 1). Regardless of vegetation complex or the kind or class of animal, number of animals occupying an area over a given period of time (stocking rate) has a profound effect on livestock production because it affects GP directly by virtue of its direct effect on FD and the subsequent effects of FD on FA. For example, at low rates of stocking, individual animal performance is maximized (Fig. 7.2) relative to the quality of forage available because GP is low. However, production/unit area is also necessarily low because number of animals/unit area is low (Fig. 7.3). But as stocking rate is increased (moderate GP), individual animal performance begins to decline because of restrictions imposed on nutrient intake by either quantitative and/or qualitative declines in the forage resource (see Chapter 2). The stocking rate at which this decline begins is commonly referred to as the **critical stocking rate** (Hart 1978), and any increase in stocking rate beyond this point normally results in a reduction in individual animal performance. Production/unit area, however, continues to increase as stocking rate is increased from low to moderate because of the increase in number of animals (Fig. 7.3). This increase continues to some maximum as stocking rate is increased, but eventually it too decreases as nutrient intake becomes progressively more restrictive.

The stocking rate at which maximum production/unit area is achieved varies as a function of rate of decline in individual animal performance. Most often it is assumed for practical reasons that rate of decline is linear (Hart 1978) because in such cases it can be shown that the stocking rate at which maximum production/unit area is achieved is precisely one-half of the stocking rate at which individual animal performance is zero. In reality, however, the relationship between stocking rate and individual animal performance is probably curvilinear (Peterson et al. 1965; Connolly 1976; Edwards 1981). The reason it is curvilinear is because a linear decline in individual animal performance is only possible if the decline in the net assimilation rate of the individual animals is uniformly linear. This is doubtful, particularly in multispecies extensive grazing systems because quantity and quality of available forage (nutrients) are seldom uniformly distributed (vertically and horizontally) over time. As a result, it is doubtful quantity and/or quality of forage consumed (nutrient intake) declines in a linear fashion as stocking rate is increased. Moreover, it is doubtful that the amount of energy expended per animal in search of nutrients (grazing) remains constant as stocking rate increases. The resulting effect is that as stocking rate is increased in extensive rangeland settings, rate of decline in individual animal performance is accelerated (Fig. 7.2).

The stocking rate at which maximum production/unit area is achieved, when individual animal gains-stocking rate responses are assumed to be curvilinear rather than linear, is near that at which individual animal performance is one-half of maximum. In other words, it is near the stocking rate that is halfway between the critical stocking rate and that at which individual animal performance is zero. This is in contrast to an assumed linear response as discussed earlier, wherein maximum production/unit area is achieved at the stocking rate that is precisely one-half of the stocking rate at which individual animal performance is zero. The relative differences in precision of estimates of optimal stocking rates between linear (precise) and curvilinear (near) individual animal-stocking rate response functions occurs because assumed linear responses are easily defined mathematically, whereas curvilinear responses are often undefinable.

But regardless of the precise rate of stocking required to maximize production/unit area, the basic problem in grazing management is that the relationships between rate of stocking and livestock production are extremely complex and highly variable in any livestock production system because GP varies widely over time and space. This variation is the result primarily of variations in quantity and quality of available forage (nutrients available) over time and space as a result of both managerially uncontrollable variables, which are primarily abiotic, and controllable variables, which are primarily biotic (see Chapter 1). Major abiotic variables are climate and the inherent productivity potential of a site as defined by such factors as soil fertility, slope, and aspect (range site). Thus, quantity and quality of available forage vary seasonally, among years and range sites, and within and among geographical regions (Sims and Singh 1978; Sala et al. 1988) regardless of the relative impact of any biotic factors.

The major biotic factor affecting quantity and quality of available forage is grazing intensity, of which stocking rate is a major determinant. Generally, as stocking rate is increased, quantity of available forage declines on both a short- and long-term basis. On a short-term basis, this decline occurs because rate of forage depletion exceeds rate of accumulation. On a long-term basis, this decline results because of the interac-

Livestock Production 165

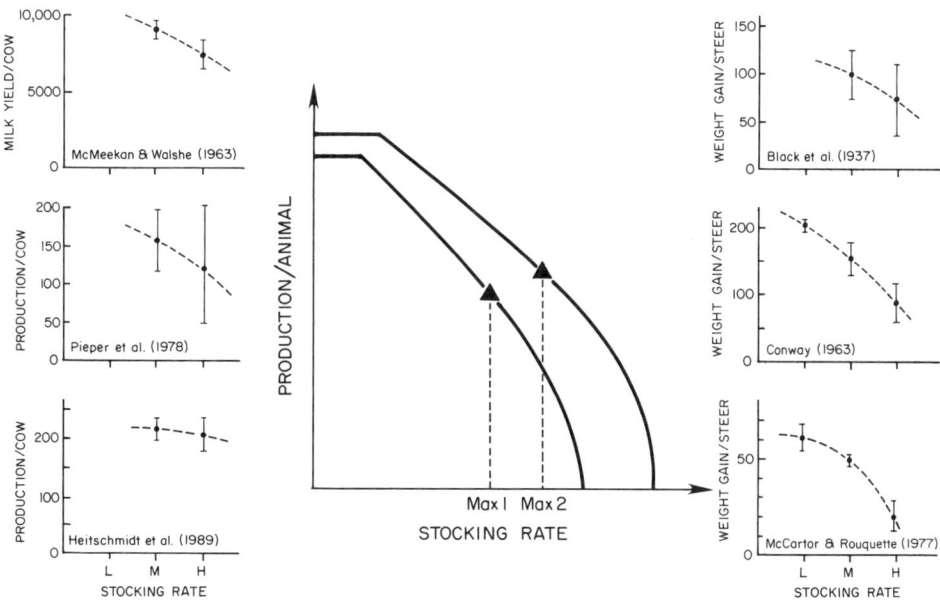

Figure 7.2. Conceptual model of functional relationship between light (L), moderate (M), and heavy (H) stocking rates and production/animal (kg) as derived from stocking rate studies by Pieper et al. (1978) and Heitschmidt et al. (1990) (year-long grazing, rangeland); McCartor and Rouquette (1977) (seasonal grazing, annual pasture); Black et al. (1937) (seasonal grazing, rangeland); Conway (1963) and McMeekan and Walsh (1963) (year-long grazing, grass-clover pasture). Vertical lines depict one 1 standard deviation of annual means. Upper line in conceptual model depicts functional relationship during periods of abundant forage production, while lower line depicts relationship during periods of limited forage production. Max 1 depicts the rate of stocking whereby production/unit area is maximized during periods of limited forage production, and Max 2 depicts the rate of stocking whereby production/unit area is maximized during periods of abundant forage production.

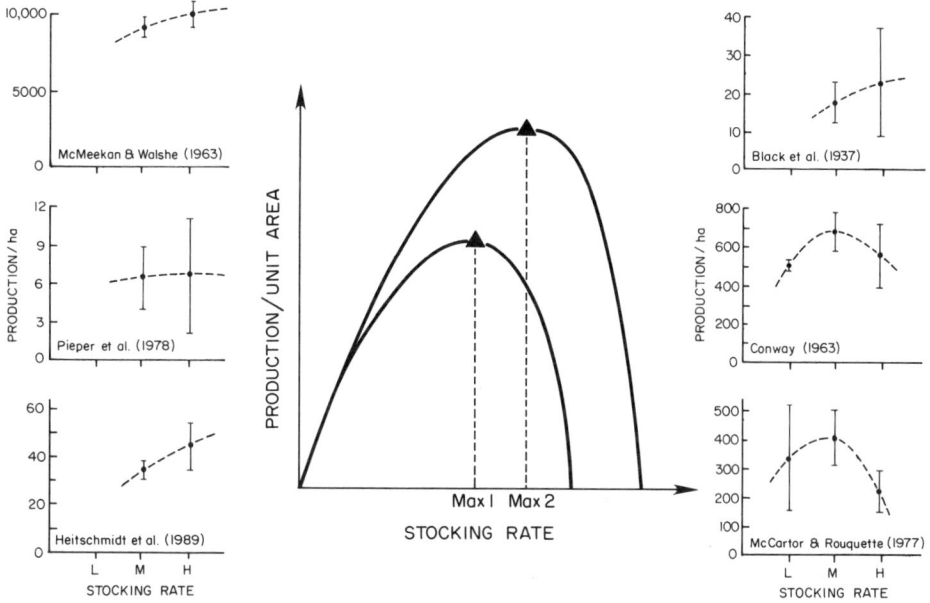

Figure 7.3. Conceptual model of functional relationship between stocking rate and production/unit area (kg) (see Fig. 7.2 for details).

tion effects of both abiotic and biotic factors on plant growth (see Chapter 4) and plant successional processes (see Chapter 5). Grazing intensity is a major factor affecting the direction, magnitude, and rate of vegetal change (range trend and condition) in rangeland ecosystems.

Stocking rate often affects quality of available forage also, but the relative effect varies over time and space depending upon the specific situation. On a short-term basis, overall forage quality often increases as grazing intensity increases because of the removal of low-quality (senesced) forage (Heitschmidt et al. 1987b, 1987c), whereas quality of available forage over the long-term varies depending upon the quality of the replacement species resulting from changes in species composition (see Chapter 5).

As a result of the integrated effects of abiotic and biotic factors on quantity and quality of available forage over time, and because of the relationship between GP and livestock production, the quantitative relationships between stocking rate and livestock production (Figs. 7.2 and 7.3) are situation specific in that they vary over time and space. As a result, optimal rates of stocking vary over time and space depending upon season, year, site, and management goals. Although such terms as under- and over-stocking are relative terms that vary as a function of management goals, "proper" stocking can be defined as that which maximizes energy capture, harvest, and conversion efficiencies within a given area on a sustained basis when assumed management goals are oriented towards maximizing livestock production over time. In this sense, under-stocking can be viewed as a tactic that generally enhances efficiency of energy capture and stability of livestock production, whereas over-stocking can be viewed as a tactic that generally reduces stability of livestock production by suppressing energy capture (see Chapters 4 and 5) and/or energy conversion efficiency (see Chapter 2).

The essence of the effects of stocking rate on livestock production is reflected in Figures 7.2 and 7.3. In both figures, the upper curve reflects the functional relationship between stocking rate and livestock production when nutrient availability/unit area of land is high (low GP), while the lower curve reflects the relationship when nutrient availability/unit area of land is low (high GP). The distance between the two curves reflects the potential effect of varying levels of nutrient availability on livestock production at various rates of stocking. At low rates of stocking, the effects are limited because differences in livestock production are a function of quality of forage. However, potential differences increase dramatically as stocking rate is increased because livestock production is affected by both quantity and quality of forage available. In other words, as forage demand (stocking rate) increases, stability of livestock production decreases. This is reflected by the general increase in the standard deviations of annual means derived from a wide array of stocking rate-livestock production studies as summarized in the margins of Figures 7.2 and 7.3.

Unfortunately, quantity and quality of available forage in dynamic, multi-species rangeland ecosystems often vary more as a function of the highly variable, managerially uncontrollable abiotic factors, as discussed earlier, rather than the managerially controlled biotic factors (Noble 1986; see Chapter 5). As a result, the optimal stocking rates for livestock production in extensive rangeland settings vary widely among seasons, years, and sites, within and among geographical regions (Morley 1966; McCown 1982). This concept is reflected in Figures 7.2 and 7.3 in that the optimal stocking rate for maximizing livestock production when quantity and/or quality of forage available is low, Max 1, is less than the optimal rate when quantity and/or quality of forage available is high, Max 2. It can be concluded also that the probability that frequently occurring abiotic events, such as drought, will become catastrophic, is related to rate of stocking. This is demonstrated in Figures 7.2 and 7.3 in that livestock production at the optimal rate of stocking during periods of reduced forage production, Max 1, would be greater than at the heavier rate of stocking required to maximize production during periods of ample forage production, Max 2. In other words, as rate of stocking increases so does frequency of occurrence of catastrophic events (Noy-Meir and Walker 1986). Thus, proper rates of stocking for sustained livestock production are necessarily moderate in virtually all range livestock production systems.

Kinds and Classes of Animals

The concept of GP is related generally only to total forage demand/availability within a given area (Hodgson 1979; Scarnecchia and Kothmann 1982). But when forage demand is defined so as to incorporate the concept of preference within the GP function (Fig. 7.1, section E), the underlying rationale for mixed animal grazing is easily grasped because:

1. Most grazing lands generally support a combination of forage classes (grasses, forbs, shrubs, and trees); and
2. Dietary preference, nutrient requirements, and foraging abilities vary among kinds and classes of animals (see Chapters 2 and 3).

Thus, GP varies among forage classes over time and space as a function of the unique assemblage of plants and animals present. By matching the forage demand of various kinds and classes of animals to the forage available, an overall increase in harvest and conversion efficiencies can often be realized, thereby increasing livestock production.

The most common practice employing this strategy is multi-species grazing. The effects of multi-species grazing on livestock production has been the subject of many studies throughout the world (Aucamp et al. 1986). For example, in the Edwards Plateau region of Texas such studies have continued without interruption for the past 38 years (Taylor 1985). Vegetation in this region is a mixture of short and mid-grasses with an abundance of forbs and a moderate overstory of browse. Grazing studies have focused on quantifying the relative effects of various combinations of cattle, sheep, and goats on livestock production. Yearlong grazing treatments stocked at equal rates of total FD include: cattle (100%); sheep (100%); goats (100%); cattle (50%) + goats (50%); and cattle (50%) + goats (25%) + sheep (25%). Results (Fig. 7.4) show that:

1. Production/animal was significantly greater for steers when grazed in combination with sheep and goats than when grazed alone;
2. Production/ewe was significantly greater when sheep were grazed with cattle and goats than when grazed alone; and
3. Production/goat was not affected by species mix.

The results from these studies reflect the relative effects of varying levels of GP on livestock production. Previous research has clearly established that cattle prefer grass, sheep prefer grass and forbs, and goats prefer browse and grass (see Chapter 2). Although forage availability was not measured in these studies, results from diet studies (Taylor 1985) suggest that when cattle were replaced by goats and/or sheep, individual cattle performance increased because FD for the grass component was reduced (lower GP). Likewise, it can be assumed production/ewe increased when some sheep were replaced by cattle and goats because GP on the forb component declined. Production/goat was presumably unaffected by species mix because rate of stocking was below the critical stocking rate necessary to cause a shift in diet selection from a high-quality browse and grass dominated diet to a lower quality diet.

Selection of the proper kinds and classes of animals is also related to environmental and economic constraints. Two examples of variations in kinds and classes of animals relative to these constraints are the preponderance of *Bos indicus* type cattle in the hot, arid regions of the world and the dominance of growing rather than breeding livestock in regions with an abundance of annual cereal grains.

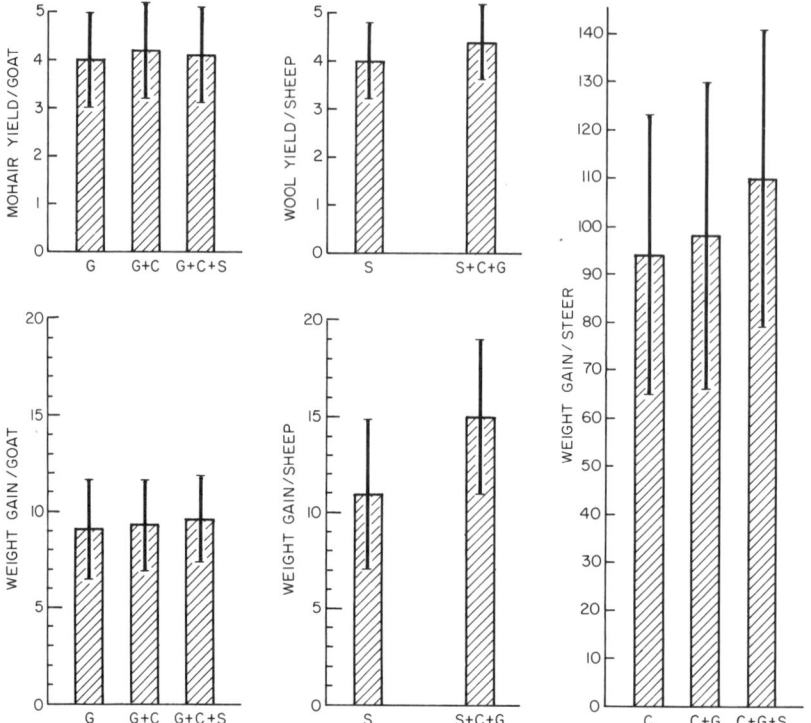

Figure 7.4. Production/animal of mutton goats (G), steers (C), and mutton sheep (S) when grazed year-long at a moderate rate of stocking in various combinations (Taylor 1985 and unpublished). Vertical lines depict 1 standard deviation of annual means.

Spatial Distribution of Animals

It is well known that livestock preferentially select various plants and plant parts during the grazing process (see Chapters 2 and 3). They also preferentially select various assemblages of plant species (plant communities) in which to graze (see Chapter 3). Because the distribution of various plant communities varies spatially as a function of such factors as toposequence, soil type, aspect, and past grazing history (see Chapter 5), GP varies among plant communities depending upon livestock distributional patterns across a landscape relative to the kinds and amounts of forage available.

Common practices utilized to enhance livestock distributional patterns are cross-fencing and strategic placement of salt, mineral, and watering facilities. There is also a strong interaction between multi-species grazing and spatial grazing patterns. This interaction effect is related primarily to the spatial distributional pattern of preferred and non-preferred plant communities/forage classes across a landscape, the innate behavioral patterns of various kinds and classes of animals, and the social interaction effects of various animal species that promote the establishment of multiple herds of grazing animals.

Studies examining the potential impact of varying livestock distributional patterns on livestock production have been reviewed previously (Squires 1981). More recently, however, the basic effects of cattle distributional patterns on livestock production have been demonstrated by Hart et al. (1988). At nearly equal rates of stocking and GP, they reported that average daily gain (ADG) of yearling heifers in Wyoming in a 518-ha pasture was significantly less than ADG in 24- and 34-ha pastures. Maximum distance to water was 5.4 km in the larger pasture and 1.4 km in the smaller pastures. Herbage utilization, a measure of harvest efficiency, averaged 50% in the smaller pastures and was similar regardless of distance from water. Utilization in the larger pasture averaged 41% and ranged from 60% near water to 25% near the back of the pasture.

Temporal Distribution of Animals

Justification for temporal variation in livestock grazing strategies is related primarily to the short- and long-term effects of defoliation on quantity of forage produced (efficiency of energy capture) and consumed (harvest efficiency), and secondarily to the quality of forage produced and consumed (efficiency of conversion). This conclusion is dictated by the first law of thermodynamics (see Chapter 1) because the presence of forage (captured solar energy) is a prerequisite to forage quality determinations. Still, the nutritional aspects of the grazing animal should be considered (Launchbaugh et al. 1978) in light of the effect of temporal variations in quantity and quality of forage available on livestock production.

There is essentially only one mechanism whereby an increase in livestock production can be expected to result as a direct function of temporal adjustment of grazing events, i.e. increased forage quality. Forage quality is seldom directly enhanced by deferment from grazing although it may be indirectly enhanced if deferment induces a desirable qualitative change in species composition (see Chapter 5). This long-term response is in contrast, however, to the potential short-term effect of grazing on forage quality. For example, Heitschmidt et al. (1987b) showed that forage quality [percentage crude protein (CP) and organic matter digestibility (OMD)] in a north Texas grassland increased over the short-term as GP increased. But the increase was the result of a decline in relative amounts of low-quality senesced forage rather than an absolute increase in amounts of high-quality, actively growing forage. This concept is the underlying mechanism justifying the use of such practices as early intensive stocking (EIS). Utilizing EIS tactics, Launchbaugh et al. (1983) showed that in certain instances, cattle production/ha increased approximately 25% over season-long grazing (SLG) when rate of stocking was increased 2-fold and length of grazing season was halved. The apparent reason for this outcome was attributed to the effect of seasonal changes in forage quality on steer conversion efficiency. This conclusion was supported by the ADG of the steers which was nearly equal in both treatments during the duration of the IES treatment (May 1–July 15) but declined thereafter in the SLG treatment (May 1–October 1).

It should be noted, however, that the response in this study was mediated through an adjustment in stocking rate in addition to manipulation of the temporal distribution of the grazing season. The strategic timing of the shortened grazing season was selected primarily in consideration of the nutritional aspects of the grazing animal; in short, the aim was to enhance conversion efficiency rather than efficiency of energy capture. This objective was realized as evidenced by the 25% increase in production/ha with IES, although rate of stocking in both treatments was equal when averaged across the 172-day duration of the SLG treatment. However, it should be noted that the benefits of IES tactics would be difficult to capture in a year-long operation stocked only with breeding animals.

A second example of a management practice designed in consideration of the nutritional needs of grazing livestock is adjustment of breeding seasons (see Chapter 2) in accordance with the cyclic dynamics of forage quality and the cyclic nutritional needs of breeding animals. Because both the nutritional needs of breeding stock and quantity and quality of available forage vary over time, livestock production can be enhanced by proper manipulation of GP over time relative to nutrient demand and nutrient availability. The relative benefits derived from this practice are similar, however, regardless of stocking rate or grazing system because temporal growth patterns of vegetation in most rangeland ecosystems are similar regardless of grazing regime.

GRAZING SYSTEMS

Grazing systems are management tools designed to balance the conflicting relationships between energy capture, harvest, and conversion efficiencies (see Chapter 1). They are designed firstly to enhance livestock production over time by either improving and/or stabilizing the quantity (efficiency of energy capture) and/or quality (efficiency of conversion) of forage produced and/or consumed (efficiency of harvest). Production improves if the benefits of rest or deferment exceed the detrimental impacts of grazing; stabilization results if the benefits of rest exactly equal the detrimental impacts of grazing; while degradation results when the benefits of rest are less than the detrimental impacts of grazing. But regardless of the effects of a grazing system on the vegetation complex, its effects on livestock production vary depending upon its direct effects on GP.

Concepts and Terminology

A **grazing system** is considered "a specialization of grazing management which defines recurring periods of grazing and deferment for 2 or more pastures or management units" (Soc. Range Manage. 1989). For the purposes of this chapter, we limit our discussion to grazing systems stocked year-long. However, the basic ecological concepts and principles presented herein apply equally well to seasonal grazing systems because they too include recurring periods of grazing and deferment. The only difference between year-long and seasonal grazing systems is the tactic utilized to attain desired periods of grazing and deferment (number pastures > number of herds vs. buy/sell seasonally).

As defined earlier, grazing pressure (GP) is the ratio of forage demand (FD) to forage available (FA) at any instant. This is a conceptually convenient definition when considering the effects of GP on livestock production in a 1-pasture, 1-herd, continuously grazed treatment. But the conceptual analysis of the functional attributes of grazing systems requires a refinement of this concept because at any instant at least one subdivision of all grazing systems is rested. Thus, GP varies as a function of resource (forage or livestock) or area (individual subdivision or total area) of interest. For example, in a 4-pasture, 1-herd system (Fig. 7.5) **livestock or subdivision GP** will be four-fold greater than **forage or total GP** because only one fourth of the total forage or area within the fenced boundaries of the system is available to the animals at any instant. In other words:

- **Livestock GP** = $\dfrac{\text{forage-demand (FD)}}{\text{forage available (FA) in grazed subdivision(s)}}$

- **Forage GP** = $\dfrac{\text{forage-demand}}{\text{forage present in all subdivisions}}$

In concert with this concept, it can be seen that forage GP will not vary among grazing systems, regardless of number of subdivisions, if total FA and FD are equal, whereas livestock GP will vary dramatically depending upon the effect each grazing system has on stock density (Fig. 7.5). Appreciation for this concept is paramount for the development of an understanding of the potential impact of various grazing systems on livestock production.

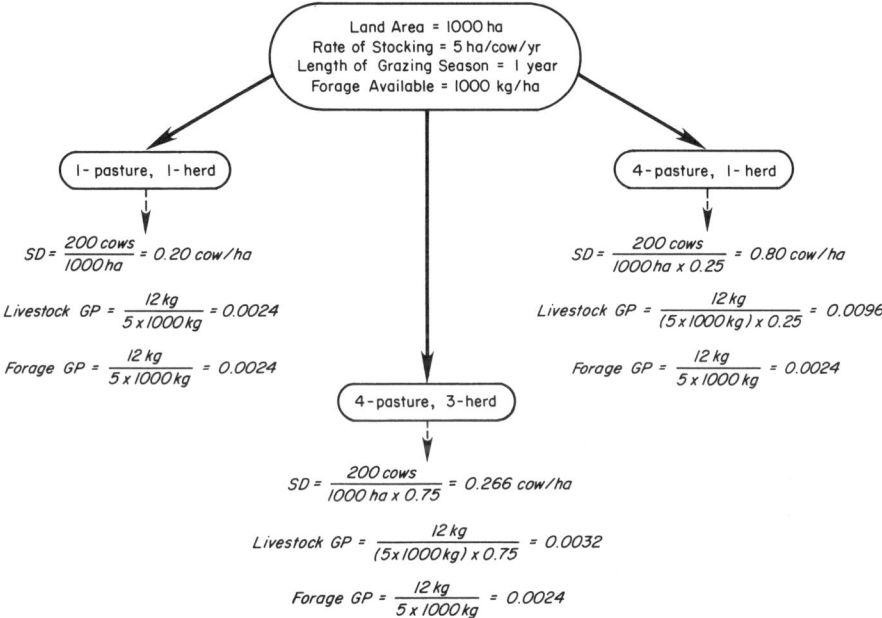

Figure 7.5. Flow chart depicting relative effects of two grazing systems on stock density (SD) and grazing pressure (GP). Broken lines are the equivalent of IF statements, and solid lines with arrows are equivalent to THEN statements (see Fig. 7.1 for details).

Effects on Livestock Production

The effects of various year-long grazing systems on livestock production have been reviewed previously by many (Heady 1961; Driscoll 1967; Shiflet and Heady 1971; Herbel 1974; Gammon 1978; Pieper 1980). Although interpretive conclusions vary among these and other reviewers, all tend to agree that the relative benefits derived from any grazing system, in terms of livestock production, vary greatly over time and space thereby making it difficult to make any definitive, universal conclusions concerning relative merits. This is not surprising, however, when the varied results are examined within the conceptual framework presented earlier relative to the effect of GP on individual animal performance. For example, it is easy to understand why individual animal performance generally declines, or at least does not improve immediately following the establishment of any grazing system (Wilson et al. 1984), unless rate of stocking is reduced (high livestock GP). Moreover, it is easy to understand why improvement in livestock production is dependent primarily on improvement in quantity and/or quality of forage available and/or consumed over time or space.

Tactics. There are two fundamental management tactics utilized to enhance quantity and/or quality of forage produced over time. These are commonly referred to as:

1. High utilization grazing (HUG); and
2. High performance or high production grazing (HPG) (Booysen and Tainton 1978).

The functional difference between these two tactics is related to the discretionary manner in which they affect the competitive interactions of preferred (high GP) and non-preferred (low GP) plant species. With HUG strategies, all plants are moderately to intensively defoliated during a grazing period, whereas with HPG strategies only the preferred plants are defoliated and then only at light to moderate intensities. Improvement with HUG tactics stems from the discretionary effects moderate to heavy levels of defoliation have on the competitive abilities (see Chapter 4) of the preferred and non-preferred species when both are defoliated. Improvement with HPG tactics stems from the varying effects lenient levels of defoliation have on the competitive abilities of the preferred plants relative to the competitive abilities of the ungrazed, non-preferred plants. The on-going (immediate or short-term) impact of these two tactics on livestock production are quite different because HUG tactics necessarily require livestock consume non-preferred forages whereas HPG tactics require consumption of only preferred forages. As a result, individual animal performance is usually less with HUG than HPG tactics whereas animal production/unit area is greater.

Implementation of a grazing system may also enhance livestock production if additional fencing is required whereby improved livestock distribution patterns are realized. This effect has been demonstarted by Hart et al. (1988) in Wyoming. At equal rates of stocking, they reported that ADGs of cattle were equal in an 8-pasture, 1-herd rotation system (24 ha/pasture) and a continuously grazed treatment if the continuously grazed pastures (24 and 34 ha) were small, but were greater in the rotational system when the continuously grazed pasture was large (518 ha).

Types of Grazing Systems. Functionally, there are four basic types of grazing systems (Fig. 7.6):

1. Deferred rotation (DR);
2. Rest rotation (RR);
3. High intensity-low frequency (HILF); and
4. Short duration (SD).

The functional aspects of each system center around the employment of either HPG or HUG tactics. The major factor affecting such is stocking rate in that high rates of stocking insure HUG tactics are employed rather than HPG regardless of type of system. However, assuming rate of stocking is moderate the general features of each are as follows.

Deferred rotation (DR) systems are multi-pasture, multi-herd systems designed to maintain or improve range condition utilizing HPG tactics. Stock density is moderate, length of graze long, and length of rest moderate. An example of a DR system is

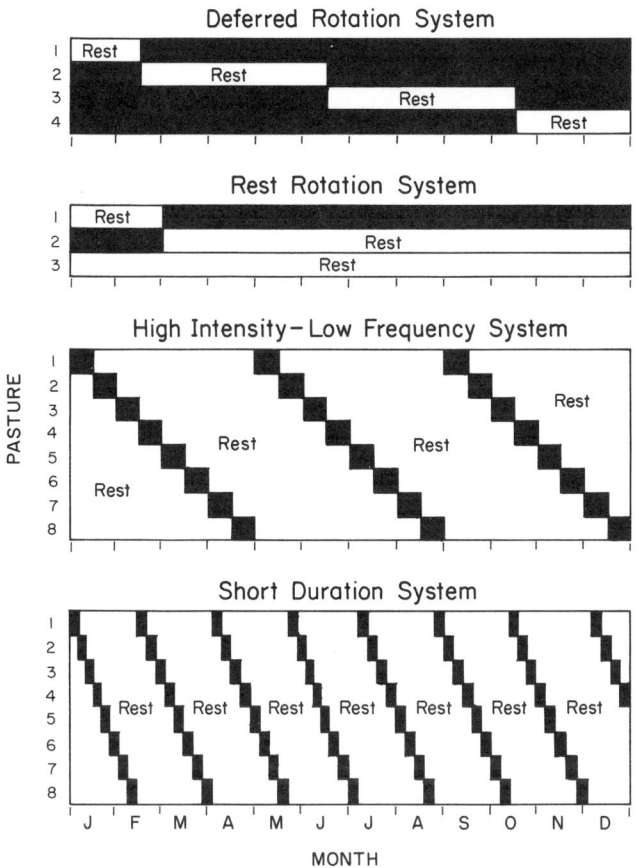

Figure 7.6. Conceptual model of sequential schedule of graze-rest periods for deferred rotation (DR), rest rotation (RR), high intensity-low frequency (HILF), and short duration (SD) type grazing systems (after Kothmann 1980).

the 4-pasture, 3-herd system developed by Merrill (1954) in the southern mixed-grass prairies of the U.S. (Fig. 7.6).

Rest rotation (RR) systems are either multi-pasture, multi-herd or multi-pasture, single herd. They are designed to maintain or improve range condition utilizing a combination of HPG and HUG tactics. Stock density ranges from moderate (multi-herd) to heavy (single herd); length of graze ranges from short (HPG tactics) to long (HUG tactics), and length of rest from long to short. An example of a RR system is the 3-pasture, 1-herd Santa Rita system developed by Martin (1978) for the arid desert grasslands of the southern U.S. (Fig. 7.6).

High-intensity, low-frequency (HILF) systems are multi-pasture systems usually stocked with a single herd of livestock (Fig. 7.6). They are designed to maintain or improve range condition utilizing HUG tactics. Stock density is high, length of graze moderate, and length of rest long.

Short duration (SD) systems are similar to HILF systems (Fig. 7.6) except HPG rather than HUG tactics are employed to maintain or improve range condition. This is achieved by reducing length of graze and rest period while maintaining high stock densities.

Livestock Production. The relative effects of the four types of grazing systems discussed above on livestock production are presented in Figure 7.7. These generalized

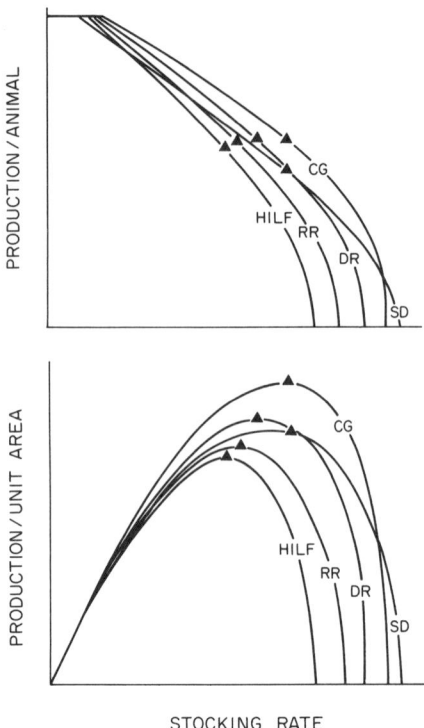

Figure 7.7. Conceptual model of the theoretical on-going effects of stocking rate on livestock production in continuous (C), deferred rotation (DR), rest rotation (RR), high intensity-low frequency (HILF), and short duration (SD) type grazing systems. Triangles (▲) reflect point on livestock production-stocking rate curve whereby production/unit area is maximized. Generally, HPG tactics are operative at stocking rates less than that which maximize production/unit area, and HUG tactics are operative at greater rates of stocking.

response curves are based on five important assumptions.

1. Vegetation complex is a multi-species, temporally variable grassland grazed year-long.
2. All systems are implemented using existing pastures.
3. Quantity and quality of available forage (i.e., range condition, our definition) is equal in all systems.
4. Management skill is sufficient to insure that livestock production in the more intensively managed type systems, particularly SD, is not unduly suppressed by improper rotation schedules.
5. There are no kinds or classes of animal by grazing system interaction effects.

These assumptions are prerequisites for the development of livestock-stocking rate response curves because of the differential effects that various grazing systems may elicit relative to the four principles of grazing management. A major problem in grazing system research in extensive rangeland settings lies in our inability to effectively employ multi-factor experimental designs to elucidate single-factor effects of a multitude of factors that affect livestock production. For example, livestock production in a 1-herd, multi-paddock RG system may vary as a function of a number of factors including stocking rate, number of paddocks (i.e., stock density), and rate of rotation (i.e., length of graze/rest periods). Thus to properly evaluate such a system requires at least a twice replicated 3 × 3 factorially designed experiment. Although such studies are desirable, they are often not feasible because of resource limitations relative to land, labor, and capital. Thus, any interpretive summary of the effects of various grazing systems on livestock production, such as that presented in Figure 7.7, must be based on a general understanding of the potential impact that the four types of grazing systems have on livestock GP over time and space.

Assumption #1 is made to eliminate any confusion that may arise concerning the potential impact of grazing systems on livestock production in tame pasturage and/or seasonal grazing environments. For example, it may be argued that continuous grazing is superior to rotational grazing when the early intensive stocking (EIS) strategy, discussed earlier, is employed on an area of land. However, for the present purposes, we would simply argue that seasonal grazing strategies such as EIS are functionally equivalent to year-long grazing HILF type systems (i.e., long graze, long rest).

Assumption #2 is necessary to eliminate the differential impact that various systems may have on livestock production as a result of redistribution of animals from increased subdivisions. This does not mean, however, that grazing systems do not enhance livestock distribution patterns as a result of their effect on stock density, as hypothesized by Kothmann (1980), Savory and Parsons (1980), and Malechek and Dwyer (1983). Certainly, the data of Hart et al. (1988), as discussed earlier, show stock density does not greatly impact livestock performance when pastures are small. Likewise, studies by Gammon and Roberts (1978), Kirby et al. (1986), and Walker et al. (1989c) tend to support the findings of Hart et al. (1988). Still, there seems little doubt that such would be the case if pastures were extremely large. Unfortunately, we know of no studies that clearly address this hypothesis.

Assumption #3 is necessary to eliminate potential differences in livestock production that may arise over time as a result of a change in quantity and/or quality of forage available (i.e., range condition). As discussed earlier, the potential impact of various grazing systems on range condition, and subsequently on livestock production, is a question that has not been critically addressed (Gammon 1978; Malechek 1984; Hart and Norton 1988; Holechek et al. 1989). We question, however, the value of addressing this issue on a broad scale because desired condition varies depending upon management goals. Moreover, because of regional differences in abiotic conditions and functional differences between grazing systems, no one grazing system is universally the best. Optimal length and sequential scheduling of graze-rest periods vary as a function of the interaction effects of abiotic and biotic factors on plant growth (see Chapter 4) and successional processes (see Chapter 5). Of major importance is the temporal precipitation pattern in a region because the benefits derived from rest or deferment, in terms of range improvement, are greatest during periods of ample rainfall. To improve the probability of encountering favorable growing conditions during a period of deferment, length of rest must be lengthened with increasing aridity. Thus, RR and HILF systems are designed characteristically for regions having extended periods of drought, such as desert grasslands, whereas DR and SD type systems are designed for areas where extended periods of drought are uncommon, such as temperate grasslands. As a result, universal conclusions as to the relative merits of various grazing systems, based on data from a wide array of studies conducted across a broad array of regions, are inappropriate and of little practical value.

Assumption #4 is made to eliminate the impact that varying levels of management skill can have on livestock production in any given system. Certainly, the greater the livestock GP (Fig. 7.5) the greater the precision of judgment necessary to attain satisfactory levels of livestock production (Gammon 1978).

Assumption #5 is made because of potential differences among grazing systems in their impact on livestock behavior. For example, there is little doubt that the probability of a grazing system affecting offspring survivability is greater in SD than any other type of system because of interference with the "mothering" process of breeding animals as a result of frequent rotation.

In light of these assumptions and a basic understanding of the potential effects of livestock GP on individual animal performance (HPG and HUG tactics), we believe Figure 7.7 approximates the short-term impacts of DR, RR, HILF, and SD type grazing systems on livestock production at various rates of stocking. We believe generally that critical rates of stocking (Hart 1978) will decrease in direct proportion to the manner in which each system affects livestock GP (DR<RR<HILF) and/or energy expenditures (SD>others). In theory, there is little reason to assume that high livestock GP will decrease individual animal performance in an SD type system if rate of rotation is proper (Assumption #4). However, previous research in SD type systems (Walker and Heitschmidt 1989) has shown that as rate of rotation is accelerated, time spent trailing and distance walked increase, which leads us to believe the lowest critical rate of stocking will most often be in SD type systems.

We also suggest that individual animal performance will be less in DR, RR, and HILF type systems than under continuous grazing, at all rates of stocking in excess of the critical rate. Depicted relative differences between these systems at any given stocking rate are based on how frequently we suspect restrictive performance levels will be encountered as a result of increased livestock GP.

The slower rate of decline in individual animal performance in the SD systems as rate of stocking is increased reflects the potential interaction effect that frequency and/or intensity of encounter of high livestock GP events may have on livestock performance. This hypothesis centers around the assumption that frequent encounters of high livestock GP events of short duration are cumulatively less restrictive to animal performance than infrequent encounters of long duration (see Chapter 2). In other words, how forage is rationed in situations where livestock GP is high can affect individual animal performance. Although we know of no data in direct support of this hypothesis, we believe such is the case based strictly on our personal experiences and observations (Taylor 1989; Heitschmidt et al. 1990).

This generalization does not support, however, the claim that sustainable rates of stocking in SD type systems are well above those under continuous grazing in yearlong grazing environments. This matter is currently the subject of lively debate throughout the world (Savory and Parsons 1980; Heitschmidt and Walker 1982; Gammon 1984; Skovlin 1987; Pieper and Heitschmidt 1988; Bryant et al. 1989; Taylor 1989), but the argument has little merit unless it is assumed that SD type grazing systems consistently enhance rate of forage production between grazing events. Although this is an exciting and attractive hypothesis and probably true in certain situations (McNaughton 1978; Detling 1988; Heitschmidt 1990), it is also a wellresearched, refuted hypothesis (see Chapter 1) relative to the general effects of grazing on plant growth in multi-species, arid, and semi-arid rangelands (Belsky 1986), and particularly in SD type grazing systems (Heitschmidt et al. 1987a, 1987c).

Based on the above generalization, it is easy to see why the results from various grazing system studies are relevant only when examined within the designed objectives of the study. Universal conclusions as to the relative merits of any particular grazing system are generally of limited value and inappropriate. An understanding of the functional aspects of various grazing systems suggests, however, that perhaps the major benefit derived from the employment of a grazing system is directly related to the forced use of moderate rates of stocking. Moreover, it seems reasonable to assume that the use of a mix of various grazing systems, in conjunction with continuous grazing, may be more appropriate for achieving management goals rather than the continuous use of any single system strategy (Walker et al. 1986a).

CONCLUSIONS

Livestock production is an integrated measure of energy capture, harvest, and conversion efficiencies. The principal factor affecting livestock production is grazing pressure (GP), which varies as a function of both abiotic and biotic factors. The principal abiotic factors are climate and the inherent productivity potential of a site. The principal biotic factor is grazing intensity, which varies as a function of the temporal and spatial distribution of various kinds and numbers of grazing animals. Grazing systems are management tools designed to enhance and stabilize livestock production over time. They are designed firstly to enhance efficiency of solar energy capture as evidenced by the universal incorporation of a period of deferment in the design of all grazing systems. The nutritional needs of livestock are necessarily of secondary consideration in the design of all grazing systems, the effects of which are mediated through the use of conservative rates of stocking. Research has shown that the ongoing effects of various grazing systems on livestock production vary widely over time and space. As a result, no grazing system has been shown to be universally superior to any other in terms of its ability to enhance livestock production.

CHAPTER 8

WILDLIFE

T. G. Barnes, R. K. Heitschmidt, and L. W. Varner

INTRODUCTION

Historically, abundance of wildlife has been an important factor affecting humans' ability to meet the basic goal of survival. But as civilizations developed through the acquisition of knowledge, such as that which led to the development of a livestock industry, interest shifted from hunting to farming and ranching; thus, interest in wildlife declined. This decline continued in most developing societies until social and economic technological advances provided more leisure time and economic opportunities for pursuing such secondary goals as tranquility (see Chapter 9). Evidence of this increased interest is reflected in the relatively recent establishment of numerous wildlife preserves, game ranches, and fee hunting institutions.

Today, most wildlife species residing in regions supporting technologically advanced societies are considered to be of commercial, recreational, biological, aesthetic, and/or scientific value (King 1966). Currently, the most common measure of the value of any given wildlife species is related to sport hunting and/or aesthetic viewing. Land owners in many developed societies now find a substantial portion of their income is derived from hunting and aesthetic viewing enterprises (Glover and Conner 1988). Thus, a basic understanding of the potential interaction effects of a large number of sympatric herbivores is critical for the development of ecologically and economically sound livestock and/or wildlife enterprises.

Wildlife, in the broadest sense, includes all undomesticated animals. However, for the purposes of this chapter we focus on the broad ecological aspects of wildlife management as they relate to the management of large herbivores in grazed systems. Treatment is appropriately brief because the basic ecological principles and concepts of grazing management (see Chapter 1), animal nutrition (see Chapter 2), and grazing behavior (see Chapter 3) are equally applicable to all large herbivores whether they be domesticated livestock or undomesticated wildlife species. Likewise, at least the potential impacts of large herbivorous wildlife species on plant growth and development (see Chapter 4), ecological succession (see Chapter 5), and watershed condition (see Chapter 6) are similar to those of livestock. The major difference between the management of large herbivorous wildlife species and livestock in grazed ecosystems is related to differences in animal behavior. Generally, the temporal and spatial distribution of various kinds and numbers of livestock (see Chapter 7) are much easier to control than wildlife. This is because aggressive livestock breeding practices have historically focused on suppression of undesirable behavioral attributes. This is in contrast to most wildlife populations wherein natural selection processes have favored selection for a combination of traits that insure survival of the population over time. As such, some level of unconstrained behavioral attributes has been maintained in most wildlife populations. This is the major reason the management tactics employed in livestock-dominated enterprises differ from those employed in wildlife-dominated enterprises.

BASIC HABITAT NEEDS OF WILDLIFE

Wildlife and domestic livestock share the same four basic needs: food, water, cover and space. Although the relative importance of these needs to the survival of an individual animal and/or population may vary, survival is a function primarily of the frequency that a limiting factor is encountered and its magnitude. A **limiting factor** is defined as a basic requirement that limits the size, growth, and/or quality of a wildlife population (Bailey 1984). Limiting factors vary over both time and space. For example, water may be the limiting factor when food is available and water is not, whereas food may be the limiting factor when water is available and food is not. In other words, any hierarchy of needs is dynamic in that it varies as a function of the changing attributes of the resident habitat relative to its ability to meet the changing needs of the animals present (Krebs and Davies 1984; Mangel and Clark 1986).

The major difference between the hierarchy of needs of wildlife and that of livestock is related to humans' ability to reduce the frequency of occurrence and magnitude of limiting factors. For example, in intensively managed livestock production systems, food requirements are met by adjusting livestock numbers (i.e., food demand; see Chapter 7) and/or by providing supplemental feeds. Likewise, water and cover requirements are met by constructing watering facilities and shelters, whereas space requirements are altered through redistribution of food, water, and/or cover over time and space. But these tasks are much more difficult to execute in extensively managed wildlife systems primarily because of the innate behavioral characteristics of most species. For example, number of animals is much more difficult to control because of the innate desire of most wildlife species to avoid detection and/or to escape from predators following detection. This is true regardless of whether the predator is human or some carnivorous wildlife species.

Food

Ruminant wildlife species are selective consumers whose degree of selectivity varies depending upon morphological and physiological adaptations (see Chapters 2 and 3). In this sense, wildlife, like livestock, are commonly categorized as **grass/roughage** (e.g., bison, mouflon, oryx, and cattle), **intermediate** (e.g., elk, chamois, impala, and goat), and **concentrate** (e.g. white-tailed deer, moose, giraffe) feeders (see Chapter 2; Hofmann 1989).

The major factor affecting diet selection in most wildlife species is quantity and quality (i.e., nutritional value; see Chapter 2) of food available. This is evidenced by the fact that diet composition of any wildlife species varies over time (season) and space (location) in response to variations in quantity and quality of food available. For example, mule deer diets in north Texas have been shown to vary from 2–43% browse, 0–5% grass, and 0–33% forb depending upon location and season (Fig. 8.1). This basic concept applies also to carnivorous wildlife species as evidenced in the Serengetti Plains wherein Bertram (1979) reported that "although lions feed on a number of the very large resident herbivores, such as buffalo and giraffe, which no other predator can take, they feed mainly on zebra and wildebeest when these migratory species are within their pride's area." Even the trophic level that wildlife species occupy in an ecosystem varies as a function of their morphological and physiological adaptations in concert with the quantity and quality of food available. For example, Rio Grande turkey in south Texas have been shown to act both as primary and secondary consumers (Fig. 8.2) with relative amounts of forage (i.e., primary production) and insects (i.e., secondary production) consumed varying as a function of seasonal availability.

In summary, it can be seen that the major factor affecting diet selection in most wildlife species is the quantity and quality of food available. Furthermore, it can be seen that the underlying ecological and biological principles that govern diet selection processes of wildlife species are similar to those governing livestock diet selection (see Chapters 2 and 3).

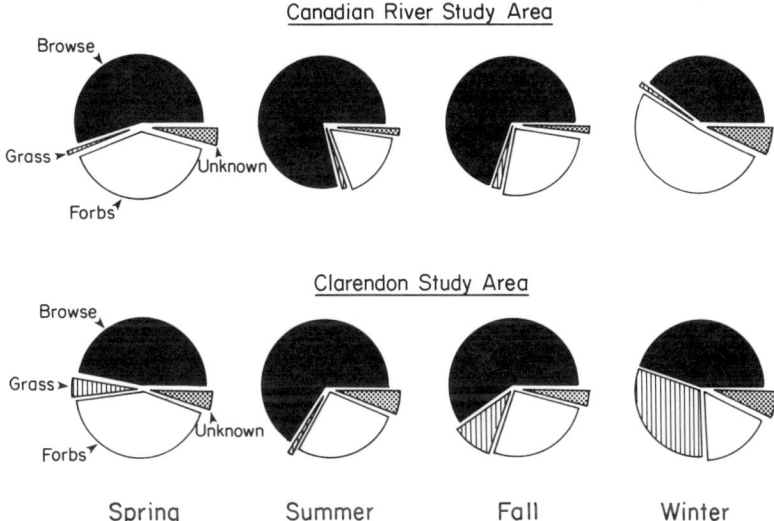

Figure 8.1. Seasonal composition of mule deer diets at two locations in north Texas (after Sowell et al. 1985).

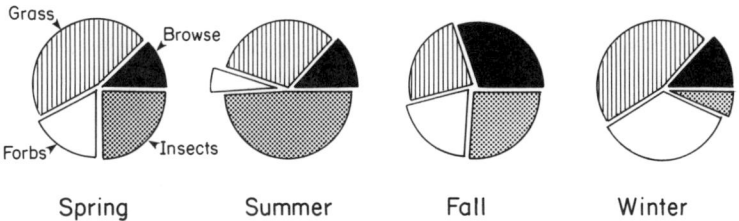

Figure 8.2. Seasonal composition of Rio Grande turkey diets in south Texas (after Litton 1977).

Water

Water is an essential compound for the sustenance of all living organisms (see Chapter 6). Thus, availability of water over time and space is a critical factor affecting the growth and survival of all animal populations including both free-ranging wildlife and fenced livestock.

Water requirements of wildlife, in terms of both quantity and quality, vary widely among species. Differences are related to behavioral (e.g., nocturnal lifestyle), morphological (e.g., storage capacity, pelage type, thickness, and color), and physiological (e.g., proportional use of metabolic water) differences. As a result of these differences, some large herbivores can survive extended periods without consuming water (e.g., camel), whereas other species cannot. Moreover, because of these adaptations, required sources of water vary widely among species ranging from free-standing (i.e., streams, lakes, and reservoirs) to that pooled on the surface of plants (i.e., dew), and/or internally incorporated into vegetation (i.e., succulence).

Cover

The structural attributes of a habitat are embodied in the concept of cover as it pertains to the functional needs of animals (Dasmann 1981). This is evidenced in that some wildlife species use vegetative cover for escape purposes whereas others prefer habitats with little or no vegetative cover. For example, mule and white-tailed deer, javelina, impala, bushbuck, and mountain reedbuck often use moderate to dense stands of brush to either avoid detection by predators or to escape from predators following detection. This is in contrast to the cover requirements of such species as pronghorn antelope and blue wildebeest which prefer open grassland habitats because the structural attributes of such habitats (low or no vegetative cover) enhance these species' ability to descry predators from afar and to escape following detection if necessary.

Cover is perceived generally to be of greater importance to large herbivorous wildlife species than livestock because of its value as **screening** and **escape habitat**. It may be argued, however, that livestock would have a similar need for cover were it not for humans' continued efforts to protect livestock from natural predators through the employment of such tactics as lethal predator suppression and the penning and herding of vulnerable species. Evidence of domestic livestock's inherent but latent need for cover is the tendency of feral livestock to preferentially inhabit dense stands of brush to avoid detection by their primary predator (i.e., humans) and/or to escape following detection.

Cover also benefits many wildlife species by virtue of its inherent ability to modify environmental conditions by providing shade during periods of high daytime temperatures and insulation during periods of low temperatures and high winds (**thermal cover**). In this sense, cover also benefits livestock particularly in the absence of man-made shelters (see Chapter 3).

There is also a strong interaction between vegetation cover and food in certain instances. For example, cover may directly enhance a herbivore's food supply if the primary cover species is a preferred forage (e.g., pricklypear cactus) of the resident herbivore (e.g., javelina) (Inglis 1985). Cover may also indirectly benefit certain species of herbivorous wildlife by creating a micro-environment that enhances the establishment and growth of certain preferred forages and/or by providing a structure that impedes the utilization of such species by competing herbivores.

Space

Although animals require space to survive, amount of space required varies depending on the spatial distribution of food, water, and cover across a landscape and evolved behavioral attributes. The amount of inter- and intra-species space required by domestic livestock is almost exclusively a function of the space required to meet food, water, and cover needs. This is evidenced by the high level of livestock performance (i.e., growth) that is achieved in crowded finishing pens. In other words, space barriers do not appear to hinder livestock performance if sufficient food, water, and cover are available. A similar argument may be made for many wildlife species in terms of individual animal growth and survival as evidenced by their performance in zoos and wildlife parks. However, many of these same species require considerably greater amounts of space to achieve acceptable levels of reproductive performance whereby survival of a population is assured. This need may be viewed primarily as an evolved behavioral response, wherein space requirements (i.e., isolation) are linked to physiological function.

LIVESTOCK-WILDLIFE INTERACTIONS

Much has been written about livestock-large herbivorous wildlife interactions in grazed systems (see e.g., Peak and Dalke 1982; Nelson 1984; Kie and Thomas 1988; Holechek et al. 1989), and it is not our intent herein to provide a detailed summary of this large, often site-, time-, and species-specific volume of knowledge. Rather, our objective is to present a conceptual model to assist in the development of an understanding of livestock-wildlife interactions regardless of location, time, or interacting species.

The model is structured around the basic premise, as presented earlier in this chapter, that livestock and wildlife have the same basic needs (i.e., food, water, cover, and space), and that relative performance in grazed systems varies in part as a function of the temporal and spatial distribution of various kinds and numbers of animals (see Chapter 7). By linking these two general concepts with specific knowledge about the ecology and biology of the interacting species, a management-oriented, decision-analysis model emerges (Fig. 8.3). The model follows the generalized decision model

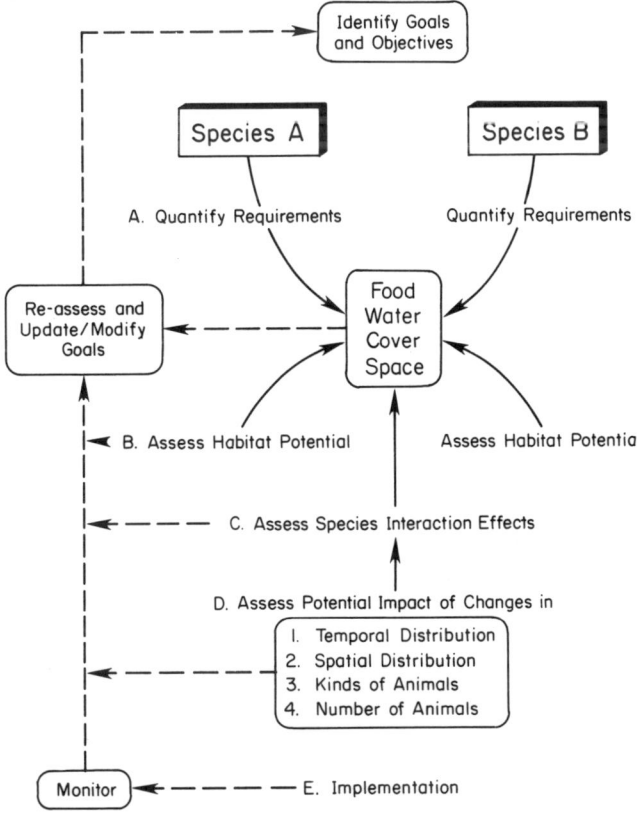

Figure 8.3. Generalized decision-analysis model for assessment of potential livestock-wildlife interactions (after Norton and Mumford 1984 and Scifres 1987).

of Norton and Mumford (1984) as presented by Scifres (1987). It is composed of four broad components: goal setting, evaluation (steps A, B, C and D), action (step E), and outcome.

Goals and Objectives

Goals reflect the desires of management (see Chapter 9), whereas evaluation assesses resource potential relative to meeting established goals (see Chapter 10). The establishment of realistic goals is often difficult because of basic misconceptions by management personnel (see Chapter 10). A common example of such in the management of livestock-wildlife grazing enterprises is the goal of maximizing both individual animal performance (e.g., trophy bucks, production per cow) and production per unit area of land. This is not an attainable goal in either wildlife or livestock production systems (see Chapter 7). Establishment of realistic goals is paramount in the development of any livestock-wildlife production enterprise, and the functional aspects of this planning model depend upon the establishment and continual assessment of management goals.

Species Requirements

The objective of this step is to collectively garner all available information relative to the basic requirements of any particular species in terms of food, water, cover, and space. Such information is normally garnered from scientific studies and/or expert opinion based on observation and experience.

Habitat Potential

The objective of this step in the planning model is to assess the potential of the habitat of interest relative to meeting the four basic needs of the species of interest. This assessment is made separately for each species assuming no interaction effects. It requires an understanding of the temporal impact of various management tactics on ecological succession (see Chapters 4 and 5). The central question is: can the basic food, water, cover, and space requirements of the targeted species be adequately met within the habitat of interest in light of species requirements and established goals?

Species Interaction Effects

The central objective of this step is to identify the magnitude of competitive overlap between two or more species relative to their ability to garner adequate resources to maintain desired population levels. Magnitude of competitive overlap varies among species depending upon a number of factors. For example, research in Oregon on pronghorn antelope, cattle, and feral horses (Fig. 8.4) has shown dietary overlap is greater between cattle and feral horses than between pronghorn antelope and cattle or pronghorn antelope and feral horses. Similarly, research in the *Acacia, Condalia, Prosopis* dominated shrub regions of south Texas has shown dietary overlap between white-tailed deer and cattle is much greater than between either jackrabbits and white-tailed deer or jackrabbits and cattle (Fig. 8.5). In similar studies, Bryant et al. (1979) found dietary overlap among white-tailed deer, sheep, and goats in the *Prosopis, Juniperus, Quercus* dominated savanna of central Texas was greater than between white-tailed deer and cattle, whereas Henke et al. (1988) showed dietary overlap between several exotic ruminants (i.e., axis deer, fallow deer, and blackbuck antelope) and cattle was much greater than between white-tailed deer and cattle. Moreover, all these studies showed that magnitude of dietary overlap varied over time (seasonally).

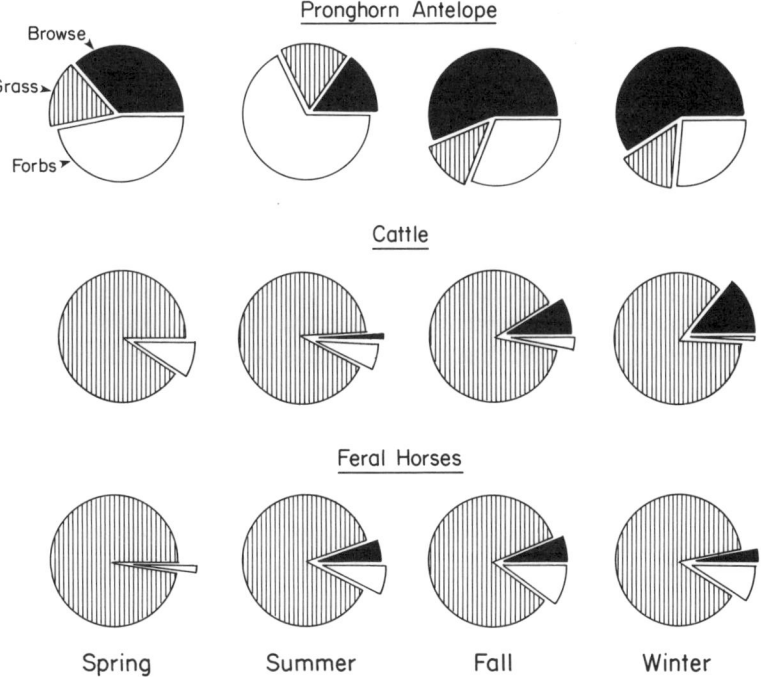

Figure 8.4. Seasonal composition of pronghorn antelope, cattle, and feral horses in Oregon (after McInnis and Vavra 1987).

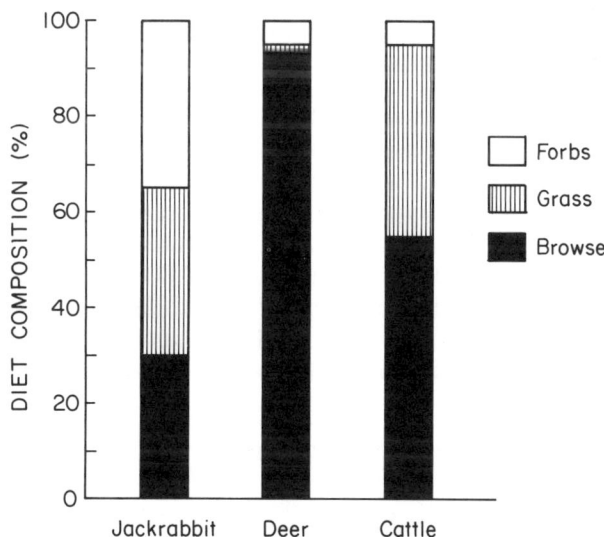

Figure 8.5. Diet composition during winter of co-existing black-tailed jackrabbits, white-tailed deer, and cross-bred cows in south Texas (from Blankenship et al. 1985).

These example studies, as well as the results from a broad array of studies throughout the world, serve to emphasize that competition for food between livestock and herbivorous wildlife varies over time and space as a function of demand relative to the quantity and quality of forage available. In other words, the **grazing pressure** concepts as presented in Chapter 7 are as functionally sound in terms of wildlife management as they are in terms of livestock management. A similar argument may be made for water, cover, and space in that demand for each resource varies depending upon each species' requirements and resource availability.

Multi-Species Grazing Tactics

The complexity of the model increases several fold at this step because it incorporates the concept that grazing management involves "the manipulation of grazing and browsing animals..." (Soc. Range Manage. 1989). It focuses on the direct, interaction effects of temporal and spatial distribution of various kinds and numbers of animals (see Chapter 7) on habitat attributes as they relate to the resource needs of the target herbivores. In essence, this step is designed to blend the functional aspects of steps A–C into an ecologically sound management plan in pursuit of established goals.

Number of Animals. Number of animals is the principal factor affecting livestock-wildlife interactions just as it is the principal factor affecting livestock production (see Chapter 7). This is so because of the direct impact number of animals has on total food, water, cover, and space demands and their subsequent availability. The magnitude of this impact varies depending upon degree of competitive overlap among the targeted herbivores and their associated behavioral traits. For example, research in the savanna regions of central Texas has shown that as livestock stocking rate increases, number of white-tailed deer decreases (Merrill et al. 1957; McMahan 1964; Reardon et al. 1978). Moreover, the magnitude of decline has been shown to vary depending upon the mix of livestock species present in that at equal rates of stocking, decline in deer numbers is greater in pastures stocked with sheep and goats than those stocked with cattle. This decline is largely attributed to greater dietary overlap between sheep, goats, and white-tailed deer than between cattle and white-tailed deer (Bryant et al. 1979; Blankenship et al. 1985).

Temporal Distribution. Temporal distribution of livestock also affects herbivorous wildlife populations. For example, research has shown livestock grazing at moderate intensities can, in some instances, enhance the quality and/or architectural structure of the forage available for subsequent grazing by wild herbivores such as elk and mule deer in North America (Anderson and Scherzinger 1975; Urness 1982) and Thompson's gazelle in Kenya (Blankenship and Overton 1974). Similar relationships have been shown between sympatric wildlife species such as black-tailed prairie dogs and bison (Coppock et al. 1983) and blue wildebeest, zebra, and Thompson's gazelle (McNaughton 1988), to name a few.

Spatial Distribution. Spatial distribution of livestock is an important factor affecting herbivorous wildlife populations. This has been demonstrated several times from studies showing white-tailed deer (Hood and Inglis 1974; Kramer 1979; Allred 1980; Cohen et al. 1989), elk (Skovlin et al. 1976), bighorn sheep (Morgan 1971; Gallizioli 1977), and moose (Denniston 1956; Schladwieler 1974) tend to vacate localized areas following introduction of livestock. Such behavior by these wild ungulates is interpreted generally to reflect an innate social intolerance to livestock since most species tend to return to the vacated areas shortly after the livestock are removed.

Implementation

Firm level goals and objectives must be continually updated and modified as each step in the planning model is completed to insure the established goals are biologically realistic and attainable. Once the biological based planning phase of the analysis (i.e., two species) or series of analyses (i.e., more than two species) is completed (Fig. 8.3, steps A–D), the selected management tactics must be evaluated in light of availability of firm level resources such as capital, facilities, and labor (see Chapter 9). Likewise, it is critical that key output parameters be continually monitored following implementation to chart progress towards established goals (see Chapter 10). Such monitoring should be quantitative whenever possible (Kei and Thomas 1988) to insure that feedback into the goal assessment process accurately reflects movement towards or away from established goals.

Monitoring

A number of documents provide detailed descriptions of established techniques for monitoring key response variables in wildlife dominated ecosystems (Schemnitz 1980; Hays et al. 1981; Chambers and Brown 1983; Cook and Stubbendieck 1986; Kie and Thomas 1988). Described techniques are generally less precise than those utilized in livestock dominated enterprises because level of managerial control of the target wildlife species is far less than that exercised in livestock production enterprises. Moreover, there is a need in most wildlife systems to monitor a greater number of interacting variables than in livestock systems. For example, the major factors affecting livestock production are food and water, whereas the successful establishment, contined propagation, and survival of most wildlife populations are often closely linked to cover and space as well as food and water. Furthermore, the relative impact of each factor on different wildlife species varies; thus the desired monitoring technique varies depending upon species. The step-wise ecological knowledge base derived from the employment of our suggested planning model is a prerequisite for selecting appropriate monitoring techniques.

The difficulty associated with the monitoring process in natural resource based systems is compounded further by the incorporation of the effects of time on the resource base (i.e., succession; see Chapters 4 and 5). Although the interaction effects of livestock on wildlife can be examined independently, they are in reality dynamic, multi-facet relationships that vary over time depending upon the targeted species and the habitat of residency. Two examples serve to emphasize this complexity. The first study centers on the observed increase in white-tailed deer populations in the savanna region of south Texas during the past century (Inglis et al. 1986). This increase is attributed primarily to a vegetational shift in the regional habitat from an open, grass dominated savanna complex to a woody shrub dominated complex. This seral shift is attributed generally to the interaction effects of intensive livestock grazing and suppression of fire (Scifres 1980) against a backdrop of subtle climatic changes (see Chapter 5). As a result of this shift, white-tailed deer habitat has improved over time as measured by the increase in number of white-tailed deer.

Although all the precise reasons for this increase in the white-tailed deer population are unknown, there is little doubt all are related to changes in habitat quality in terms of resource availability. For example (Table 8.1), it has been shown generally that quality of habitat for cattle has declined in terms of quantity and quality of available forage, remained neutral in terms of cover and space, and improved in terms of water. This is in contrast to white-tailed deer in that quality of habitat has improved in terms of food, water, and cover, and declined in terms of spatial needs. The changes in food and water are related to the shift from a grass dominated savanna to a shrub dominated scrubland. Water development has enhanced the habitat for both cattle and white-tailed deer, whereas humans' physical presence, within itself (i.e., popula-

Table 8.1. Generalized effects of change over past 50 years in south Texas habitat relative to basic needs of cattle and white-tailed deer (negative, positive, and neutral). Adapted from Inglis et al. (1986).

Attribute	Species	
	Cattle	White-tailed Deer
Food	−	+
Water	+	+
Cover	0	+
Space	0	−

tion density), is perceived to have reduced the overall quality of the habitat in terms of meeting the spatial needs of the white-tailed deer. However, when the effect of human presence is evaluated relative to the role played in altering the food, water, and cover attributes of the regional habitat, an entirely different perception emerges. Recognition of such simply serves to emphasize the complexity associated with the management of any natural resource including livestock-wildlife enterprises.

Another classic example of a time-related interaction effect has been reported by Walker et al. (1987). In this study, they reported water development on two South Africa game reserves generally enhanced wild ungulate populations during periods of average to above average rainfall. This increase in ungulate populations was the result of an increase in forage availability through spatial redistribution of the animals. However, they subsequently reported that during drought animal death losses were much greater than would have been expected in the absence of "improved" water development because of a general pre-drought depletion of forage reserves.

These two studies singularly and in combination reveal something of the complexity involved in any analysis of livestock-wildlife interactions. Both serve to emphasize that for any action there is a corresponding, multi-facet reaction, the value of which varies (i.e., positive, negative, neutral) over time depending upon management goals (see Chapter 10). These studies also reveal the universal applicability of the proposed planning model (Fig. 8.3) in terms of species and location of interest. The model is as applicable to the management of avian species as it is to the management of wild ungulates. For example, bobwhite quail are a much studied (e.g., Rosene 1969; Lehmann 1984; Guthery 1986), economically important gamebird in many regions of the U.S., and specific knowledge of the basic needs of bobwhite quail (Fig. 8.3, step A), their preferred habitat (step B), and the interaction effects of such (step C) and other wildlife and/or livestock species on bobwhite quail populations (step D) is a prerequisite for developing ecologically and economically sound management tactics to meet firm level goals. For example, bobwhite quail require a wide array of cover types for a wide array of activities including **nesting** (moderately dense grass-forb mixture), **roosting** (sparse, short-statured, open canopy), **loafing** (bare ground under elevated canopy), **screening** (low growing shrubs), **escape** (dense ground cover), **dusting** (bare ground), and **thermal** protection (dense ground cover under woody overstory). Although such needs are in some respects similar to the needs of many domestic and wild ungulates, they are also unique to bobwhite quail thereby demonstrating the need for specific knowledge about the needs of each species of interest.

Similarly, it is imperative that knowledge of habitat attributes be specific. For example, the spatial distribution of various habitats across a landscape is often a critical factor affecting wildlife populations. Knowledge of such effects is imbedded in the concept of **edge effect** which broadly refers to the impact that the spatial arrangement of various habitats may have on certain wildlife populations. Generally, the greater the number of habitat types or habitat boundaries (edges) present within an area, the greater the diversity and abundance of wildlife. Such is the case because many wildlife species require several types of habitat to meet their needs. As a result, the potential impact of a broad array of juxtapositioned habitats and their associated edges on targeted wildlife populations must be evaluated when implementing any landscape level management tactic.

CONCLUSIONS

Wildlife are important consumers in grazed systems. Their perceived role and subsequent value varies among societies depending upon societal goals. Wildlife and livestock share the same basic needs (food, water, cover, and space). Food habits vary among large herbivorous wildlife species in the same manner as livestock depending on morphological and physiological adaptations (see Chapter 2), behavioral attributes (see Chapter 3), and quantity and quality of food available (see Chapter 7). Similarly, because water is an essential compound for the sustenance of all organisms (see Chapter 6), all wildlife species require water, amount of which varies among species as a function of habitat of occupancy, activity, and evolved behavioral, morphological and physiological adaptations. Cover requirements vary among wildlife species depending upon desired use. However, cover requirements of large herbivorous wildlife species are functionally similar to livestock as evidenced by the manner in which feral livestock use cover. All wildlife require space with the amount dependent primarily on the spatial distribution of food, water, and cover and secondarily on evolved behavioral attributes. Knowledge of the basic needs of targeted wildlife and livestock species in concert with knowledge of habitat attributes provides a basic foundation for implementing sound livestock-wildlife grazing management tactics.

CHAPTER 9

SOCIAL AND ECONOMIC INFLUENCES ON GRAZING MANAGEMENT

J. R. Conner

INTRODUCTION

Grazing management is defined as the manipulation of livestock to accomplish a desired result (Soc. Range Manage. 1974). The desired result varies, however, as a function of many social factors of which economics is usually dominant. The objective of this chapter is to examine the impact that social factors in general, and economics in particular, have on the adoption of various grazing management strategies and tactics.

Societal goals vary widely as a function of a multitude of factors. This variation in goals is related to the variation among societies in the goals of each individual. Maslow (1954) argues that there is a basic ordering or hierarchy of psychological needs or goals common to all humans (Fig. 9.1). In this conceptual model, biological survival is the most basic goal, and spiritual tranquillity the most advanced goal. The basic thesis presented by Maslow is that the attainment of higher goals is only attempted after the more basic goals have been met. For example, in primitive societies with relatively unsophisticated institutions and technologies, humans' predominate focus is on achieving individual and/or group survival. However, in highly developed societies with sophisticated institutions and technologies, humans' biological survival becomes more certain and its place as the predominate goal is overtaken by needs for financial security, social prominence, and intellectual stimulation.

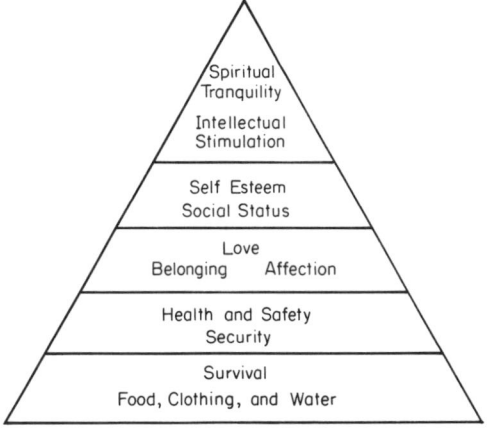

Figure 9.1. Hierarchy of man's needs (goals), adapted from Maslow (1954).

The resulting effect of this variation in societal goals is that grazing management strategies also vary since human culture and the knowledge base embodied in it affects how we use ecosystems to accomplish goals. Lack of sophisticated institutions and technologies dictate an opportunistic approach to ecosystem use. Thus migratory herding is characteristically the primary grazing management practice utilized in primitive societies. But as sophisticated institutions are established and expanded knowledge bases developed, more complex livestock production practices are adopted, such as those incorporating use of supplemental feeds and cultivated forages, feed grains in finishing yards, and grazing systems. This added complexity is necessary because the more intricate underlying institutions, such as property rights, restrict such practices as migratory herding. This increased complexiity is also a reflection of humans' expanded knowledge base and their focus on higher goals.

Unfortunately, however, the functional aspects of grazed ecosystems remain constant regardless of social and economic factors. The ecological principals which, for example, regulate the rate and extent of deterioration in ecological condition class (Chapter 5) are functionally constant regardless of whether over-grazing occurs in a primitive or sophisticated society. The human consequences, however, are quite different in the two societal complexes in that ecological imbalances in primitive societies are frequently corrected only through the death of beast and/or human, whereas in more developed societies corrections are most commonly reached through economic convulsions. In other words, humans are better able to survive temporary ecological imbalances in the more developed societies thanks to the existence of more sophisticated institutions and technology, and human suffering is expressed in the form of financial losses and reduced social status and self esteem.

ECONOMICS AND GRAZING MANAGEMENT

Economics is the study of how people, individually and collectively, use scarce resources to satisfy their wants and needs (goals) (Robbins 1932; Samuelson 1964; Ferguson 1966). Resources, in this context, are the things used to accomplish human goals and objectives. In this book, the resources of primary interest are grazing land ecosystems used for livestock and wildlife production. Because the primary goals in most sophisticated societies are related to economics, and because livestock/wildlife production in these ecosystems vary primarily as a function of the temporal and spatial distribution of various kinds and numbers of livestock (Chapter 6), an understanding of the potential economic impact of these four grazing tactics is critical.

The Effect of Number of Animals on Economic Goals

Stocking rate (number of animals) is the major factor affecting the potential magnitude of profits realized by a ranching enterprise. The importance of stocking stems from the relationships presented in Figures 7.2 and 7.3 which show that as stocking rate is increased, production/individual animal decreases while production/unit-area of land increases to some maximum and then declines. Because of these basic relationships the incentive to maintain high stocking rates as a means of achieving financial goals is tremendous. However, because of temporal and spatial variation in quantity and quality of forage produced in rangeland environments, the optimal rates of stocking required to continually maximize production/unit-area of land varies. Moreover, this variation is acutely reflected in terms of profits, because at high rates of stocking production costs generally increase at a somewhat faster rate than the rate of

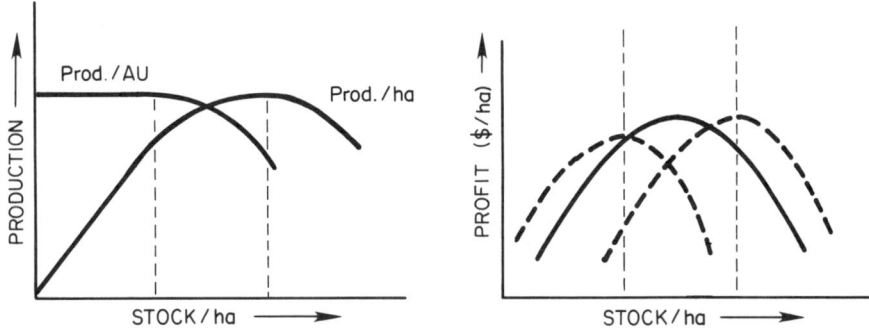

Figure 9.2. Relationship between production/animal, production/unit area of land, profit/unit area, and animal numbers/unit area.

increase in gross returns. Thus, as stocking rate is increased beyond a moderate level, profit levels begin to decline (Fig. 9.2). Moreover, this reduction in profits reduces rate of increase in net worth and increases the probability of encountering a catastrophic loss.

These concepts have been adequately demonstrated from studies at the Texas Experimental Ranch as presented in Figures 6.2 and 6.3. During the period from 1982 through 1987; average residual returns (i.e., net returns to land, management, and profit)/cow from year-long continuous grazing stocked at moderate and heavy rates were about $70 and $61, respectively, while average net returns/hectare were $11 and $13, respectively (Heitschmidt et al. 1990). However, returns/cow over the 6-year period ranged from about $14 to $109 in the moderately stocked treatment, and from about $-18 to $96 in the heavily stocked treatment while net returns/hectare ranged from $12 to $17 and from $-3 to $21, respectively. These data revealed two important points:

1. the optimal stocking rate for maximizing economic returns varied dramatically among years; and
2. the probability of encountering catastrophic losses was much greater at the heavy than moderate rate of stocking.

Moreover, it is anticipated that over time the production potential of the heavily stocked treatment will continue to decline as range condition declines. Thus, economic stability will decline and financial risks will increase substantially.

The Effects of Species and Class of Animals on Economic Goals

Species and classes of grazing animals also affect a rancher's ability to meet financial goals because annual profits may be enhanced by selecting a species or class of grazing animal that can most efficiently utilize the particular forage available (Chapter 2). For example, if browse makes up a large component of the forage base, the rancher might be able to safely stock more animal units of goats than cattle/unit area because goats utilize browse more efficiently than cattle. Often, particularly if the forage base is diverse, a combination of two or more types of grazing animals can utilize the forage more efficiently than a single species. In this case, the total animal units/unit area can be increased with a combination of livestock over the number of animal units that could safely be stocked with either species alone.

Combinations of livestock may also enhance a rancher's ability to avoid catastrophic losses because the probability of simultaneously suffering economic losses in any given year in two or more diverse enterprises is usually much less than the probability of experiencing losses in any given year for a single enterprise. This risk management strategy, known as investment diversification, has long been an accepted business practice.

Choosing an alternative class or species of grazing animal, or a combination of grazing animals, is often difficult because of differences in production requirements, costs, and product prices associated with the different animals. Fences, for example, that may be well suited for cattle production may be totally inadequate for goats. In this case, changing from a single cattle enterprise to a goat enterprise or a combination of cattle and goats requires a significant increase in fencing costs which might more than off-set any gains in income resulting from the more efficient utilization of the forage base.

Effects of Spatial Distribution of Livestock on Economic Goals

The effects of spatial distribution of livestock on profits and risk avoidance are difficult to assess because the potential impact varies tremendously among enterprises. Although it is well known that an increase in livestock production can be attained through enhanced livestock distributional patterns (Chapter 6), increased costs may limit or totally eliminate profit potentials. Major cost factors include fencing, water development, and labor. These costs are further confounded because labor costs may either increase or decrease over time as a function of the specific situation. Labor savings may be realized as a result of enhanced livestock gathering and supplemental feeding regimes, whereas labor costs may increase as a result of increased time spent gathering and feeding livestock due to a greater number of herds.

Effects of Temporal Distribution of Animals on Economic Goals

Temporal distribution of animals can have a dramatic effect on a rancher's ability to increase profits and, ultimately, increase net worth and reduce risk. For example, Launchbaugh (1986) reports net returns/hectare increased two-fold when an intensive early stocking (IES) strategy was used in place of a season-long grazing (SLG) strategy (see Chapter 6). At normal rates of stocking, net returns/hectare over a 4-year period from SLG (May 1–October 1) averaged $20.54, while returns/hectare from IES (May 1–July 16) averaged $21.31 (normal stocking rate), $41.87 (double normal stocking rate), and $46.62 (triple normal stocking rate). Major cost savings were related to death losses, health costs, equipment fuel and repairs, mineral supplements, and interest.

In terms of variation in net returns among years, there was little difference between the two normal stocked treatments as net returns/hectare ranged from $7.14 to $32.41 in the SLG, and from $8.64 to $35.59 in the IES treatment. However, as stocking rate was increased variation in returns increased also. Net returns/hectare ranged from $19.22 to $67.28 at the 2× rate of stocking, and from $14.36 to $78.42 at the 3× rate of stocking. Thus, based on these data, risk avoidance was greatest in the double normal stocking rate IES treatment (minimum return equals $19.22) and least in the normal stocking rate SLG treatment (minimum return equals $7.14). From this example it can be seen that profits and risk avoidance were both increased through a temporal shift in grazing. But it is important to note that the positive response was related in part to temporal patterns of forage growth in that rate of forage production in this region is consistently greatest in spring and early summer (Sims and Singh 1978b). A major factor affecting the relative success of any grazing strategy used in rangeland environments must be related to climatic rainfall patterns particularly with regards to the temporal distribution of animals.

It should also be noted that the use of such practices as IES are closely tied to ranchers' goals other than strict economic gains. For example, IES would most likely not be an acceptable practice for a rancher who prefers to operate a cow/calf enterprise. In this instance, the difference in profits and risk avoidance between stocker IES and cow/calf year-long grazing enterprises can be viewed as a cost to the rancher to run the enterprise of his/her choice. For example, if a rancher continues to operate a cow/calf enterprise although a $5/hectare-greater profit could be realized with a stocker IES strategy, the cost of this personal preference is $5/hectare.

Effects of Grazing System on Economic Goals

The effects of any given grazing system on the attainment of economic goals is complex and difficult to assess. The complexity arises because there is essentially an infinite number of grazing systems, and their effects on economic goals vary as a function of current numbers and configuration of pastures, labor constraints, managerial ability, and personal preference. Moreover, implementation of a grazing system generally involves certain time constraints because the primary justification for the establishment of all grazing systems is related to their effect on range condition. Thus, a grazing system may be viewed by the rancher as a tool contributing to the goal of sustained growth in profits and net worth over time as garnered through improved ecological condition and range productivity. To the extent that a grazing system allows for maintenance of range condition with higher stocking rates than would otherwise be possible, it may be viewed by the rancher as contributing to the goal of maximizing annual profits.

The rancher's decision regarding the establishment of a grazing system hinges on the tradeoff between the benefits he/she expects to gain in the form of current or future carrying capacity and the costs of additional fences, water facilities, and operating labor required to implement and maintain the grazing system. Like most other decisions the rancher makes, the optimal combination of pasture size, length of rest, or deferment period and stocking rate is unique to his/her individual goal hierarchy and set of resources. In addition, since the goals and resources are likely to change through time, so will the optimal grazing strategy.

Generalities based on economic contrasts between grazing systems are difficult to formulate because of the presence of confounding factors such as differences in fencing, development of water and working facilities, labor costs, costs of destocking and restocking, managerial ability, etc. For example, Heitschmidt et al. (1990) contrasted economic returns from 1982 through 1987 from cow/calf performance in heavily (HC) and moderately (MC) stocked year-long continuous grazing treatments (see stocking rate section), a moderately stocked 4-pasture, 3-herd deferred rotation (DR) system, and a very heavily stocked 16-paddock, 1-herd rotational grazing system (RG). Net returns/hectare over the 6 years averaged about $13, $11, $16, and $16, respectively, for the HC, MC, DR, and RG treatments. Range in returns was from about $−3 to $21 in the HC treatment, $2 to $17 in the MC treatment, $9 to $22 in the DR treatment, and from $5 to $20 in the RG treatment. The analyses were based on the assumption that annual labor, fuel, and equipment costs were equal among treatments. These assumptions were made so as to reduce the level of specificity to the research location. Based on these analyses, it was concluded that profit potential was greatest in the DR and RG treatments, and risk avoidance was greatest in the DR treatment.

LAND OWNERSHIP AND GRAZING MANAGEMENT

In most modern capitalistic societies, the resource owner may use the resource to satisfy his or her own goals. In most of these societies land is either owned by individual citizens, business firms, or organizations, i.e., privately owned; or it belongs to a societal group such as a state or the nation, i.e., publicly owned. Thus, if the land is privately owned it may be used, with few restrictions, as its owner wishes to satisfy his or her goals and objectives. If it is publicly owned, the land must be used to satisfy the goals of society.

Grazing Management on Privately Owned Land

Most agricultural land in modern societies is used in the production of marketable products. In areas where climatic, topographic, and/or soil characteristics, or desires of the owner render the land unsuited for cultivation, it is used to produce livestock and/or wildlife and related products through grazing. These livestock/wildlife production operations, commonly called ranches, are generally operated as businesses. In the ranching business the forage grown on the land supplies a significant portion, if not all, of the feed for the animals. The grazing land and animals are used in conjunction with labor, capital, management expertise, etc., to induce animal reproduction and/or growth which can be sold for currency. The currency can then be used to maintain and/or replace the resources used in the production process and provide income to the rancher which he/she can then use to meet personal goals.

But management goals and objectives vary among ranching enterprises as a result of numerous social and economic factors. Still, in the context of a business firm, ranches included, the major goal is continuous survival of the business enterprise. If this goal is achieved, then the rancher is assured of achieving at least some nonfinancial goals such as maintaining his/her chosen occupation and lifestyle (spiritual tranquillity) (Conner 1984b). Survival as a business generally requires that other supportive goals also be achieved. That is, to survive the ranch must produce a profit in most years, and in years when losses occur they must not be of sufficient magnitude to eliminate the ranch owner's net worth. Over the long term then, the goal of firm survival requires that the rancher also achieve the goals of obtaining profits and avoiding catastrophic losses.

To insure that these goals are achieved most ranchers adopt yet another supportive goal, i.e., sustained growth in profits and net worth over time because these goals provide an ever increasing cushion or margin of safety against business failure due to catastrophic losses. Attainment of these goals also insures the rancher's standard of living will improve along with retirement security and/or his/her family's inheritance.

A rancher attempts to meet goals by managing the available set of resources. He/she accomplishes this by selecting a set of livestock production enterprises and grazing strategies believed to provide the best opportunity to meet these goals.

In summary, it can be seen that a rancher's goals in a developed society generally relate to long-term security, self-esteem, and spiritual tranquillity (Fig. 9.1), and that the level of economic profits attained from various grazing management tactics is a major factor affecting goal achievement. Thus, a fundamental knowledge of the potential impact of the four broad principles of grazing management (see Chapter 6) on economic profits is essential.

Grazing Management on Publicly Owned Land

A frequently stated goal of most modern societies is to maintain and/or improve the general welfare of its citizens. The problem with this goal is that it masks the problem of establishing the "correct" ordering of secondary goals that contribute to improvement in the general welfare and avoids the issue of how the welfare is distributed among the individuals and subgroups of citizens making up the society. In the final analysis, the ordering or ranking of society's goals, and the allocation of resources to meet them, is accomplished through the political processes and institutions that the society has established. Whether they are equitable and/or efficient is beyond the scope of this volume.

Why would a modern capitalistic society choose public ownership of a part of its land? Would not a capitalistic society encourage private ownership of productive resources? The answer is yes, private ownership would be preferred unless, through public control, the society can achieve goals that could not be achieved, or that would be achieved less efficiently, if the land were privately owned.

And what are these goals that can be best achieved through public ownership of land? The goals most often cited as justification for public ownership include conservation of natural resources, preservation of environmentally unique and/or fragile areas, and creation of opportunities for present and future citizens of the society to observe and enjoy unique wildlife and "natural" areas. Public ownership is necessary to achieve these goals because if the areas were privately owned, the resources (rangeland) would be subject to ecological deterioration through use in the production of readily marketable goods and/or services.

Goals relating to preservation of resources for the enjoyment of future generations are particularly hard to achieve under private ownership since the current private owners have no way of marketing the service they are expected to provide to future generations. Thus, a private owner uses the land to produce a product or service for which current markets exist, even if the process results in irreversible degradation of the resource.

Most publicly owned land is capable of supporting more than one use goal, some of which are not mutually exclusive, but all of which will be held as "most important" by some of the different individuals and groups in the society. Thus, the government agencies charged with the responsibility of allocating public lands among these different possible uses must constantly reevaluate the "correctness" of their allocations relative to the equity among different interest groups and the efficiency of the land in its different uses.

One traditional use of a large portion of publicly owned land throughout the world is to furnish forage for privately owned livestock through grazing. In the absence of government regulation, the grazing land is often abused by overgrazing which results from each livestock owner trying to maximize his use of the land. In such cases, herd owners are provided no incentive to conserve forage for future use and/or range improvement because forage not used by one individual's herd will likely be used by another's.

In some cultures, this problem is exacerbated by the individual herd owner's primary goal being the maximization of herd size. In these instances, the owner is willing to sacrifice production per animal and sometimes total herd production of meat or milk because numbers of animals may be more important than quality or total herd production.

This type of use of publicly owned land might be well suited for primitive societies where individuals are free to migrate with their herds over vast areas. Also, in primitive societies over-use problems are usually resolved by war or starvation which reduces numbers to levels that the land can support. In more developed societies, where the public land is more limited and bounded by privately owned land, unregulated use of public land by private herd owners usually results in land abuse and degradation. The remedy for this problem is either privatization of the land or government regulation which restricts the number and/or time period that livestock can be grazed on a given area.

One way of regulating grazing is by issuing "grazing permits". Most public land on which grazing permits are issued has other simultaneous uses, otherwise, society's interest would be better served by selling the land into private ownership. Most commonly, the "other uses" include providing habitat for wildlife which in turn provide recreational and aesthetic benefits to members of society other than ranchers. Thus, the government agencies charged with allocating the land among its alternative uses must carefully regulate the grazing use so that it does not impinge on, or unduly interfere with, the other uses.

The execution of this task is a constant source of concern to ranchers, resource agency personnel, and others interested in public grazing because there are inevitable use conflicts; particularly on the more productive range sites like riparian areas. Riparian areas are sources of lush vegetation and drinking water for livestock, which are enticed to utilize these sites more frequently than more arid sites. These same areas, however, are also critical hunting, breeding, nesting, and feeding grounds for wildlife; critical areas for stream bank erosion and silting; and important fisheries. Excluding livestock entirely from riparian areas in many cases reduces the value of the rangeland for grazing to the point that it is not worth the fee that the rancher must pay for the permit. On the other hand, restricting the use of riparian areas by livestock to selected specific sites requires fencing which is expensive and detracts from the recreational and aesthetic benefits that can be derived from the area. In addition, overgrazing and erosion problems are usually intensified for the specified livestock use areas because of the increased concentration of livestock on the smaller areas.

Grazing management problems on public lands are not, however, restricted to those related to leased livestock grazing permits. Wildlife and feral horse and burro populations are generally protected and encouraged to increase on most publicly owned rangelands. In many instances the populations of these wild animals have increased to the point that their forage demand has exceeded what the area can supply. To prevent deterioration in the ecological trend of the rangeland and/or starvation of part of the wild animals, the managing agencies have had to resort to reducing the number of the wild animals through hunting and/or capturing and relocating them to new areas with less grazing pressure.

SUMMARY

The ecological aspects of a grazed ecosystem are functionally constant regardless of the socio-cultural aspects of the human population interacting with it. Thus, regardless of how sophisticated the society, manipulation of temporal and spatial distribution and kinds and numbers of grazing animals are the only means by which humans can manage grazing land to achieve their desired goals.

The roles that grazing management play in the use of grazing land to achieve goals depend on whether the land is privately or publicly owned. Privately owned rangeland is principally used as a resource in a productive process. The purpose of the productive process is to help the land owner achieve goals such as firm survival which in turn usually requires achieving profits and avoiding catastrophic losses. Grazing management, in this case, generally consists of striving to balance the desire for maximum annual profits/unit area with the necessity of leaving some reserve forage for emergency use and allowing for improvement in the productive potential of the rangeland. This role of grazing management is, however, often complicated through social and institutional influences on the use of privately owned rangeland.

Publicly owned grazing land is characterized by its intended role in meeting non-market goals of conservation, preservation, and equity of opportunity to enjoy unique recreational and aesthetic experiences. The role of grazing management is expanded to include facilitating many simultaneous but different land uses and users and usually amounts to keeping livestock from damaging the resource or from impinging on the other uses and users.

CHAPTER 10

THE DECISION-MAKING ENVIRONMENT AND PLANNING PARADIGM

J. W. Stuth, J. R. Conner, and R. K. Heitschmidt

INTRODUCTION

The sheer complexity of the knowledge presented in the previous chapters provides insight as to why grazing management is largely a heuristic art rather than a science. In the first nine chapters of this book, we attempt to sequentially present the basic concepts that we believe are critical to transforming grazing management from a heuristic to an analytical decision process. Chapter 1 focuses on principles and concepts of grazing management relative to ecosystem structure and function, with special emphasis on energy flow. Chapter 2 focuses on nutritional aspects of livestock production as affected by various management tactics, while Chapter 3 provides insight as to how animals react to their environment and subsequently adjust their foraging tactics to maintain their health and well-being. The impact of herbivory on growth and development of plants is examined in Chapter 4, and its integrated effects on various plant assemblages is examined in Chapter 5. In Chapter 6, specific attention is focused on the impact of herbivory on quantity and quality of water and soil. The integrated short- and long-term effects of herbivory on livestock and wildlife production are examined in Chapters 7 and 8. Finally, in Chapter 9 the potential impact of social and economic factors on humans' ability to meet goals through grazing management is discussed. In this chapter our objective is to provide insight as to how humans can most effectively use this broad knowledge base to meet personal goals.

THE DECISION-MAKING ENVIRONMENT

System

It is impossible to separate grazing management from ranch or firm management since resources utilized and derived are integral parts of the ranch firm. Grazing management must be viewed in the context of a system comprised of interacting components which can be and are manipulated (Fig. 10.1). Interrelationships among ranch resources, such as people, finances, land, vegetation, climate, animals, and time, as well as activities and external influences must be understood and taken into account by the decision maker. The impact of each decisive action must be evaluated in advance and the outcome monitored. The manager must also be able to anticipate

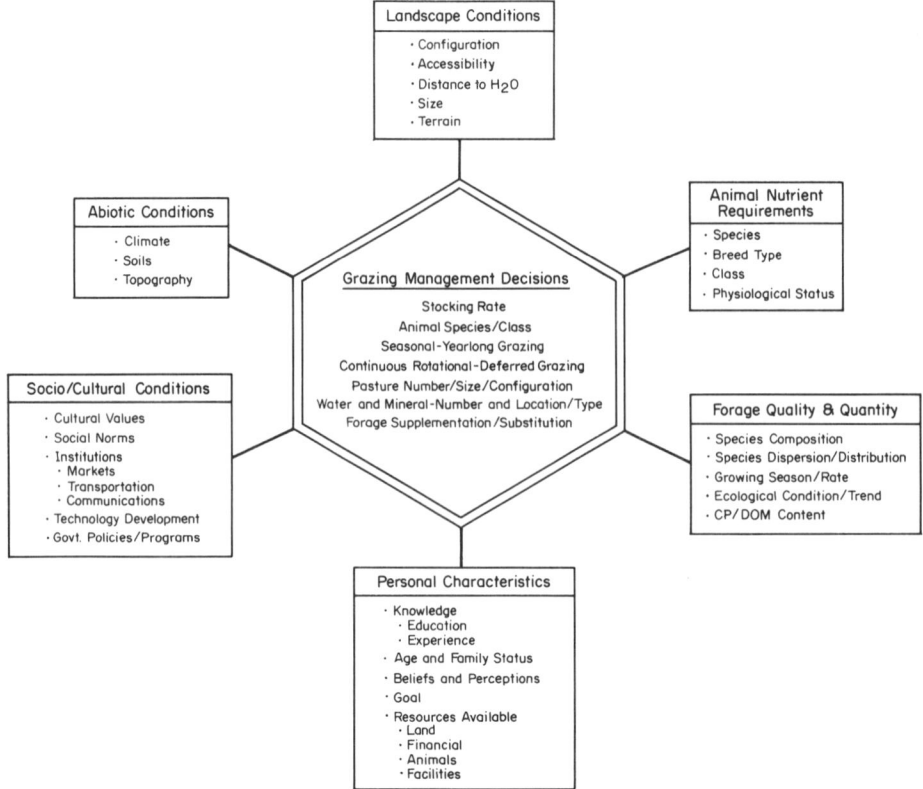

Figure 10.1. Interrelationships among resources, activities, and external influences which constitute the grazing management environment.

and implement timely changes to optimize decisional outcome. As Wilcox (1982) stated, "the successful rancher is that one who can, firstly identify the different factors which will affect operation of the ranch, and secondly, can anticipate the changes in them that will influence his success. This successful rancher is the one who avoids crisis in the running of his enterprise ... No operator should allow himself to get into the crisis situation, but should arrange his management style to anticipate the changes which will be necessary in the operation and make those changes affecting in a timely manner." Likewise, White et al. (1988) stated, "This [anticipating change] is an impossible task if the ranch has not developed a logical and practical approach for analyzing information, evaluating plans, and directing daily operations ... It is doubtful that any person can accurately assimilate the mass of information and predict the overall ranch outcome without detailed planning and evaluation."

The Nature of Expertise

The effectiveness of a manager in using available information (i.e., expertise) has a psychological basis deriving from memory, perception, and problem-solving skills and exists only in the social context in which the manager functions. The level of acquired expertise relates to learning and assimilation of knowledge in the cognitive processes, judgmental behavior, social behavior backed by social knowledge, creative behavior, analytical behavior, and ability to establish and pursue firm practices when necessary (Greenwell 1988).

Decision making in grazing management, as in any other management occupation, requires a strong foundation of technical (cognitive) knowledge enveloped in a functional social knowledge of his/her cultural environment. The quality of the decision depends on the analytic and judgmental skills of a manager. Analysis is a particular type of thinking which transforms one representation into one or more different representations making it easier to perceive particular meanings which were not readily evident in a previous representation (Greenwell 1988).

In an attempt to reflect the relative importance of mental states and behavioral components of decision making, we have constructed a matrix which ranks the relative importance of each of these factors in their relation to grazing management (Fig. 10.2). Judgmental skills coupled with the analytical and creative capacities of the individual manager have a critical impact on the quality of the decisions made relative to the grazing manager, while social skills are of lesser importance. Deep cognitive skills are also of importance in planning and implementing other closely associated actions such as herd health programs and livestock handling techniques.

Even though a great deal is often known that would assist managers in the making of quality decisions, many livestock production decisions are inappropriate and/or incorrect because the information base of the manager is incomplete and/or inaccurate. Unhappily the functional base is often incomplete because the manager has failed to search out and/or assimilate all the available information on a given subject, or in some cases the needed information is simply not available.

The grazing manager's perception of the world is shaped by personal characteristics, experience, and education operating within the context of age, family status, beliefs, and socio/cultural environments. Perception can be viewed as a mental filter that permits only certain facts to pass through to the analytical centers of the brain (Fig. 10.3). Thus, perception is a major factor affecting a manager's decision-making ability because it is the mechanism through which the manager builds internal information bases, whether accurate or inaccurate but generally incomplete. Perception plays a particularly strong role in the complex decision-making environment with which grazing management must work.

Behavioral Components of Expertise	Mental State of Expertise		
	Deep Cognitive Skills	Judgemental Skills	High Level Social Skills
Highly Creative	Occasionally Important	Very Important	Seldom Important
Analytical	Occasionally Important	Important	Occasionally Important
Strict Procedural	Seldom Important	Occasionally Important	Not Important

Figure 10.2. Relative importance of knowledge and behavioral factors in relation to grazing management.

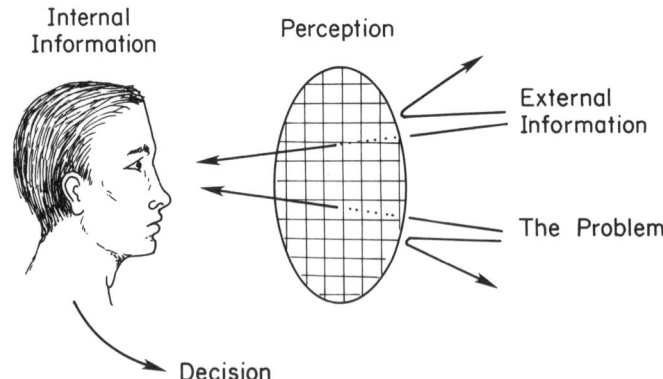

Figure 10.3. Relation of individual perception to the planning process.

FUNCTIONS OF MANAGEMENT

Several good books on agricultural business management (Koontz and O'Donnell 1972; Duft 1979; Kay 1981) are available which clearly describe and define in great detail managerial functions. All generally agree upon three to five managerial functions, depending on how the categories are defined. For our purposes, we use three categories: planning, operating, and monitoring (see Fig. 10.4) (Conner 1984a).

Planning

Planning is the primary function of management. If done correctly, it is an almost continuous process. It is a function that must be performed by everyone involved in management, from the land owner to the laborer. It is often helpful to think of planning as a stepwise process. The steps, in the order in which they must be performed, are:

1. Establish and prioritize goals.
2. Inventory and assess resources.
3. Identify and analyze alternatives.
4. Select the alternative(s) that most nearly achieves all goals and objectives.

Operations

Operations include those managerial functions relating to getting the job done. Most of the sub-functions in this category relate to organizing, selecting, directing, and motivating people to implement and carry out the plans selected in the planning process. As in other aspects of the ranching business, there are alternative organizational structures, personnel incentive programs, etc. Thus, the selection of the best operating practices for any given ranch business should also grow out of the planning process.

The operation of a ranch business, like any other business, benefits from a well-defined organizational structure that delineates the areas of responsibility of all the people involved in the organization. A good organizational structure eliminates duplication and overlapping responsibilities, makes it easier to select the "right" person(s) for the jobs to be done, and facilitates communication and motivation. The larger the business, the more formal the organizational structure should be. However,

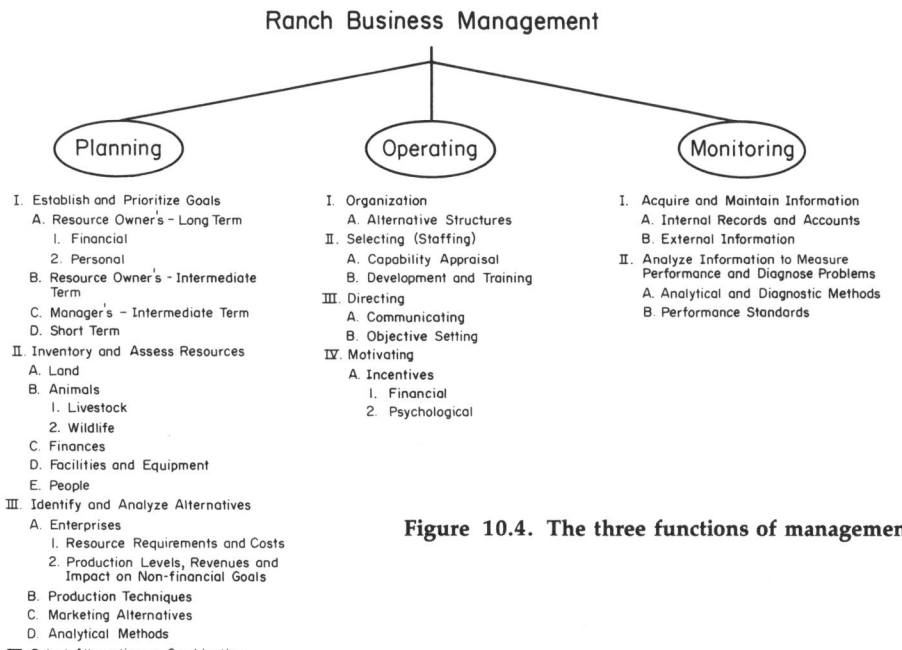

Figure 10.4. The three functions of management.

even the small ranch will usually benefit from a clear delineation of areas of responsibility.

Securing and retaining personnel is traditionally a troublesome function for ranchers. How frequently we all hear the complaint that "you just can't hire good hands any more." Most often, however, it can be shown that the problem results from poor, or non-existent, staffing and personnel programs, poor communication between the various levels of management and/or between management and labor, and the failure of management to provide adequate motivation for employees. In many cases, lack of motivation is not the result of low monetary benefits, but rather stems from the failure of the employer to provide a means for the employee to attain an acceptable level of personal satisfaction from his/her work.

Monitoring

Monitoring includes those managerial functions which are necessary to measure the performance of the ranch operation with respect to its progress toward achieving the goals and objectives delineated in the planning function. There are two general sub-functions included in monitoring. The first sub-function is the acquisition and maintenance of information. The acquisition and maintenance of information involves the maintenance of internal records and accounts such as expenditures and receipt ledgers, bank account balances, and production records by pasture, herd, and/or individual cow. It is also necessary to acquire and maintain information from external sources such as livestock and feed prices, market outlook reports, import/export regulations, etc.

The second sub-function is the use of the information gathered using analytic and diagnostic techniques to measure performance. Typical analytical and diagnostic techniques include annual income and net worth statements and financial ratios such as debt-equity, annual rate of return on investment, etc. Also included are actual cash flow reports, production measures per hectare and per head, and comparisons of prices received to monthly and annual average prices. All of these analytical and diagnostic methods are helpful in determining where adjustments are needed to more nearly achieve all the ranch goals and objectives.

By now, it should be obvious that none of these three functions, planning, operating, and monitoring, go forward in isolation. In fact, they are all part of a continuous process with the monitoring function providing new information as to availability and capability of resources and the relative feasibility of alternative plans. In turn, each change in plans requires operating adjustments to implement and carry out those plans. In addition, since humans are intelligent, emotional animals, goals and objectives often change with time, thus necessitating reassessment of plans.

THE PLANNING PROCESS

Now, let us turn to the practical steps in the planning process and examine each of them in more detail.

Goals

The role of goals is discussed extensively in Chapter 9, but another look, from a planning perspective, is warranted. The totality of ranch goals and objectives must be defined to serve as the measure against which success is judged. Goals establish the milestones that warrant achievement. They delineate the differences between where and what the ranch is now, and where and what the manager wants it to be in the future. Yet, as absolutely essential as they are, failure to define and reevaluate goals and objectives is one of the most common failures and cause of failures in the ranching industry today. Many ranchers who do define goals fail to prioritize, or rank them as to importance, leading to other kinds of problems.

Goals and objectives can be grouped in many ways: long-term -short-term, personal-financial, general-specific, etc. However, the most important and useful grouping should be based on order of importance. The most important goals are those of the resource owners, i.e., the land owner, livestock owner, etc. If all of these resources are owned by the ranch firm then this task is much easier than if the resources are owned by different entities, because there is less likelihood that goals will conflict. The important point, regardless of the nature of resource ownership, is that if the resource owner(s) do not clearly identify and communicate their goals, then there is little chance that the other people associated with the business can be very effective in helping the resource owner(s) achieve them.

Resource Assessment

Resources are the things used to achieve goals and objectives. Types of resources vary greatly, but in the ranching business it is useful to group them into five broad categories: lands, animals (livestock and wildlife), finances, facilities, and people. The organization and operation of a ranch involves all of these resources in some measure. But the specific combination of resources available to any one ranch differs from the combination available to any other. Thus, the resources available to each ranch must be carefully evaluated if the unique combination available is to be organized and used to best meet the totality of ranch goals and objectives.

The assessment process must be thorough enough to provide clear indications of the inherent capabilities and limitations of the resources. Most people in the ranching business are familiar with many of these assessments and find capability assessments easy. For example, most ranchers are able to estimate the "proper" stocking rates for their ranches and can generally estimate the productivity of their cow herd under normal conditions. Alternatively, many ranchers overlook, or incorrectly assess, the ranch business's financial capability to survive annual operating losses brought about by drought or prolonged low prices, or the capability of the labor and management of the ranch to adapt to and efficiently operate a new grazing system.

The important point in any resource assessment is that it must thoroughly evaluate all of the resources available to the firm. They are all important and the failure to use any one of them efficiently can be detrimental to the achievement of some or all of the ranch goals and objectives.

Identification and Analysis of Alternatives

The identification and analysis of alternative ways that a set of resources might be utilized to accomplish goals is the most difficult step in the planning process. The difficulty stems from the multitude of alternatives. That is, alternatives exist for not only enterprises like wildlife, cow-calf, mutton goats, ewe-lamb, and combinations of all of these, but also production techniques such as range improvements, grazing systems, and supplemental feeding. In addition, there are marketing alternatives, such as selling direct for cash, forward contracting, or sending to custom feedlots and then selling. To further complicate the formulation of a good plan, various combinations of all of these enterprises, production, and marketing alternatives must be examined.

For each alternative, or unique combination of alternatives, the planner must first evaluate resource requirements and costs to determine if the ranch's resources are sufficient to implement a practice, or if, for example, some of the financial resources would need to be used to acquire other resources, such as facilities or livestock. Next, the expected production levels, revenues, and impacts on non-financial goals (free time, peace of mind, overreaching societal needs, etc.) likely to result from each alternative should be assessed.

Several analytical methods are available which the rancher may use in analyzing alternatives. Some of the more common methods used are enterprise budgeting, gross margin analysis, break-even analysis, cash flow analysis, investment analysis, and linear programming. Whether the analytical methods chosen to be used in any given planning situation are sophisticated or simple depends on several factors, including the accuracy and reliability needed for the estimates, the length of time over which the plan will be in effect, and how quickly the information is needed.

Alternative Selection

Finally, after all the alternative combinations are analyzed, the combination of alternatives which most nearly meets the totality of ranch goals and objectives is selected. Care should be taken to ensure that personal goals are not given undue emphasis over financial goals, and vice-versa, in the alternative selection process.

PROBLEM HIERARCHY

In practice, it is usually helpful to use the time dimension to establish hierarchical order to planning. First, long-term or strategic planning concerns the achievement of goals of the resource owners over several years. Long-term plans generally affect major changes in the way resources are used to achieve goals. Once implemented, plans at this level cannot be altered quickly. In terms of organizational structure they are usually under the control of resource owners and/or upper level management personnel.

Intermediate time periods require tactical planning. Most often tactical planning is the responsibility of middle- to upper-level management and is more closely related to the operating function than long-term planning. Intermediate term planning results in decisions of how to best effect or implement long-term plans. Decisions resulting from tactical planning are likely to be repeated at least annually.

Short-term or operational planning is characterized by those many decisions that must be made in a short time-frame. Plans for the day's or week's activities are examples of operational plans commonly made.

A Model Plan

In the following sections we model some of the practical aspects of ranch planning relative to grazing management. We use the 2900-ha Texas Experimental Ranch (Fig. 10.5) as the resource base. We chose to use the ranch in this work example because of our familiarity with the land resource as an ecological unit and the availability of long-term vegetation supply and livestock performance records.

Our analyses is developed around two broad ownership goals:

1. Maintain the quality of the natural resource in perpetuity.
2. Retain ownership.

Strategic Planning

Strategic planning for grazing management requires an inventory of resources, the analysis and selection of optimum mixes of enterprises, and an investment analysis for the improvements required by the given mix of enterprises.

Resource Inventory. Resources consist of vegetation (forage), animals (domestic/wild), land, facilities, finances, and people. Forage inventory methods vary depending on the nature of the animal enterprises, entity doing the planning, and country where the planner resides. In the United States, the most pervasive resource inventorying system is that developed by the USDA Soil Conservation Service which provides a comprehensive analysis of the ecological state and trend of the vegetation complex (see Chapter 5), rangeland condition, pasture/hayland suitability, grazeable woodland suitability, and grazeable cropland suitability as related to a soil series. A massive knowledge base has been assembled during the past 40 years relative to production potential of various soil series/range sites of a given ecological condition. Because this knowledge base has expanded rapidly over this period of time, reliance upon expert opinion and local knowledge (lore) has markedly declined.

Animals are generally the resource best understood by grazing managers as they can be counted and their response observed in a relatively short period of time. However, characterization of individual performance and nutrient inputs relative to individual output so critical to their management is rarely surveyed and so understood by the manager. This shortcoming can readily be remedied through the use of such relatively simple procedures and recording practices as individual animal identification, records of performance (conception, weight, fiber intake), and monitoring by way of pregnancy palpation, body condition scoring, and weighing scales.

An inventory of facilities is critical for effective proportioning of fixed costs by enterprise and identification of excessive overhead costs associated with maintenance of various possible grazing enterprises. So for example, labor costs are

The Decision-Making Environment and Planning Paradigm 209

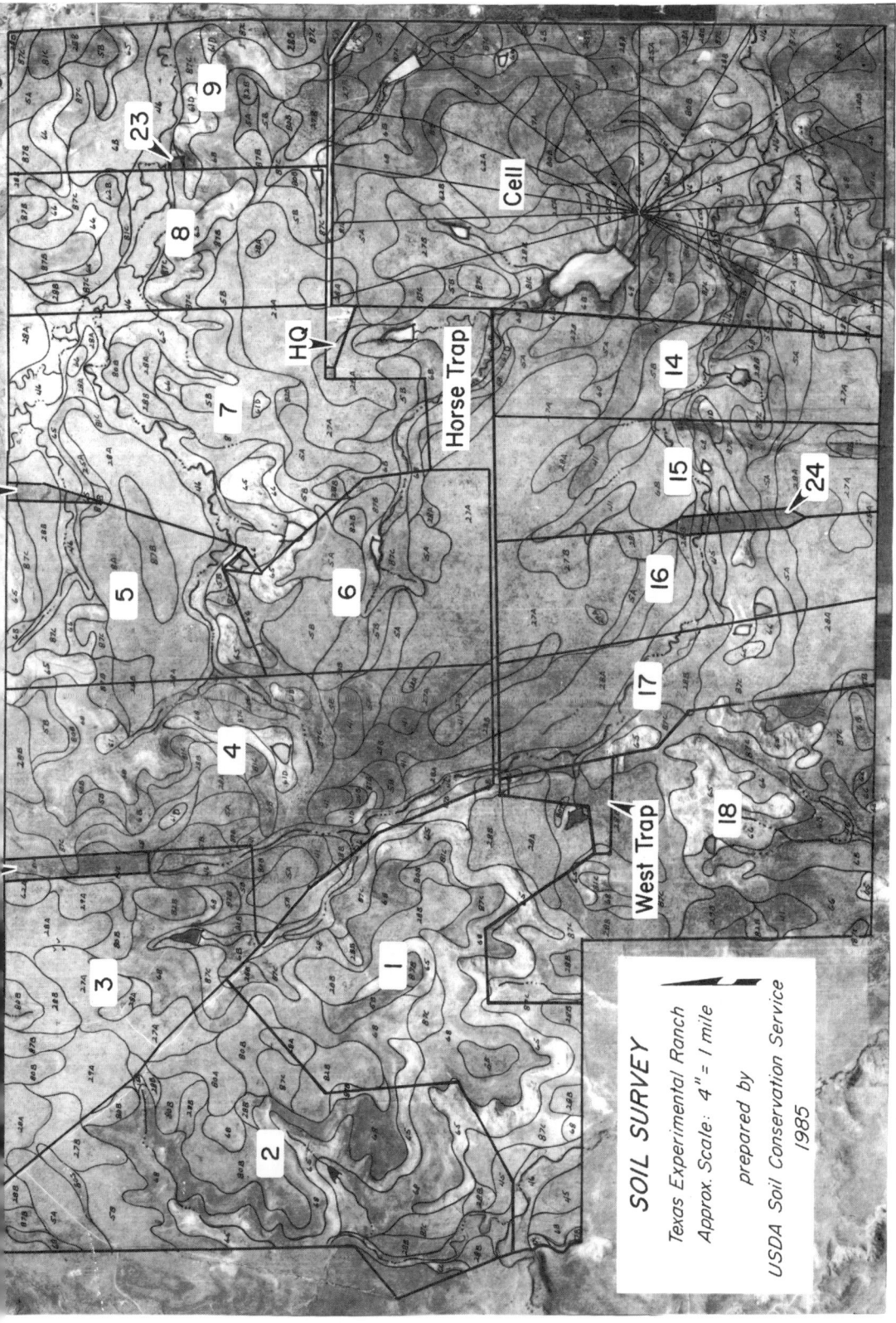

Figure 10.5. USDA-SCS Soils Map of the Texas Experimental Ranch.

often intertwined with facilities when on-ranch housing is provided. Therefore, selection of enterprises and investment strategies which affect grazing management must consider the division of labor and utilization of facilities which have both fixed and variable costs.

Finally, an inventory of financial and human resources is important because they impact the amount of capital, labor, expertise, etc., that can be effectively mobilized/utilized in the enterprise.

The first step in our working example analyses is to examine the forage resource relative to its suitability as a grazeable resource. Previous research at the ranch has shown the area is dominated by a mixture of highly palatable, warm-season short and midgrasses and Texas wintergrass, a highly preferred cool-season perennial. These grasses grow under a light to moderately dense stand of honey mesquite, an unpalatable woody tree (Heitschmidt et al. 1985). Annual and perennial forbs are not a major component other than occasional infestation of the highly unpalatable annual forb, Texas broomweed. The region is particularly suited for cow/calf enterprises because:

1. The warm-season grasses provide a suitable forage complex during late spring, summer, and early fall.
2. Texas wintergrass provides a creditable forage during late fall, winter, and spring, often of sufficient abundance to eliminate the need for winter feed supplementation for the mature cows.
3. Climatic extremes are well within the tolerance levels of cattle.

Thus, to meet our established ownership goals, the major enterprises we will consider are (1) cow/calf and (2) stocker cattle.

Figure 10.5 is an aerial photo of the ranch delineating fence lines, water impoundments, working pens, roads, etc., and associated soils. Table 10.1 is a pastures summary by soil series. Table 10.2 is a pastures summary of potential carrying capacities. In this instance, estimated carrying capacities are based on USDA Soil Conservation Service recommendations, historical records, and experience in light of the ownership goal of maintaining quality of resources in perpetuity.

An inventory of facilities reveal adequate headquarters, working pens, equipment, etc., are available for proper management and care of the livestock. Analyses also indicate adequate capital, labor, personnel, etc., to meet management goals.

Enterprise Optimization. Once an accurate resource inventory has been completed, the manager must determine which animal production enterprises are feasible (i.e., cow/calf, stocker, or a combination), the relative efficiency of each, and the net contribution each might make to overall goals. One tool for accomplishing this task is an enterprise budget which provides detailed information on resources required, costs and estimates of production levels, and product prices (Conner and Chamberlain 1985).

Figure 10.6 exhibits the data requirements and the classification of annual costs and net returns information that can be obtained. The enterprise budget is developed to project the expected costs and net returns for a typical one-year period within a planning horizon of several years. It is analogous to an annual profit and loss statement for a business firm except that it typically represents only one production enterprise and includes only the annual costs and returns attributable to that single enterprise. Capital costs represent the opportunity cost to the rancher of having his operating and investment capital tied up in cattle, fences, pens, etc., instead of having it invested in other investment opportunities, such as a savings account, where it would

Table 10.1. Acres of mapped soil series at Texas Experimental Ranch, spring 1985.

			Pasture																						Horse Trap	West Trap	House Trap	Roads	Total
Series / site	Slope (%)	Map Symbol	1	2	3	4	5	6	7	8	9	Cell	14	15	16	17	18	21	22	23	24					
...loam																										
...de clay loam	0–1	5A		27	5	38		49	25	2	21	60	47	11	20	1				2	16					324
...de clay loam	1–3	5B	30	33	13	50	3	46	45	28	16	18	25	8		3	2					11	3			334
...on clay loam	0–1	25A							5			50														55
...on clay loam	1–3	25B										3														3
...a clay loam	0–1	27A	15	9	50	19		92	61	39		29	64	82	83	1										563
...a clay loam	1–3	27B		14	12	8					13	42	13	2	21	3										128
...clay loam	0–1	29A			49										2		17									68
...en clay loam	0–1	62A			7							70						2								79
...en clay loam	1–3	62B							4			24			2					1						31
Total			45	83	136	115	3	187	136	73	50	296	149	103	128	8	19	2		3		40	3		3	1585
...ey Upland																										
...m clay	1–3	6B	15		15	11	3	8	12	3	45	113	3	50	T	2	62						2			344
...clay	0–1	28A	31	3	55	31	90	5	94	16		55	6	48	60	95	15			3		10	4		3	624
...clay	1–3	28B	93	41	26	92	25	15	20	10	17	127	18	10	8	40	60		5	2		38	20			667
Total			139	44	96	134	118	28	126	29	62	295	27	108	68	137	137		5	5		48	26		3	1635
...Slopes																										
...Rock Outcrop complex	1–8	68	139	207	40	98	1		7	19	22	162	12	14	4		11		1	2		16				755
...silty clay	1–3	87B	16	15	16	27	27	8	6	38	16			15	T			1								185
...silty clay	3–5	87C	72	46	9	66	41	29	67	60	52	76	15	15	24	45	142	2	2		5	4	6			778
Total			227	268	65	191	69	37	80	117	90	238	27	29	43	45	153	2	3	1	7	20	6			1718
...ow Clay																										
...clay	3–8	61D				22			3		10	1	6	2								2				46
...Redland																										
...clay loam	1–3	24B				3																				3
...ow																										
...stoney clay	0–1	80A		22																						22
...stoney clay	1–3	80B	11	87	34	17	1		7	3	4	70												T		234
Total			11	109	34	17	1		7	3	4	70												T		256
Shallow																										
...stoney clay loam	1–5	41				48						24	24	29	3	6	23									157
...uders complex	1–8	60											6													6
...y loam	1–3	82B	22	9	19	2		6	4		6	2					5									75
Total			22	9	19	50		6	4		6	26	30	29	3	6	28									238
...ow Clay—Shalely Hills																										
...Harpersville complex	1–8	66		2			20	12	16	11	13	8			3	8	77									170
...y Hills—Shalely Hills																										
...Harpersville complex	3–30	65	88	103			12	35	14		52	35	42			10	98		1	1						491
...y Upland																										
...to stoney clay	0–5	81C	6		9	30			4	4	11	45				4	16									129
...oney clay loam	1–8	89										26	5													31
Total			6		9	30			4	4	11	71	5			4	16									160
...y Bottomland																										
...ty clay, occasionally flooded		45	17	6								8		7					2							40
...ty clay, frequently flooded		46	36	36	12	37	38	6	112	51	32	112	15	12	18	38	6		2	2		8	1			574
Total			53	42	12	37	38	6	1122	51	32	112	23	12	25	38	6		2	4		8	1			614
			2	2	2	2	1	2	3	1	3	27	2	1	2	1	1				2					
pits, etc.			T	T								4				3									18	25
			593	662	373	613	285	292	543	324	323	1148	269	284	272	253	526	20	11	2	19	120	36	9	18	6995

Table 10.2. Estimated carrying capacity by pastures (see Fig. 10.5 and Table 10.1) at Texas Experimental Ranch.

Pastures	Size (ha)	Carry Capacity	
		(ha/AU/yr)	(AU/yr)
1	240	7	34
2	268	7	38
3	151	6	25
4	248	6	41
5	115	6	19
6	118	6	20
7	220	6	37
8	131	7	19
9	131	7	19
Cell	465	6	77
14	109	6	18
15	115	6	19
16	110	6	18
17	102	6	17
18	213	7	30
Horse Trap	49	6	8
West Trap	15	6	2
House Trap	4	0	0
Exclosures	21	0	0
Roads	7	0	0
TOTAL	2881		441

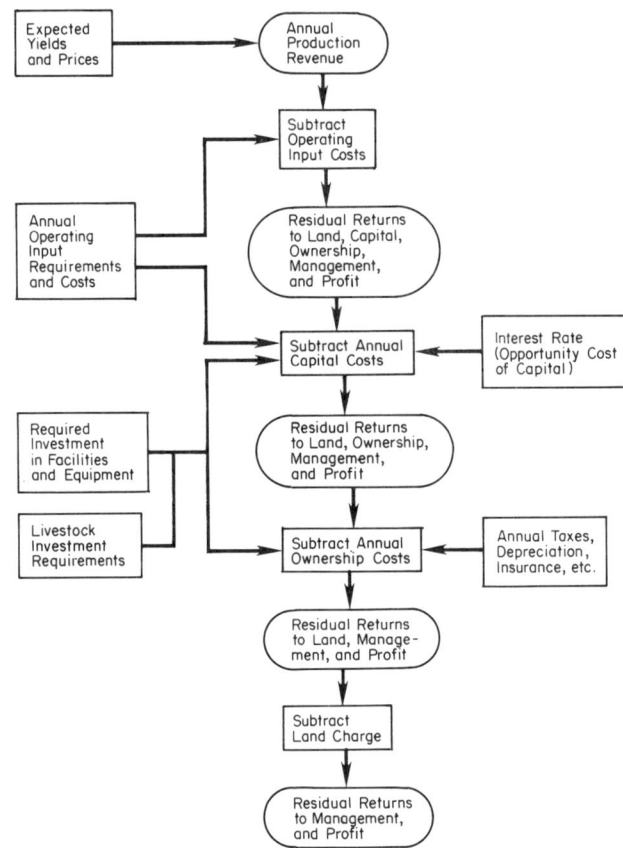

Figure 10.6. Data requirements and classification of annual costs and returns in the development of an enterprise budget.

earn an annual return in the form of interest. Ownership costs reflect the annual depreciation expenses of the purchased assets. Land is excluded because its productive potential is not used up in the annual production process, and raised cows are excluded because their replacement cost is reflected by not selling a portion of the heifer calves and putting them back in the herd each year. Annual taxes and insurance costs are also included in ownership costs. Residual, or net, returns to the unpaid resources are calculated after each of the major categories of cost have been developed and deducted from gross income. Residual returns to land, management, and profit represent what remains for the rancher to pay the opportunity cost of land use, his/her management time and expertise, and to claim as profit.

For example, Table 10.3 is an enterprise budget for a 1215-ha cow-calf operation at a moderate rate of stocking. The budget is based on performance data collected at the ranch under a year-long continuous grazing strategy (Heitschmidt et al. 1990). Table 10.4 is a similar budget for a stocker operation. Similarly, budgets can be developed for other livestock and/or wildlife enterprises as well as other grazing strategies. By comparing the net returns per acre for each, the rancher can select the one providing the highest return.

For example, Table 10.5 is a summary of cow/calf enterprise budgets for 4 grazing strategies based on livestock performance data collected at the ranch (Heitschmidt et al. 1990). Development of such budgets is appropriate in most grazing management analyses because of the potential impact of various rates of stocking and grazing systems on rangeland resources (Chapters 5 and 6), livestock (Chapter 7), and wildlife (Chapter 8) production and economic stability (Chapter 9).

A major tool utilized for identifying the effect of various enterprise mixes on management goals is linear programming. Linear programming (LP) may be used to select the level and mix of enterprises to maximize the rancher's goals subject to the specific set of resources available (Glover and Conner 1988). For example, Whitson et al. (1982), using livestock performance data from the Texas Environmental Ranch (Heitschmidt et al. 1982), determined that although year-long continuous grazing at a heavy rate of stocking resulted in 27% more net income/ha than a 4-pasture, 3-herd moderate stocked, deferred rotation system, it also increased economic risk (see Chapter 9).

More detailed information on the inventory of resources and resource requirements of the various animal enterprises can be effectively used in the LP analysis to better match animal needs to available resources. For example, the resource inventory and the enterprise budgets could be supplemented with detailed information on classes of forage (grasses, forbs, and browse) available by season (summer, fall, winter, and spring) and intake of the same forage classes by season for each animal enterprise. These data, combined with other resource parameters, enterprise requirements, and product prices in a LP solution algorithm, could provide explicit information on the best combination of enterprises to use with the available resources (Fig. 10.7). Furthermore, if product prices change, or if the forage resource were to be modified by a resource improvement practice, the model could be re-formulated and an optimal solution obtained for the new situation.

Table 10.3 Enterprise budget for cow-calf operation on 1215 hectares in rolling plains of Texas with moderate stocking rate and continuous year-long grazing strategy.

Livestock Investment Requirements

	Number	Unit	$/Unit	Total Value
Horse	2	Head	1,000	2,000
Beef bull	7	Head	1,250	8,750
Beef cow raised	167	Head	600	100,200
Beef heifer raised	22	Head	500	11,000
		Total Livestock Investment		$121,859
		$730.24/Cow		
		100.41/ha		

Production Revenue

Production Revenue	Total Produced	Wgt. Each	Unit	$/Unit	Total Units Produced	Total
Steer calves	72	255.9	kg.	1.41	18,424.8	26,044.38
Heifer calves	50	231.4	kg.	1.33	11,570.0	15,333.62
Cull cows	18	409.1	kg.	0.80	7,363.6	5,913.00
			Total Production Revenue			$47,291.00
			$283.18/Cow			
			38.93/ha			

Operating Costs

Operating Inputs	Quantity	Unit	$/Unit	Total/Cost
Range cubes	455.50	kg.	0.24	107.72
Vet. & med. exp	1,040.41	Dol.	1.00	1,040.41
Salt	2,277.30	kg.	0.15	350.70
Miscellaneous	55.11	Dol.	1.00	55.11
Marketing	140.00	Head	10.00	1,400.00
Fence repair	45.00	Dol.	1.00	45.00
Equipment fuel and lube	416.25	Dol.	1.00	416.00
Equipment repair	575.45	Dol.	1.00	575.45
Operating Labor	846.00	Hr.	10.00	8,460.00
	A Total Operating Cost			$12,450.64
	$74.55/Cow			
	10.25/ha			
	Residual Returns to Land, Capital, Ownership, Management and Profit			$34,840.36
	$208.62/Cow			
	28.68/ha			

Capital Costs

Capital Requirement	Quantity	Unit	Rate of Return	Total Cost
Annual operating cap.	12,450.64	Dol.	.05	622.53
Facil.& equip.invest.	24,479.62	Dol.	.10	2,447.96
Livestock investment	121,950.00	Dol.	.10	12,195.00
	Total Capital Cost			$15,265.49
	$91.41/Cow			
	12.57/ha			
	Residual Returns to Land, Ownership, Management and Profit			$19,374.87
	$117.21/Cow			
	16.01/ha			

Table 10.3. Continued.

Ownership Costs	
Dep., Ins., Taxes	Total Cost
Facilities and equipment	$3,517.90
Livestock	1,237.69
Total Ownership Costs $28.48/Cow 3.93/ha	$4,753.59
Residual Returns to Land, Management and Profit $88.74/Cow 12.20/ha	$14,819.28

Table 10.4. Enterprise budget for stocker grazing operation on 1215 hectares in rolling plains of Texas with moderate stocking rate and winter-spring grazing.

Livestock Investment Requirements

	Number	Unit	$/Unit	Total Value
Horse	2	Head	1,000	2,000
Total Livestock Investment $5.67/Head 1.65/ha				$2,000

Production Revenue

	Total Produced	Wgt. Each	Unit	$/Unit	Total Units	Total Revenue
Stockers	342	340.9	kg.	1.32	116,590.9	153,900
Total Production Revenue $435.98/Head 126.67/ha						$153,900

Operating Costs

Input item	Quantity	Unit	$/Unit	Total Cost
Stockers	353 @ 231.8	kg.	1.44	$117,920
Range Cubes	15,812.5	kg.	0.24	3,795
Salt	2,473.30	kg.	0.15	371
Vet. & med. exp.	1,765	Dol.	1.00	1,765
Marketing	342	Head	10.00	3,420
Fence repair	45	Dol.	1.00	45
Equipment fuel & Lube	416	Dol.	1.00	416
Equipment repair	575	Dol.	1.00	575
Operating Labor	846	Hr.	10.00	8,460
Total Operating Cost $387.44/Head 112.57/ha				$136,767
Residual Returns to Land, Capital, Ownership, Management and Profit $48.54/Head 14.10/ha				$17,133

Table 10.4. Continued.

Capital Costs

	Quantity	Unit	Rate of Return	Total Cost
Annual operating capital	136,767	Dol.	.05	1,828
Facil. & equip. Invest.	24,480	Dol.	.10	2,448
Livestock Investment	2,000	Dol.	.10	200
	Total capital Cost $12.71/Head 3.69/ha			$4,486
	Residual Returns to Land, Ownership, Management and Profit			$12,647
	$35.83/Head 10.41/ha			

Ownership Costs

Dept. Ins. & Taxes		Total Costs
Facil. & equip.		3,518
Livestock		100
	Total ownership cost $10.25 head 2.98/ha	$3,6188
	Residual Returns to Land, Management and Profit	$9,029
	$25.58/Head 7.43/ha	

Table 10.5. Summary of cow/calf enterprise budgets for four grazing strategies in the rolling plains of Texas.

Residual Returns to Land, Management and Profit	Grazing Strategy			
	Heavily stocked year-long contin.	Moderately stocked year-long continuous	3-herd 4-pasture deferred rotation	1-herd 14-pasture rotational grazed
$ Per hectare	13.21	11.02	15.98	16.38
$ Per cow	60.81	69.57	93.12	62.72

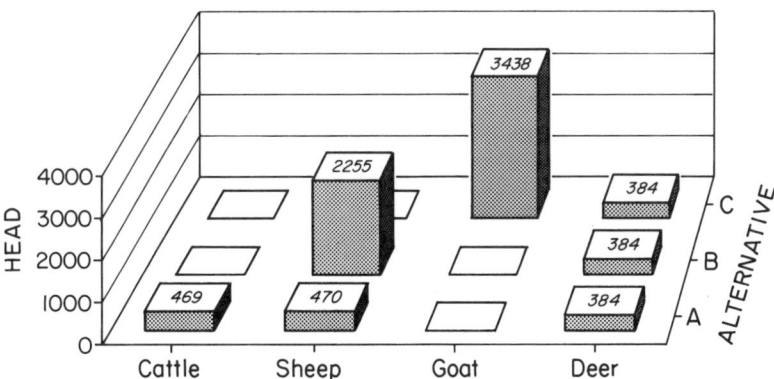

Figure 10.7. Linear programming solutions of optimal livestock enterprise lines and resulting net return under three restrictions on livestock.

Investment Analysis. Another strategic planning factor involves the question of making major investment and/or operational changes such as the installation of new livestock water facilities or cross fences necessary to implement an alternative grazing system. The appropriate analytical technique for determining whether such practices will result in significant positive contributions toward the rancher's goals is investment analysis.

Brush control practices, fences, water facilities, etc. are all potential investments which require a relatively large initial capital outlay and produce variable annual returns over several years. Other investment opportunities always exist so money invested in a particular improvement practice or program cannot be used for such alternatives. Any resource manager must insist that any improvement practice earn a return on the money invested in it at least as large as the return that could be earned in alternative investments (i.e., opportunity cost).

One way to compare the annual income earning potential of different investments is to use the "average annual rate of return" which is the average annual net earnings or profits divided by the total amount of the investment with the quotient expressed as a percentage. One example, and a standard against which to measure an investment, is an investment in a savings account earning 9% interest which is an average annual rate of return of 9%.

To calculate the addition to total annual net profit derived from a specific rangeland improvement practice, one must calculate the changes in annual costs of production and revenues from products and/or service produced solely as a result of the resource improvement practice. Since costs and revenues derived from improvement practices differ drastically in the years after the practice is implemented, these changes must be estimated for each year for several years into the future (Scifres et al. 1985).

Because the costs and returns associated with the resource improvement practice occur at different times in the planning period they must be adjusted to reflect their present value before being compared. Present value is the worth today of a sum of money that is to be available sometime in the future. The idea that there is a difference is based on the time value of money. Revenue received a year from now is not worth a dollar today because the use of it is foregone for one year. To equate the two, i.e., to estimate the present value of currency that is to be received a year from now, simply multiply the currency by the discount factor.

The discount factor for each year into the future is

$$\frac{1}{(1+r)^n}$$

where
 r = the "real" rate of return from next best alternative investment, plus differences for risk
 n = the number of each year into the future.

If r = 10%, then the discount factor for year 1 is

$$\frac{1}{(1+.10)^1} = .909\text{---}$$

and the discount factor for year 2

$$\frac{1}{(1+.10)^2} = .826\text{---}.$$

The investment analysis procedure simply accounts for the added costs and revenues associated with the resource improvement practice occurring in each year in the planning period. It discounts the investment costs and revenues with the appropriate discount factor, then sums the discounted costs and revenues, and finally subtracts the summed discounted costs from the summed discounted revenues. The difference is the net present value (NPV) of the investment in the resource improvement practice.

1. If the NPV is zero, the investment in the improvement practice is estimated to earn exactly the same average annual rate of return as the alternative investment, i.e., the selected discount rate.
2. If the NPV is greater than zero, the investment in the improvement practice is estimated to produce an average annual rate of return that is greater than the alternative.
3. If the NPV is negative, the range improvement practice is estimated to produce an average annual rate of return less than the specified rate of the alternative and is economically unwise based on the stated criterion.

The investment analysis procedure also produces an additional investment indicator: the internal rate of return (IRR). Internal rate of return is defined as the discount rate that results in a NPV of zero. Thus, it is a direct estimate of the average annual rate of return the investment in the improvement practice will produce over the entire planning period. This indicator is particularly useful in comparing the relative merit of alternative investments which may require different time periods and/or different initial outlays of capital because it reports returns as an average annual percentage. Net present value, on the other hand, only reveals the present value of net returns in dollars over or under the specified discount rate.

In summary, it can be seen that a combination of analytical tools must be utilized to identify the effect of various grazing management strategies on goal attainment. For example, recent analyses at the Texas Experimental Ranch show moderate rates of stocking are necessary to maintain the range resource (Goal 1) relative to watershed (Pluhar et al. 1987) and range condition (Heitschmidt et al. 1985) regardless of grazing strategy. Moreover, the analyses also show moderate rates of stocking are necessary to reduce (Goal 2) economic risks (Whitson et al. 1982; Heitschmidt et al. 1990), although heavier rates are necessary to maximize average profit/ha on an annual basis. Thus, selection of the "best" strategy requires owners to reevaluate goals and objectives and adopt acceptable compromises.

Tactical Planning

Once a mixture of enterprises and alternative practices has been selected, the manager begins the implementation stage of planning. That is, the animals are bought and sold, the fences erected, and/or water or land developed. Once in place, the manager must plan for the upcoming year or production cycle (Fig. 10.3). It is difficult to return animals or undo fences and water developments because funds have already been spent. The manager must develop grazing and husbandry tactics to best recapture the value of the funds expended which involves a projection of responses for the upcoming year.

This process calls for assessment of current conditions relative to expected cycles in forage production. Most managers have acquired rules of thumb relative to setting stocking from one year to the next. Generally, the planning position for the next year is based on heuristic knowledge of rainfall probabilities in relation to the risk-taking tendencies of the manager. Because rainfall patterns and probabilities vary around the globe and within regions, it is difficult to provide a general set of rules that matches all planning environments.

Typical planning tools required for tactical planning include forage supply/demand balance analysis, annual grazing plan, nutritional balance and mediation analyses, and partial budget analysis. Often managers are faced with surpluses/deficits in forage supply in their normal grazing regime. By comparing the monthly distribution of forage production, regardless of scheduled use, to the physiologically weighted monthly livestock demand on a whole-firm basis, the manager is able to obtain a clearer picture of whether forage deficits or excesses are a function of current grazing practices or characteristics of the forage base of the ranch (Table 10.6). Grazing tactics can often be altered so stored feeds/forage can be strategically fed to help alleviate forage balance problems.

Grazing planning tools require that the manager schedule the use of pastures in terms of the differences between animal demand based upon their physiological profile against the growth cycle of the available forage (Fig. 10.8). This exercise cannot be undertaken in isolation from management goals.

Perhaps the most difficult management objective to plan for is the mediation of animal nutritional deficiencies within a grazing regime to meet performance goals of livestock (see Chapter 2). Managers are at present utterly dependent upon experience in feeding food supplements of differing kinds and amounts at different times of the year to maintain the nutritional status of the animal(s). These heuristic feeding programs are typically founded upon empirical feeding studies conducted under constantly controlled conditions which can seldom be correlated with the situation faced by the manager. In this case, as in many others, the manager must internalize such scientific information and then apply it to his/her unique set of resources in order to determine if such studies meet his/her needs. The major limiting factor to moving this largely heuristic planning process to a more quantitative practice is the lack of suitable analytical tools for assessing the nutritional well-being of an animal.

The economic analytical tools, including linear programming and partial budgeting, have now been adapted for tactical planning in the stock raising industry.

Pasture	Area	Capacity	% Capacity	Desired Rest Period (d)												# Graze Cycles	Days Grazed	Graze/Rest Ratio
				120	120	90	75	75	90	90	90	75	90	120	120			
				J	F	M	A	M	J	J	A	S	O	N	D			
1	240	34	.146		7+13		11		13		2+11			5	57	.185
2	268	38	.163		14		12+10		10+4		19			4	59	.193
3	151	25	.107		4+5		7		8			13		4	37	.113
4	248	41	.176	3		13		13+3		13			18+		4	63	.209
7	115	19	.082	10	6		7		6					4	29	.086
6	118	20	.086	10	6		8			8				4	32	.096
7	220	37	.158	8+11		12		13+1		14				4	59	.193
8	131	19	.082	10		6		7		7				4	30	.090
Total	1491	233															

Herd Structure (#Hd)

	J	F	M	A	M	J	J	A	S	O	N	D
Cows (Mature)	215	215	215	215	215	215	215	215	182	182	182	182
Bulls (Mature)				10	10	10						
Repl. Heifers (24-36 mo)									33	33	33	33
Calves	65	155	220	220	220	220	220	220				

Forage Balance (AUM)

	J	F	M	A	M	J	J	A	S	O	N	D
Forage Supply	84	56	195	391	531	447	280	84	280	224	84	140
Forage Demand	215	226	236	260	271	281	280	301	169	179	190	200
Monthly Balance	-131	-170	-41	131	260	166	0	-217	111	45	-106	-60
Accumulated Balance	199	29	-12	131	391	558	558	341	452	497	391	330

Figure 10.8. Grazing control chart of an 8-pasture, 1-herd rotational grazing system and associated monthly forage balance analysis.

LP models are commonly used to determine the most efficient combination of foodstuffs to correct nutritional deficiencies in animals if those deficiencies are accurately characterized. Partial budgeting permits the manager to identify only those revenues and costs contained in an enterprise budget that change when a particular management practice is discontinued or altered. Positive benefits are reflected in increased revenues and/or reduced costs while negative consequences are reflected in reduced revenues and/or increased costs. The difference between positive and negative effects helps the manager determine whether a change in operations will improve or reduce profits. Clearly, partial budget analysis depends upon having developed good enterprise budgets to help guide the manager through the analytical process. Typical grazing management problems that lend themselves to partial budget analysis include the question of purchasing or raising the replacement brood animals or retention of growing animals as "stockers" in anticipation of changing forage and/or market conditions. For example, a dramatic change in supplemental feed costs in the budgets presented in Table 10.5 would also dramatically alter residual returns. Such changes must be anticipated and their impacts evaluated to minimize economic risk in conjunction with optimizing profit.

Operational Planning

Once a grazing plan has been implemented and feedstuffs purchased for the upcoming feeding season, the manager is faced with responding quickly to the cycle of plant production that actually occurs during course of the operating year. Due to conflicting logistical and market forces, most grazing managers cannot readily and quickly stock/destock animals as growth rates of forage increase or decrease in response to variable climatic conditions. However, the manager can establish key decision points and contingency plans for reacting to forage/market conditions should they deviate markedly from pre-planned levels (Anderson and Hardaker 1973).

For instance, if two management objectives are 1) to improve the ecological condition of the range, and 2) to increase grazing capacity over the long term, a criterion for establishing decision points to avoid overgrazing demands a clear understanding of the percentage of forage production available each month, the minimum forage residues required to maintain the hydrological properties of the soils and vigor of the plants, the probability of rainfall, the husbandry calendar, and market projections. On ranges located in areas having dual peak rainfall patterns, a major percentage of the forage is produced in one or the other of these two periods, e.g., 70% grazing capacity in season one. Not infrequently, managers are faced with forage conditions poorer than normal near the end of their major production cycle, so are forced to assess their chances of surviving until the next rainfall period. In arriving at a decision, they must also assess the probability of receiving at least median levels of rainfall during the next production cycle. This assessment can then be weighed against the minimum forage residue standards set by the long-term objectives. The manager is in this case at a decision point where he/she must assess the alternative actions relative to logistic and market constraints available.

The kinds of analytical methods being urged here require that managers have access to long-term weather records and current market projections as well as the recently developed techniques for monitoring forage supply and animal condition. This information coupled with current animal condition and inventory information allows the manager to budget forage and determine cost-effective tactics to dampen seasonal swings in stock numbers.

FUTURE DIRECTIONS OF PLANNING

In the previous sections of this chapter, we have focused on characteristics of grazing management expertise and analysis that can be performed to assist managers in their decision process at all levels of management. Many of these decisions involve complex biological processes, a high degree of uncertainty about future weather, and perceptions of market trends. As stated previously, these decisions are made in a climate of uncertainty depending on the degree of understanding the manager possesses relative to biological relationships and the amount of information that can be acquired to access market trends and weather risks. When information deficiencies exist, the traditional source for enlightenment has been trusted producers, trade journals, and action agency personnel who service knowledge requirements of the resource manager. However, the advent of low-cost microcomputers has given rise to a new dimension in information technology which improves transfer of new technology and information as well as analysis of complex systems.

Moving the planning process from local heuristic to scientifically based planning requires development of computerized decision support systems which allow the manager to input a broad array of parameters unique to his planning environment and analyze those parameters with a set of scientifically proven algorithms. Such systems have emerged in recent years which facilitate planning and analysis of grazed systems (Stuth et al. 1990). Most of these systems have focused on strategic planning relative to investment analysis, forage inventorying, herd inventory projections, and enterprise analyses. A new wave of tactical planning systems are emerging to address such issues as grazing planning and control, nutritional mediation, herd transactions, and fodder management.

Operational planning of grazing involves assessment of current conditions, determination of probable future response, and analysis of potential solutions (Stuth et al. 1990). In order to move operational decision making into computerized decision support for grazing requires increased resolution and amount of data relative to current conditions and probable responses. To meet these requirements, the planner must implement monitoring systems which assess forage supply, animal demand and nutritional status, and weather events leading to a decision point. Currently there are few methodologies available to efficiently assess forage supply on a scale necessary for ranch firms. A major limitation is the inability of many managers to identify plant species and estimate standing crop. New hydrologic-based forage production models linked to long-term weather records could assist managers or advisors of managers in projecting probable responses of current forage supply in the near term. However, action agencies will have to place information on-line relative to information on physical characteristics of soils, maximum potential response of major species, and long-term weather records. High-speed communications technology, satellite relay, and mass storage devices are advancing at such a high rate that resource information should be common place in those countries willing to commit the necessary resources to organize their resource knowledge.

Recent advances in fecal profiling via near-infrared spectroscopy allows rapid assessment of protein and energy status of free-roaming animals (Stuth et al. 1990). This technology should offer the resource manager the ability to monitor nutritional status of their livestock in the same manner that farmers monitor soil fertility. Requirements for nutritional mediation could then be carefully assessed and optimum supplement programs established.

As pointed out in Chapter 9, advanced technology is a product of organized societies. Obviously, therefore, sophisticated decision-making aids such as discussed here have little relevance to subsistence societies unless monies diverted from more highly organized societies are made available to organizations possessing knowledge required for planning at the firm/property level in such societies. Such aid, however, should be less costly in the future as the cost/computing power continues to decline rapidly while ease of use and sophistication of software tools increases. The world is rapidly moving toward a situation in which resource managers can access advanced planning systems designed for the increasing analytical power of microcomputers, download planning applications, resource data and other kinds of information from centrally built and maintained data bases; conduct spatial analyses of resources; run simulation models of biological and hydrologic processes; and access expert opinion using artificial intelligence methodologies. The fact that resource managers have not widely used these technologies in the past has in part been due to technology limitations, the failure of researchers and extension personnel to demonstrate the value of such systems to producers, as well as the unwillingness of producers to undertake the learning activities necessary to effectively use this planning technology. Planning, as previously noted, is a personal process which depends primarily on internalized knowledge. Scientific information is typically passed by action agency personnel, such as extension officers to ranchers, but such agents are external sources of information to the land manager. When the manager is faced with a crisis decision, external information is discounted while a premium is placed on traditional second-hand knowledge which has been internalized. Therefore, the decision is entirely or nearly entirely driven by perceptions based on a more or less ill-founded, scientifically insupportable knowledge basis.

Scientifically defensibly, knowledge-based decision support systems offer the manager a means of blending his knowledge with that of experts in a planning environment which he controls. He can access external information and develop "what if" scenarios reflecting his personal perception of the current situation and future responses (Fig. 10.9). Once in control, the value of external information can be better assessed and personal perceptions tested. This process allows for technology to be evaluated in relation to unique planning environments. The value of these systems will be proved only after a critical mass of managers access the system and weigh the results. If properly configured, delivered, and maintained, decision-support systems should usher in the information age for the grazing manager. These tools, coupled with a greater understanding of biological, ecological, economic, and socio/political forces impinging on the livestock producer, should advance decision making to a level whereby sustainability of the resource and the enterprise is better assured.

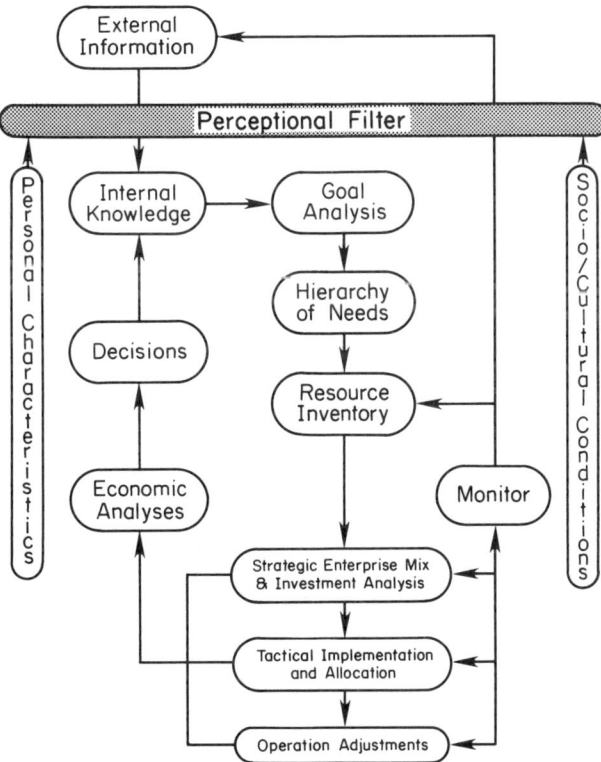

Figure 10.9. Hierarchical view of decision making which allows for internalization of external information in the context of a rancher's unique set of resources and managerial environment.

APPENDIX A

ANIMAL SPECIES LIST

COMMON NAME	SCIENTIFIC NAME
Axis Deer	*Cervus axis*
Bighorn Sheep	*Ovis canadensis*
Bison	*Bison bison*
Black-tailed Jackrabbit	*Lepus* spp.
Black-tailed Prairie Dog	*Cynomys ludivicianus*
Blackbuck Antelope	*Antilope cervicapra*
Blue Wildebeest	*Connocheates taurinus*
Brahman Cattle	*Bos indicus*
Bushbuck	*Tragelaphus scriptus*
Camel	*Camelus* spp.
Cape Buffalo (African)	*Synceros caffer*
Cattle	*Bos* spp.
Chamois	*Rupicapra rupicapra*
Coyote	*Canis latrans*
Dik Dik	*Madoqua kirki*
Elk	*Cervus canudensis*
Fallow Deer	*Cevis dama*
Feral Horse	*Equus caballus*
Giraffe	*Giraffa cameleopardis*
Goat	*Capra hircus*
Horse	*Equus caballus*
Impala	*Aepyceros melamus*
Javelina	*Tayassu tajacu*
Lion	*Felis leo*
Moose	*Alces alces*
Mouflon	*Ovis musimon*
Mountain Reedbuck	*Reduna fulvorufula*
Mule Deer	*Odocoileus hemonius*
Oryx	*Oryx beisa*
Pig	*Sus* spp.
Pronghorn Antelope	*Antilocapra americana*
Rio Grande Turkey	*Meleagris gallapavo*
Sheep	*Ovis aries*
Thompson's Gazelle	*Gazella thomsonii*
White-tailed Deer	*Odocoileus virginianus*
Zebra	*Equus* spp.

APPENDIX B

SUPPLEMENTAL PLANT SPECIES LIST

COMMON NAME	SCIENTIFIC NAME
Acacia	*Acacia* spp.
	Acacia karroo Hayne
	Acacia hockii DeWild
Algarrobo	*Prosopis juliflora* (= glandulosa Torr. var. Torreyana [L. Benson, M C. Johnst.])
Baccharis	*Baccharis pilularis* var. *consanguinea* (DC.) C. B. Wolf
Baobab	*Adansonia digitata* L.
Barley	*Hordeum vulgare* L.
Bermudagrass	*Cynodon dactylon* (L.) Pers.
Big Sagebrush	*Artemisia tridentata* Nutt.
Black Grama	*Bouteloua eriopoda* (Torr.) Torr.
Blue Grama	*Boutelous gracilis* (H. B. K.) Lag. ex Steud.
Bluebunch Wheatgrass	*Agropyron spicatum* (pursch). Scribn. & Smith.
Browntop	*Agrostis* spp.
Buffalograss	*Buchloe dactyloides* (Nutt.) Engelm.
Canarygrass	*Phalaris tuberosa* L.
Cane Bluestem	*Bothriochloa barbinodis* (Lag.) Herter
Cheatgrass	*Bromus tectorum* L.
Common Curlymesquite	*Hilaria belangeri* (Steud.) Nash
Condalia	*Condalia* spp.
Creosote Bush	*Larrea tridentata* (DC.) Cov.
Crested Wheatgrass	*Agropyron cristatum* (L.) Gaertn.
Eucalyptus	*Eucalyptus* L'Heritier spp.
Hard fescue	*Festuca ovina* L.
Honey Mesquite	*Prosopis glandulosa* Torr. var. *glandulosa*
Hooded Windmillgrass	*Chloris cucullata* Bisch.
Huisache	*Acacia farnesiana* (L.) Willd.
Juniper	*Juniperus* spp.
Little Bluestem	*Schizachyrium scoparium* (Michx.) Nash
Little Bluestem	*Schizachyrium scoparium* var. *frequens* (Hubb.) Gould
Live Oak	*Quercus virginiana* Mill.
McCartney Rose	*Rosa bracteata* Wendl.
Medusahead	*Taeniatherum caput-medusae* (= asperum) Simonkai) Nerski
Mesquite	*Prosopis* spp.
Mimosa	*Mimosa pigra* L.
Narrow-leaved Ironbark	*Eucalyptus crebra* F. Muell.
Oak	*Quercus* spp.
One-seeded Juniper	*Juniperus monosperma* (Engelm.) Sarg.

COMMON NAME	SCIENTIFIC NAME
Orchardgrass	*Dactylis glomerata* L.
Perennial ryegrass	*Lolium perenne* L.
Pines	*Pinus* Lindl. spp.
Pinyon Pine	*Pinus monophylla* Torr. and Frem.
Ponderosa Pine	*Pinus ponderosa* Laws
Poplar Box	*Eucalyptus populnea* F. Muell.
Prickly Acacia	*Acacia* Mill. sp.
Pricklypear Cactus	*Opuntia* spp.
Quackgrass	*Agrophyron repens* L. Beauv.
Red threeawn	*Aristida longiseta* Steud.
Rubber Rabbitbrush	*Chrysothamnus nauseosus* (Pall.) Brit.
Sagebrush	*Artemsia tridentata* Nutt.
Scarlet Gilia	*Ipomopsis aggregata*
Sideoats Grama	*Boutelous curtipendula* (Michx.) Torr.
Sun Sedge	*Carex eleocharis* Bailey
Tall Fescue	*Festuca arundinacea* Schreb.
Tarbrush	*Flourensia cernua* DC.
Teosinte	*Zea mexicana* (schrad.) Reeves & Mangeldorf
Texas Broomweed	*Xanthocephalum dracunculoides* (DC.) Shinners
Texas Cupgrass	*Eriochloa sericea* (Scheele) Munro
Texas Wintergrass	*Stipa leucotricha* Trin and Ruph.
Tobosa grass	*Hilaria mutica* (Buckl.) Benth.
Umbrella Thorn	*Acacia tortilis* (Forsk.) Hayne subsp. *spirocarpa*
Western Wheatgrass	*Agropyron smithii* Rydb.

LITERATURE CITED

Adams, D.C. 1985. Effect of time of supplementation on performance, forage intake and grazing behavior of yearling beef steers grazing Russian wild ryegrass in the fall. J. Anim. Sci. 61:1037–1042.

Adams, D.C., R. C. Cochran, and P. O. Currie. 1987a. Forage maturity effects on rumen fermentation, fluid flow, and intake in grazing steers. J. Range Manage. 40:404–408.

Adams, D.C., T. C. Nelsen, W. L. Reynolds, and B. W. Knapp. 1986. Winter grazing activity and forage intake of range cows in the Northern Great Plains. J. Anim. Sci. 62:1240–1246.

Adams, D.C., R. E. Short, and B. W. Knapp. 1987b. Body size and body condition effects on performance and behavior of grazing beef cows. Nutr. Rep. Int. 35:269–277.

Akin, E. D., and D. Burdick. 1977. Rumen microbial degradation of starch-containing bundle sheath cells in warm-season grasses. Crop Sci. 17:529–533.

Akiyama, T., S. Takahashi, M. Shiyomi, and T. Okubo. 1984. Energy flow at the producer level. Oikos 42:129–137.

Alden, W. G., and I. A. McD. Whittaker. 1970. The determinants of herbage intake by grazing sheep: The interrelationship of factors influencing herbage intake and availability. Aust. J. Agr. Res. 21:755–766.

Alderfer, R. B., and R. R. Robinson. 1947. Runoff from pastures in relation to grazing intensity and soil compaction. Am. Soc. Agron. J. 29:948–958.

Allen, T. F. H., R. V. O'Neill, and T. W. Hoekstra. 1984. Interlevel relations in ecological research and management: some working principles from hierarchy theory. USDA For. Serv. Gen. Tech. Rep. RM-110. Fort Collins, CO. USA.

Allen, T. F. H., and T. B. Starr. 1982. Hierarchy: Perspectives for ecological complexity. Univ. Chicago Press, Chicago.

Allison, C. D. 1985. Factors affecting forage intake by range ruminants: a review. J. Range Manage. 38:305–311.

Allison, F. E. 1973. Soil organic matter and its role in crop production. Elsevier, New York.

Allred, K. L. 1980. Effects of a short duration grazing system on white-tailed deer (*Odocoileus virginianus*) in the Rio Grande Plain, Texas. M.S. Thesis, Texas A&M Univ., College Station. USA.

Anderson, D.C. 1987. Below-ground herbivory in natural communities: a review emphasizing fossorial animals. Quart. Rev. Biol. 62:261–286.

Anderson, D.C., K. T. Harper, and R. C. Holmgren. 1982. Factors influencing the development of cryptogamic soil crusts in Utah deserts. J. Range Manage. 35:180–185.

Anderson, E. W., and R. J. Scherzinger. 1975. Improving quality of winter forage for elk by cattle grazing. J. Range Manage. 28:120–125.

Anderson, J. R., and J. B. Hardaker. 1973. Management decisions and drought, p. 220–224. *In*: J. V. Lovett (ed.), The environment, economics and social significance of drought. Angus and Robertson, London.

Anderson, S. J., T. J. Klopfenstein, and V. A. Wilkerson. 1988. Escape protein supplementation of yearling steers grazing smooth brome pastures. J. Anim. Sci. 66:237–242.

Andrew, M. H. 1988. Grazing impact in relation to livestock watering points. Trends Ecol. Evol. 3:336–339.

Anslow, R. C. 1966. The rate of appearance of leaves on tillers of the Gramineae. Herb. Abst. 36:149–155.

Archer, S. 1989. Have southern Texas savannas been converted to woodlands in recent history?. Am. Nat. 134:545–561.

Archer, S., and J. K. Detling. 1984. The effects of defoliation and competition on regrowth of tillers of two North American mixed-grass prairie graminoids. Oikos 43:351–357.

Archer, S., and J. K. Detling. 1986. Evaluation of potential herbivore mediation of plant water

status in a North American mixed-grass prairie. Oikos 47:287–291.
Archer, S., M. G. Garrett, and J. K. Detling. 1987. Rates of vegetation change associated with prairie dog (*Cynomys ludovicianus*) grazing in North American mixed-grass prairie. Vegetatio 72:159–166.
Archer, S., C. J. Scifres, C. R. Bassham, and R. Maggio. 1988. Autogenic succession in a subtropical savanna: rates, dynamics and processes in the conversion of a grassland to a thorn woodland. Ecol. Monogr. 58:111–127.
Archer, S., and L. L. Tieszen. 1980. Growth and physiological responses of tundra plants to defoliation. Arct. Alp. Res. 12:531–552.
Archer, S., and L. L. Tieszen. 1986. Plant responses to defoliation: hierarchical considerations, p. 45–59. *In*: O. Gudmundsson (ed.), Grazing research at northern latitudes. Plenum, New York.
Arnold, G. W. 1981. Grazing behavior, p. 79–104. *In*: F. H. W. Morley (ed.), Grazing animals. Elsevier, New York.
Arnold, G. W., and M. L. Dudzinski. 1966. The behavioural responses controlling the food intake of grazing sheep, p. 367–371. *In*: A. G.G. Hill (ed.), Proc. 10th Int. Grassl. Congr. Valtioneuvofton Kirjapaino, Helsinke. FINLAND.
Arnold, G. W., and M. L. Dudzinski. 1967. Studies on the diet of the grazing animal. II. The effect of physiological status in ewes and pasture availability on herbage intake. Aust. J. Agr. Res. 18:349–359.
Arnold, G. W., and M. L. Dudzinski. 1978. Ethology of free-ranging domestic animals. Elsevier, New York.
Arnold, J. F. 1955. Plant life-form classification and its use in evaluating range conditions and trend. J. Range Manage. 8:176–181.
Aspinall, D. 1961. The control of tillering in the barley plant. Aust. J. Biol. Sci. 14:493–505.
Atkinson, C. J., and J. F. Farrar. 1983. Allocation of photosynthetically-fixed carbon in *Festuca ovina* L. and *Nardus stricta* L. New Phytol. 95:519–531.
Aucamp, A. J., J. E. Danckwerts, and N. M. Tainton. 1986. Management of integrated grass/browse systems in semi-arid South African savannas, p. 393–394. *In*: P. J. Joss, P. W. Lynch and O. B. Williams (eds.), Rangelands: A resource under seige. Aust. Acad. Sci., Canberra.
Aucamp, A. J., J. E. Danckwerts, W. R. Teague, and J. J. Venter. 1983. The role of *Acacia karroo* in the False Thornveld of the Eastern Cape. Proc. Grassl. Soc. S. Afr. 18:151–154.
Austin, M. P., O. B. Williams, and L. Belbin. 1981. Grassland dynamics under sheep grazing in an Australian mediterranean-type climate. Vegetatio 46/47:201–212.
Azevedo, J., and D. L. Morgan. 1974. Fog precipitation in coastal California forests. J. Ecol. 55:1135–1141.
Baile, C. A., and M. A. Della-Fera. 1981. Nature of hunger and satiety control systems in ruminants. J. Dairy Sci. 64:1140–1152.
Baile, C. A., and J. M. Forbes. 1974. Control of feed intake and regulation of energy balance in ruminants. Physiol. Rev. 54:160–214.
Bailey, D. W., L. R. Rittenhouse, R. H. Hart, and R. W. Richards. 1988. Cattle memory for food resource level and locations. Proc. West. Sec. Am. Soc. Anim. Sci. 39:8–11.
Bailey, J. A. 1984. Principles of wildlife management. John Wiley, New York.
Baker, R. D., F. Alvarez, and L. P. Le Du. 1981. The effect of herbage allowance upon the herbage intake and performance of suckler cows and calves. Grass Forage Sci. 36:189–199.
Barbour, M. G., J. H. Burk and W. D. Pitts. 1987. Terrestrial plant ecology. 2nd ed. Benjamen/Cummings, Menlo Park, CA. USA.
Barth, R. C., and J. O. Klemmedson. 1978. Shrub-induced spatial patterns of dry matter, nitrogen, and organic carbon. Soil Sci. Soc. Am. J. 42:804–809.
Beckmann, G. G., and K. J. Smith. 1974. Micromorphological changes in surface soils following wetting, drying and trampling, p. 832–845. *In*: G. K. Rutherford (ed.), Soil microscopy: Proc. 4th Int. Working-meeting on Soil Micromorphology. Limestone Press, Kingston, Ontario. CAN.
Begon, M., J. L. Harper, and C. R. Townsend 1986. Ecology: individuals, populations and communities. Sinauer, Sunderland, MA. USA.
Belovsky, G. E. 1978. Diet optimization in a generalist herbivore: the moose. Theor. Pop. Biol. 14:105–134.
Belovsky, G. E. 1981a. Food plant selection by a generalist herbivore: the moose. Ecology 62:1020–1030.
Belovsky, G. E. 1981b. Optimal activity times and habitat choice of moose. Oecologia (Berlin) 48:22–30.

Belovsky, G. E. 1984a. Herbivore optimal foraging: A comparative test of three models. Am. Nat. 124:97–115.
Belovsky, G. E. 1984b. Moose and snowshoe hare competition and a mechanistic explanation from foraging theory. Oecologia (Berlin) 61:150–159.
Belovsky, G. E. 1986a. Generalist herbivore foraging and its role in competitive interactions. Am. Zool. 26:51–69.
Belovsky, G. E. 1986b. Optimal foraging and community structure: implications for a guild of generalist grassland herbivores. Oecologia (Berlin) 70:35–52.
Belovesky, G. E., and J. B. Slade. 1986. Time budgets of grassland herbivores: body size similarities. Oecologia (Berlin) 70:53–62.
Belsky, A. J. 1983. Small-scale pattern in grassland communities in the Serengeti National Park, Tanzania. Vegetatio 55:141–155.
Belsky, A. J. 1986. Does herbivory benefit plants? A review of the evidence. Am. Nat. 127:870–892.
Belsky, A. J., R. G. Amundson, J. M. Duxberry, S. J. Richa, A. R. Ali, and S. M. Mwonga. 1989. The effects of trees on their physical, chemical and biological environments in a semi-arid savanna in Kenya. J. Appl. Ecol. 26:1005–1024.
Bennett, H. H. 1939. Soil conservation. McGraw-Hill, New York.
Berg, R. T., and R. M. Butterfield. 1976. New concepts of cattle growth. Sydney Univ. Press, Sydney.
Bertram, B. 1979. Pride of lions. Charles Scribner's Sons, New York.
Bines, J. A., S. Suzuki, and C. C. Balch. 1969. The quantitative significance of long-term regulation of food intake in the cow. Brit. J. Nutr. 23:695–704.
Biondini, M. E., C. D. Bonham, and E. F. Redente. 1985. Secondary successional patterns in a sagebrush (*Artemisia tridentata*) community as they relate to soil disturbance and soil biological activity. Vegetatio 60:25–36.
Biswell, H. A., and J. E. Weaver. 1933. Effect of frequent clipping on the development of roots and tops of grasses in prairie sod. Ecology 14:368–390.
Black, J. L., and P. A. Kenney. 1984. Factors affecting diet selection by sheep II. Height and density of pasture. Aust. J. Agr. Res. 35:565–578.
Black, W. H., A. L. Baker, V. I. Clark, and O. R. Mathews. 1937. Effect of different methods of grazing on native vegetation and gains of steers in Northern Great Plains. USDA Tech. Bull. No. 547.
Blackburn, W. H. 1975. Factors influencing infiltration and sediment production of semiarid rangelands in Nevada. Water Resour. Res. 11:929–937.
Blackburn, W. H. 1983. Influence of brush control on hydrologic characteristics of range watersheds, p. 73–88. *In*: K. C. McDaniel (ed.), Proc. Brush Manage. Symp. Texas Tech Press, Lubbock. USA
Blackburn, W. H. 1984. Impacts of grazing intensity and specialized grazing systems on watershed characteristics and responses, p. 927–983. *In*: Developing strategies for rangeland management. Westview Press, Boulder, CO. USA.
Blackburn, W. H., T. L. Thurow, and C. A. Taylor, Jr. 1986. Soil erosion on rangeland, p. 31–39. *In*: Proc. Use of Cover, Soils and Weather Data in Range. Monitor. Symp. Soc. for Range Manage., Denver, CO. USA.
Blackburn, W. H., and P. T. Tueller. 1970. Pinyon and juniper invasion in black sagebrush communities in east-central Nevada. Ecology 51:841–848.
Blankenship, L. H., and S. A. Overton. 1974. Resource management on a Kenya ranch. J. S. Afr. Wildl. Manage. Assoc. 4:185–190.
Blankenship, L. H., L. W. Varner, T. J. Fillinger, and G. Wampler. 1985. Competition for food among herbivores on Texas rangelands—preliminary report, p. 1030–1055. *In*: I. Harta Sanchez, E. Jardel and L. H. Blankenship (eds.), Proc. 1st Int. Wildl. Symp., Vol. II. SEDUE, Mexico City.
Booysen, P. De V., 1967. Grazing and grazing management terminology in Southern Africa. Proc. Grassl. Soc. S. Afr. 2:45–57.
Booysen, P. De V. and Neil M. Tainton. 1978. Grassland management: principles and practice in South Africa, p. 551–554. *In*: D. N. Hyder (ed.), Proc. 1st Int. Range. Congr. Soc. Range Manage., Denver, CO. USA.
Booysen, P. De V., N. M. Tainton, and J. D. Scott. 1963. Shoot-apex development in grasses and its importance in grassland management. Herb. Abst. 33:209–213.
Botkin, M. J. M., and L. S. Y. Wu. 1981. How ecosystem processes are linked to large mammal population dynamics, p. 373–387. *In*: C. W. Fowler and T. D. Smith (eds.), Dynamics of large mammal populations. John Wiley, New York.

Boyle, M., W. T. Frankenberger, Jr., and L. H. Stolzy. 1989. The influence of organic matter on soil aggregation and water infiltration. J. Prod. Agr. 2:290–299.

Brady, N.C. 1974. The nature and properties of soils. 8th ed. Macmillan, New York.

Branson, F. A. 1953. Two new factors affecting resistance of grasses to grazing. J. Range Manage. 6:165–171.

Branson, F. A. 1985. Vegetation changes on western rangeland. Range Monogr. No. 2. Soc. Range Manage., Denver, CO. USA.

Branson, F., and J. E. Weaver. 1953. Quantitative study of degeneration of mixed prairie. Bot. Gaz. 114:397–416.

Braunack, M. V., and J. Walker. 1985. Recovery of some soil surface properties of ecological interest after sheep grazing in a semi-arid woodland. Aust. J. Ecol. 10:451–460.

Briske, D. D. 1986. Plant response to defoliation: morphological considerations and allocation priorities, p. 425–427. *In*: P. J. Joss, P. W. Lynch and O. B. Williams (eds.), Rangelands: a resource under siege. Aust. Acad. Sci., Canberra.

Briske, D. D., and J. L. Butler. 1989. Density-dependent regulation of ramet populations within the bunchgrass *Schizachyrium scoparium*: Interclonal versus intraclonal interference. J. Ecol. 77:963–974.

Brotherson, J. D., and S. R. Rushforth. 1983. Influence of cryptogamic crusts on moisture relationships of soils in Navajo National Monument, AZ. Great Basin Nat. 43:73–78.

Brown, B. J., and T. F. H. Allen. 1989. The importance of scale in evaluating herbivory impacts. Oikos 54:189–194.

Brown, B. J., and J. J. Ewel. 1988. Response to defoliation of species-rich and monospecific tropical plant communities. Oecologia (Berlin) 75:12–19.

Brown, J. R., and S. Archer. 1987. Woody plant seed dispersal and gap formation in a North American subtropical savanna woodland: the role of domestic herbivores. Vegetatio 72:159–166.

Brown, J. R., and S. Archer. 1989. Woody plant invasion of grasslands: establishment of honey mesquite (*Prosopis glandulosa* var. *glandulosa*) on sites differing in herbaceous biomass and grazing history. Oecologia (Berlin) 80:19–26.

Brown, J. R., and S. Archer. 1990. Water relations of a perennial grass and seedling versus adult woody plants in a subtropical savanna, Texas. Oikos 57:366–374.

Bryant, F. C., B. E. Dahl, R. D. Pettit, and C. M. Britton. 1989. Does short duration grazing work in arid and semiarid regions? J. Soil Water Conserv. 44:290–296.

Bryant, F. C., M. M. Kothamnn, and L. B. Merrill. 1979. Diets of sheep, Angora goats, Spanish goats and white-tailed deer under excellent range conditions. J. Range Manage. 32:412–417.

Bucher, E. H. 1987. Herbivory in arid and semi-arid regions of Argentina. Revista Chilenda de Historia Natural 608:265–273.

Buffington, L. C., and C. H. Herbel. 1965. Vegetational changes on a semidesert grassland range from 1858 to 1963. Ecol. Monogr. 35:139–164.

Burns, J. C. 1978. Symposium: forage quality and animal performance. Antiquality factors as related to forage quality. J. Dairy Sci. 61:1809–1820.

Burroughs, W., D. K. Nelson, and D. R. Mertens. 1975. Evaluation of protein nutrition by metabolizable protein and urea fermentation potential. J. Dairy Sci. 58:611–619.

Busso, C. A., R. J. Mueller, and J. H. Richards. 1989. Effects of drought and defoliation on bud viability in two caespitose grasses. Ann. Bot. 63:477–485.

Butler, J. L., and D. D. Briske. 1988. Population structure and tiller demography of the bunchgrass *Schizachyrium scoparium* in response to herbivory. Oikos 51:306–312.

Butterfield, C. H., and J. W. Stuth. 1991. Landscape constraints on foraging area selection. J. Appl. Anim. Behav. Sci. (in press).

Caldwell, M. M. 1984. Plant requirements for prudent grazing, p. 117–152. *In*: Developing strategies for rangeland management. Westview Press, Boulder, CO. USA.

Caldwell, M. M., T. J. Dean, R. S. Nowak, R. S. Dzurec, and J. H. Richards. 1983. Bunchgrass architecture, light interception, and water-use efficiency: assessment by fiber optic point quadrats and gas exchange. Oecologia 59:178–184.

Caldwell, M. M., J. H. Richards, D. A. Johnson, R. S. Nowak, and R. S. Dzurec. 1981. Coping with herbivory: Photosynthetic capacity and resource allocation in two semiarid *Agropyron* bunchgrasses. Oecologia (Berlin) 50:14–24.

Caldwell, M. M., J. H. Richards, J. H. Manwaring, and D. M. Eissenstat. 1987. Rapid shifts in phosphate acquisition show direct competition between neighbouring plants. Nature 327:615–616.

Canfield, R. H. 1957. Reproduction and life span of some perennial grasses of southern Arizona. J. Range Manage. 10:199–203.
Carlson, D. H., T. L. Thurow, R. W. Knight, and R. K. Heitschmidt. 1990. Effect of honey mesquite on the water balance of Texas Rolling Plains rangeland. J. Range Manage. 43:491–496.
Carman, J. G., and D. D. Briske. 1982. Root initiation and root and leaf elongation of dependent little bluestem tillers following defoliation. Agron. J. 74:432–435.
Carman, J. G., and D. D. Briske. 1985. Morphologic and allozymic variation between long-term grazed and non-grazed populations of the bunchgrass *Schizachyrium scoparium* var. *frequens*. Oecologia (Berlin) 66:332–337.
Casal, J. J., V. A. Deregibus, and R. A. Sanchez. 1985. Variations in tiller dynamics and morphology in *Lolium multiflorum* Lam. vegetative and reproductive plants as affected by differences in red/far-red irradiation. Ann. Bot. 56:553–559.
Casal, J. J., R. A. Sanchez, and V. A. Deregibus. 1986. The effect of plant density on tillering: the involvement of R/FR ratio and the proportion of radiation intercepted per plant. Environ. Exp. Bot. 26:365–371.
Caswell, H., F. Reed, S. N. Stephenson, and P. A. Werner. 1973. Photosynthetic pathways and selective herbivory: a hypothesis. Am. Nat. 107:465–480.
Caton, J. S., A. S. Freeman, and M. L. Galyean. 1988a. Influence of protein supplementation on forage intake, in situ forage disappearance, ruminal fermentation and digesta passage rates in steers grazing dormant blue grama rangeland. J. Anim. Sci. 66:2262–2271.
Caton, J. S., W. C. Hoefler, M. L. Galyean, and M. A. Funk. 1988b. Influence of cottonseed meal supplementation and cecal antibiotic infusion in lambs fed low quality forage. II. Serum urea nitrogen, insulin, somatotropin, free fatty acids and ruminal and cecal fermentation. J. Anim. Sci. 66:2253–2261.
Chacon, E. A., and T. H. Stobbs. 1976. Influence of progressive defoliation of a grass sward on the eating behavior of cattle. Aust. J. Agr. Res. 27:709–727.
Chambers, J. C., and R. W. Brown. 1983. Methods for vegetation sampling and analysis on revegetated mined lands. USDA For. Serv. Gen. Tech. Rep. INT-151. Ogden, UT. USA.
Chapin, F. S. 1980. The mineral nutrition of wild plants. Ann. Rev. Ecol. Syst. 11:233–260.
Chapin, F. S., and G. R. Shaver. 1985. Individualistic growth response of tundra plant species to environmental manipulations in the field. Ecology 66:564–576.
Chapman, D. F., D. A. Clark, C. A. Land, and N. Dymock. 1983. Leaf and tiller growth of *Lolium perenne* and *Agrostis* spp. and leaf appearance rates of *Trifolium repens* in set-stocked and rotationally grazed hill pastures. N.Z.J. Agri. Res. 26:159–168.
Chapman, D. F., D. A. Clark, C. A. Land, and N. Dymock. 1984. Leaf and tiller or stolon death of *Lolium perenne*, *Agrostis* spp., and *Trifolium repens* in set-stocked and rotationally grazed hill pastures. N.Z.J. Agr. Res. 27:303–312.
Charney, J., P. H. Stone, and W. J. Quirk. 1975. Drought in the Sahara: A biophysical feedback mechanism. Science 187:434–435.
Charnov, E. L. 1976. Optimal foraging: the marginal value theorem. Theor. Pop. Biol. 9:129–136.
Chatterton, N.J., P. A. Harrison, J. H. Bennett, and W. R. Thornley. 1987. Fructan, starch and sucrose concentrations in crested wheatgrass and redtop as affected by temperature. Plant Physiol. Biochem. 25:617–623.
Chew, R. M. 1982. Changes in herbaceous and suffrutescent perennials in grazed and ungrazed desertified grassland in southeastern Arizona, 1958–1978. Am. Midl. Nat. 108:159–169.
Chew, R. M., and A. E. Chew. 1965. The primary productivity of a desert shrub (*Larrea tridentata*) community. Ecol. Monogr. 35:355–375.
Christie, E. K. 1975. A study of phosphorus nutrition and water supply on the early growth and survival of buffelgrass grown on a sandy red earth from south-west Queensland. Aust. J. Exp. Agr. Anim. Husb. 15:239–249.
Church, D.C. 1969. Digestive physiology and nutrition of ruminants. Vol. 1. Oregon St. Univ. Press, Corvallis. USA.
Cinguemani, V., J. R. Owensby, Jr., and R. G. Baldwin. 1978. Input data for solar systems. US Dep. Energy Rep., Environ. Resour. Assess. Branch. Interagency Agree. No. E(49–26)–1041.
Clark, F. E. 1977. Internal cycling of [15]nitrogen in shortgrass prairie. Ecology 58:1322–1333.
Clarke, S. E., E. W. Tisdale, and N. A. Skoglund. 1947. The effects of climate and grazing practices on short-grass prairie vegetation in southern Alberta and southwestern Saskatchewan. Ministry Agr. Tech. Bull. No. 46. Saskatoon, Saskatchewan. CAN.

Clary, W. P. 1987. Herbage production and livestock grazing on pinyon-juniper woodlands, p. 440–447. *In*: R. L. Everitt (ed.), Proceedings: pinyon-juniper conference. USDA For. Serv. Gen. Tech. Rep. INT-215. Ogden, UT. USA.

Coffin, D. P., and W. K. Lauenroth. 1988. The effects of disturbance size and frequency on a shortgrass plant community. Ecology 69:1609–1617.

Cohen, W. E., D. L. Drawe, F. C. Bryant, and L. C. Bradley. 1989. Observations on white-tailed deer and habitat response to livestock grazing in South Texas. J. Range Manage. 42:361–365.

Colebrook, W. F., J. L. Black, and P. A. Kenney. 1987. A study of factors influencing diet selection by sheep, p. 85–86. *In*: M. Rose (ed.), Herbivore nutrition research. Aust. Soc. Anim. Prod., Brisbane.

Coleman, D.C., R. Andrews, J. E. Ellis, and J. W. Singh. 1976. Energy flow and partitioning in selected man-managed and natural ecosystems. Agro-Ecosystems 3:45–154.

Coleman, S. W., T. D. A. Forbes, and J. W. Stuth. 1989. Measurements of the plant-animal interface in grazing research, p. 37–52. *In*: G. C. Martin (ed.), Grazing research methods. Am. Soc. Agron., Madison, WI. USA.

Coley, P. D. 1986. Cost and benefits of defense by tannins in a neotropical tree. Oecologia (Berlin) 70:238–241.

Collins, S. L. 1987. Interaction of disturbances in tallgrass prairie: a field experiment. Ecology 68:1243–1250.

Collins, S. L., and D. E. Adams. 1983. Succession in grasslands: thirty-two years of change in a central Oklahoma tallgrass prairie. Vegetatio 51:181–190.

Collins, S. L., and S.C. Barber. 1985. Effects of disturbance on diversity in mixed-grass prairie. Vegetatio 64:87–94.

Collins, S. L., J. A. Bradford, and P. L. Sims. 1987. Succession and fluctuation in *Artemisia* dominated grassland. Vegetatio 73:89–99.

Connell, J. H. 1978. Diversity in tropical rain forests and coral reefs. Science 199:1302–1310.

Conner, J. R. 1984a. Holistic ranch management: how, when, where and what needs to be done, p. 1–10. *In*: L. D. White and D. Guynn (eds.), Proc. 1984 Int. Ranchers Roundup. Texas Agr. Ext. Serv., College Station. USA.

Conner, J. R. 1984b. Why do you own a ranch?, p. 39–48. *In*: L. D. White and D. Guyan (eds.), Proc. 1984 Int. Ranchers Roundup. Texas Agr. Ext. Serv., College Station. USA.

Conner, J. R. 1985. Range improvements: the decision process, p. 174–185. *In*: L. D. White and T. R. Troxel (eds.), Proc. 1985 Int. Ranchers Roundup. Texas Agr. Ext. Serv., College Station. USA.

Conner, J. R., and P. J. Chamberlain. 1985. Profitability analysis of ranch investments, p. 107–117. *In*: L. D. White and T. R. Troxel (eds.), Proc. 1985 Int. Ranchers Roundup. Texas Agr. Ext. Serv., College Station. USA.

Connolly, J. 1976. Some comments on the shape of the gain-stocking rate curve. J. Agr. Sci. 86:103–109.

Conrad, H. R., A. D. Pratt, and J. W. Hibbs. 1964. Regulation of feed intake in dairy cows. I. Change in importance of physical and physiological factors with increasing digestibility. J. Dairy Sci. 47:54–62.

Conway, A. 1963. Effect of grazing management on beef production. Irish J. Agr. Res. 2: 243–258.

Cook, C. W. 1970. Energy budget of the range and range livestock. Colorado Agr. Exp. Sta. Bull. TB109. Ft. Collings. USA.

Cook, C. W. 1971. Comparative nutritive values of forbs, grasses and shrubs, p. 303–310. *In*: C. M. McKell, J. P. Blaisdell and J. R. Goodin (eds.), Wildland shrubs: their biology and utilization. USDA For. Serv. Gen. Tech. Rep INT-1. Ogden, UT. USA.

Cook, C. W., and L. E. Harris. 1968. Effect of supplementation on intake and digestibility of range forage. Utah Agr. Exp. Sta. Bull. 475. Logan. USA.

Cook, C. W., and L. A. Stoddart. 1953. Some growth responses of crested wheatgrass following herbage removal. J. Range Manage. 6:267–270.

Cook, C. W., L. A. Stoddart, and F. E. Kinsinger. 1958. Responses of crested wheatgrass to various clipping treatments. Ecol. Monogr. 28:237–272.

Cook, C. W., and J. Stubbendieck. 1986. Range research: basic problems and techniques. Soc. Range Manage., Denver, CO. USA.

Cooper, H. W. 1953. Amounts of big sagebrush in plant communities near Tensleep, Wyoming as affected by grazing treatment. Ecology 34:186–189.

Cooper, S. M., and N. Owen-Smith. 1986. Effects of plant spinescence on large mammalian herbivores. Oecologia (Berlin) 68:446–455.

Coppock, D. L., J. E. Ellis, J. K. Detling, and M. I. Dyer. 1983. Plant-herbivore interactions in a North American mixed-grass prairie. II. Responses of bison to modification of vegetation by prairie dogs. Oecologia (Berlin) 56:10–15.

Costello, D. F. 1964. Range dynamics control—an ecological urgency, p. 91–107. *In*: D. J. Crisp (ed.), Grazing in terrestrial and marine environmentals. Blackwell, Oxford. ENG.

Cowan, R. T., J. J. Robinson, I. McDonald, and R. Smart. 1980. Effects of body fatness on lambing and diet in lactation on body tissue loss, feed intake and milk yield of ewes in early lactation. J. Agr. Sci. 95:497–514.

Crawley, M. J. 1983. Herbivory—The dynamics of animal-plant interactions. Univ. California Press, Berkeley. USA.

Crider, F. J. 1955. Root-growth stoppage resulting from defoliation of grass. USDA Tech. Bull. 1102.

Dahl, B. E., and D. N. Hyder. 1977. Development morphology and management implications, p. 258–290. *In*: R. E. Sosebee (ed.), Rangeland plant physiology. Soc. Range Manage., Denver, CO. USA.

Danckwerts, J. E., and A. J. Aucamp. 1986. The effect of range condition on the grazing capacity of semi-arid South African savanna, p. 229–230. *In*: P. J. Joss, P. W. Lynch and O. B. Williams (eds.), Rangeland: a resource under seige. Aust. Acad. Sci., Canberra.

Dasmann, R. F. 1981. Wildlife biology. John Wiley, New York.

Davidson, J. L., and F. L. Milthorpe. 1966. Leaf growth in *Dactylis glomerata* following defoliation. Ann. Bot. 30:173–184.

Davis, J. H., and C. D. Bonham. 1979. Interference of sand sagebrush canopy with needle and thread. J. Range Manage. 32:384–386.

Davis, M. B. 1982. Quaternary history and the stability of forest communities, p. 132–153. *In*: D.C. West, H. H. Shugart and D. D. Botkin (eds.), Forest succession: concepts and application. Springer-Verlag, New York.

Day, T. A., and J. K. Detling. 1990. Grassland patch dynamics and herbivore grazing preference following urine deposition. Ecology 71:180–189.

Dean, R., J. E. Ellis, R. W. Rice, and R. E. Bement. 1975. Nutrient removal by cattle from a shortgrass prairie. J. Appl. Ecol. 12:25–29.

Demment, M. W. 1982. The scaling of ruminoreticulum size with body weight in East African ungulates. Afr. J. Ecol. 20:43–47.

Demment, M. W., and P. J. Van Soest. 1981. Body size, digestive capacity, and feeding strategies of herbivores. Winrock Int. Livestock Res. Publ. Morrilton, AR. USA.

Demment, M. W., and P. J. Van Soest. 1985. A nutritional explanation for body-size patterns of ruminant and non-ruminant herbivores. Am. Nat. 125:641–672.

Denniston, R. H., II. 1956. Ecology, behavioral and population dynamics of the Wyoming or Rocky Mountain moose, *Alces alces shirasi*. Zoologica 41:105–118.

Deregibus, V. A., R. A. Sanchez, J. J. Casal, and M. J. Trlica. 1985. Tillering responses to enrichment of red light beneath the canopy in a humid natural grassland. J. Appl. Ecol. 22:199–206.

Deregibus, V. A., M. J. Trlica, and D. A. Jameson. 1982. Organic reserves in herbage plants: Their relationship to grassland management, p. 315–344. *In*: M. J. Recheigl (ed.), Handbook of agricultural productivity. Vol. I. Plant productivity. CRC Press, Bocona, FL. USA.

Detling, J. K. 1988. Grasslands and savannas: Regulation of energy flow and nutrient cycling by herbivores, p. 131–148. *In*: L. R. Pomeroy and J. J. Alberts (eds.), Concepts of ecosystem ecology. Springer-Verlag, New York.

Detling, J. K., M. I. Dyer, and D. T. Winn. 1979. Net photosynthesis, root respiration, and regrowth of *Bouteloua gracilis* following simulated grazing. Oecologia (Berlin) 41:127–134.

Detling, J. K., and E. L. Painter. 1983. Defoliation responses of western wheatgrass populations with diverse histories of prairie dog grazing. Oecologia (Berlin) 57:65–71.

Detling, J. K., D. J. Winn, C. Proctor-Gregg, and E. L. Painter. 1980. Effects of simulated grazing by below-ground herbivores on growth, CO_2 exchange, and carbon allocation patterns of *Bouteloua gracilis*. J. Appl. Ecol. 17:771–778.

Dirzo, R., and J. L. Harper. 1982. Experimental studies on slug-plant interaction. IV. The performance of cyanogenic and acyanogenic morphs of *Trifolium repens* in the field. J. Ecol. 70:119–138.

Doran, J. W., and D. M. Linn. 1979. Bacteriological quality of runoff water from pastureland. Appl. Environ. Microbiol. 37:985–991.

Douglass, J. E. 1983. The potential for water yield augmentation for forest management in the Eastern United States. Water Resour. Bull. 14:351–358.

Dregne, H. E. 1987. Reflections on the U.N. plan of action to combat desertification. Desert. Cont. Bull. 15:8–11.

Driscoll, R. S. 1967. Managing public rangelands: effective livestock, grazing practices and systems for national forests and national grasslands. USDA For. Serv. Agr. Info. Bull. 315.

Duft, K. D. 1979. Principles of management in agribusiness. Prentice-Hall, Reston, VA. USA.

Dyer, M. I., and U. G. Bokhari. 1976. Plant-animal interactions: studies of the effects of grasshopper grazing on blue grama grass. Ecology 57:762–772.

Dyer, M. I., J. K. Detling, D.C. Coleman, and D. W. Hilbert. 1982. The role of herbivores in grasslands, p. 255–295. *In*: J. R. Estes, R. J. Tyrl and J. N. Brunken (eds.), Grasses and grasslands: systematics and ecology. Univ. Oklahoma Press, Norman. USA.

Dyksterhuis, E. J. 1949. Condition and management of rangeland based on quantitative ecology. J. Range Manage. 2:104–115.

Edwards, P. J. 1981. Grazing management, p. 323–354. *In*: N. M. Tainton (ed.), Veld and pasture management in South Africa. Shuter and Shooter, Pietermaritzburg. S. AFR.

Eissenstat, D. M., and M. M. Caldwell. 1988. Competitive ability is linked to rates of water extraction: a field study of two arid land tussock grasses. Oecologia (Berlin) 75:1–7.

Ellis, J. E., J. A. Wiens, C. F. Rodell, and J. C. Anway. 1976. A conceptual model of diet selection as an ecosystem process. J. Theor. Biol. 60:93–108.

Ellison, L. 1960. Influence of grazing on plant succession of rangelands. Bot. Rev. 26:1–78.

Emmanuel, W. R., H. H. Shugart, and M. Stevenson. 1985a. Climatic change and the broad-scale distribution of terrestrial ecosystem complexes. Clim. Change 7:29–43.

Emmanuel, W. R., H. H. Shugart, and M. Stevenson. 1985b. Response to comment: climatic change and the broad-scale distribution of terrestrial ecosystem complexes. Clim. Change 7:457–460.

Esau, K. 1960. Anatomy of seed plants. John Wiley, New York.

Etter, A. G. 1951. How Kentucky bluegrass grows. Ann. Missouri Bot. Gard. 38:293–375.

Evans, P. S. 1973. The effect of repeated defoliation to three different levels on root growth of five pasture species. N.Z.J. Agr. Res. 16:31–34.

Ferguson, C. E. 1966. Microeconomic theory. Richard D. Irvin, Homewood, IL. USA.

Ferrell, C. L., and T. G. Jenkins. 1987. Influence of biological types on energy requirements, p. 1–7. *In*: Proc. Grazing Livestock Nutr. Conf. Univ. Wyoming, Laramie. USA.

Field, C., and H. A. Mooney. 1986. The photosynthesis—nitrogen relationship in wild plants, p. 25–55. *In*: T. J. Givnish (ed.), On the economy of plant form and function. Cambridge Univ. Press, Cambridge, MA. USA.

Finch, V. A. 1984. Heat stress as a factor in herbivores under tropical conditions, p. 81–105. *In*: F. M. C. Gilchrist and R. I. Mackie (eds.), Herbivore nutrition in the subtropics and tropics. Science Press, Craighall. S. AFR.

Floate, M. J. S. 1981. Effects of grazing by large herbivores on nitrogen cycling in agricultural ecosystems. *In*: F. E. Clark and T. Rosswall (eds.), Terrestrial nitrogen cycles. Ecol. Bull. 33:585–601.

Fogden, M. P. L. 1978. The impact of lagomorphs and rodents on the cattle rangelands of northern Mexico. Report, 1973–1977; Centre for Overseas Pest Research, College House, London.

Foran, B. D. 1986. The impact of rabbits and cattle on an arid calcareous shrubby grassland in central Australia. Vegetatio 66:49–59.

Forbes, J. M. 1971. Physiological changes affecting voluntary food intake in ruminants. Proc. Nutr. Soc. 30:135–142.

Forbes, J. M. 1980. Hormones and metabolites in the control of food intake, p. 145–160. *In*: V. Ruckebusch and P. Thivend (eds.), Digestive physiology and metabolism in ruminants. AVI, Westport, CT. USA.

Forbes, T. D. A., and S. W. Coleman. 1987. Herbage intake and ingestive behavior of grazing cattle as influenced by variation in sward characteristics, p. 141–152. *In*: Floyd P. Horn, John Hodgson, John J. Mott and Ray W. Broughman (eds.), Grazing-lands Research at the Plant-Animal Interface Symp. Winrock Int., Morrilton, AR. USA.

Forbes, T. D. A., E. M. Smith, R. B. Razor, C. J. Dougherty, V. G. Allen, L. L. Erlinger, J. E. Moore, and F. M. Rouquette, Jr. 1985. The plant-animal interface, p. 95–116. *In*: Watson, V. H. and C. M. Wells, Jr. (eds.), Simulation of forage and beef production in the Southern Region. USDA South. Coop. Ser. Bull. 308.

Fox, D. G. 1987. Physiological factors influencing voluntary intake by beef cattle, p. 193–207. *In*: F. N. Owens (ed.), Feed Intake by Beef Cattle Symp. Oklahoma St. Univ., Stillwater. USA.

Fox, D. G., C. J. Sniffen, and J. D. O'Connor. 1988. Adjusting nutrient requirements of beef cattle for animal and environmental variations. J. Anim. Sci. 66:1475–1495.

Freer, M. 1981. The control of food intake by grazing animals, p. 105–124. *In*: F. H. W. Morley (ed.), Grazing animals. Elsevier, New York.

Frischknecht, N.C. 1963. Contrasting effects of big sagebrush and rubber rabbitbrush on production of crested wheatgrass. J. Range Manage. 16:70–74.

Fulbright, J. E., and S. L. Beasom. 1987. Long-term effects of mechanical treatment on white-tailed deer browse. Wildl. Soc. Bull. 15:560–564.

Gade, A. E., and F. D. Provenza. 1986. Nutrition of sheep grazing crested wheatgrass versus crested wheatgrass—shrub pastures during winter. J. Range Manage. 39:527–530.

Gallizioli, S. 1977. Overgrazing on desert bighorn ranges. Trans. Desert Bighorn Counc. 21:21–23.

Gammon, D. M. 1978. A review of experiments comparing systems of grazing management on natural pastures. Proc. Grassl. Soc. S. Afr. 13:75–82.

Gammon, D. M. 1984. An appraisal of short duration grazing as a method of veld management. Zimbabwe Agr. J. 81:59–64.

Gammon, D. M., and R. B. Roberts. 1978. Patterns of defoliation during continuous and rotational grazing of the Matopos Sandveld in Rhodesia. 1. Selectivity of grazing. Rhod. J. Agr. Res. 16:117–131.

Gamougoun, N. D., R. P. Smith, M. K. Wood, and R. D. Pieper. 1984. Soil, vegetation and hydrologic responses to grazing management at Fort Stanton, New Mexico. J. Range Manage. 37:538–541.

Gatsuk, L. E., O. V. Smirnova, L. I. Vorontzova, L. B. Zaugolnova, and L. A. Zhukova. 1980. Age states of plants of various growth forms: a review. J. Ecol. 68:675–696.

Gifford, R. M., and C. Marshall. 1973. Photosynthesis and assimilate distribution in *Lolium multiflorum* Lam. following differential tiller defoliation. Aust. J. Biol. Sci. 26:517–526.

Glenn-Lewin, D.C. 1980. The individualistic nature of plant community development. Vegetatio 43:141–146.

Glover, M. D., and J. R. Conner. 1988. A model for selecting optimal combinations of livestock and deer lease-hunting enterprises. Wildl. Soc. Bull. 16:158–163.

Golley, F. B. 1960. Energy dynamics of a food chain of an old-field community. Ecol. Monogr. 30:187–206.

Golley, F. B. 1961. Energy values of ecological materials. Ecology 42:581–584.

Goodloe, S. 1969. Short duration grazing in Rhodesia. J. Range Manage. 22:369–373.

Gosz, J. R., R. T. Holmes, G. E. Likens, and F. H. Bormann. 1978. The flow of energy in a forest ecosystem. Sci. Am. 238:93–102.

Gould, F. W. 1975. Texas Plants: A Checklist and Ecological Summary.

Greene, L. W., W. E. Pinchak, and R. K. Heitschmidt. 1987. Seasonal dynamics of minerals in forages at the Texas Experimental Ranch. J. Range Manage. 40:502–506.

Greenwell, M. 1988. Knowledge engineering for expert systems. Ellis Horwood, Chichester. ENG.

Grovum, W. L. 1986. A new look at what is controlling food intake, p. 1–40. *In*: F. N. Owens (ed.), Proc. Feed Intake by Beef Cattle Symp., Oklahoma St. Univ., Stillwater. USA.

Guthery, F. S. 1986. Beef, brush and bobwhites: quail management in cattle country. Caesar Kleberg Wildl. Res. Inst. Press, Kingsville, TX. USA.

Hakkila, M. D., J. L. Holechek, J. D. Wallace, D. M. Anderson, and M. Cardenas. 1987. Diet and forage intake of cattle on desert grassland range. J. Range Manage. 40:339–341.

Handl, W. P., and L. R. Rittenhouse. 1972. Herbage yield and intake of steers. Proc. West. Sec. Am. Soc. Anim. Sci. 23:197–200.

Hanley, T. A. 1982. The nutritional basis for food selection by ungulates. J. Range Manage. 35:146–151.

Hanley, T. A., and K. A. Hanley. 1982. Food resource partitioning by sympatric ungulates on Great Basin rangelands. J. Range Manage. 35:152–158.

Harberd, D. J. 1962. Some observations on natural clones in *Festuca ovina*. New Phytol. 61:85–100.

Harper, J. L. 1977. The population biology of plants. Academic Press, New York.

Harrington, G. N. 1973. Brush control: a note of caution. East Afr. Agr. For. J. 39:95–96.

Harrington, G. N., A. D. Wilson, and M. D. Young (eds.). 1984. Management of Australia's rangelands. CSIRO. Melbourne.

Hart, R. H. 1978. Stocking rate theory and its application to grazing on rangeland, p. 547–550. *In*: D. N. Hyder (ed.), Proc. 1st Int. Range. Congr. Soc. Range Manage., Denver, CO. USA.

Hart, R. H., and E. F. Balla. 1982. Forage production and removal from western and crested wheatgrass under grazing. J. Range Manage. 35:362–366.

Hart, R. H., and B. E. Norton. 1988. Grazing management and vegetation response, p. 493–525. *In*: P. T. Tueller, (ed.) Vegetation science applications for rangeland analysis and management. Kluwer, Dordrecht. NETHERLANDS.

Hart, R. H., M. J. Samuel, and J. Bissio. 1988. Role of rotation and distribution in short-duration grazing on range, p. 291. (Abstr.) Ann. Meet. Soc. Range Manage., Denver, CO. USA.

Hastings, J. R., and R. M. Turner. 1965. The changing mile: an ecological study of vegetation change with time in the lower mile of an arid and semi-arid region. Univ. Arizona Press, Tucson. USA.

Havstad, K. M., and D. E. Doornbos. 1987. Effects of biological type on grazing behavior and energy intake, p. 9–15. *In*: Grazing Livestock Nutr. Conf. Univ. Wyoming, Laramie. USA.

Havstad, K. M., and J. C. Malechek. 1982. Energy expenditure by heifers grazing crested wheatgrass of diminishing availability. J. Range Manage. 35:447–450.

Havstad, K. M., A. S. Nastis, and J. C. Malechek. 1983. The voluntary forage intake of heifers grazing a diminishing supply of crested wheatgrass. J. Anim. Sci. 56:259–263.

Hawksworth, D. L. 1971. Lichens as litmus for air pollution: a historical review. Int. J. Environ. Studies. 1:281–296.

Hays, R. L., C. Summers, and W. Seitz. 1981. Estimating wildlife habitat variables. USDA Fish Wildl. Serv. FWS/OBS-81/47.

Heady, H. F. 1961. Continuous vs. specialized grazing systems: a review and application to the California annual type. J. Range Manage. 14:182–193.

Heady, H. F. 1964. Palatability of herbage and animal preference. J. Range Manage. 17:76–82.

Hegarty, M. P. 1982. Deleterious factors in forages affecting animal production, p. 133–150. *In*: J. B. Hacker (ed.), Nutritional limits to animal production from pastures. Commonwealth Agr. Bur., Farnham Royal. UK.

Heitschmidt, R. K. 1990. The role of livestock and other herbivores in improving rangeland vegetation. Rangelands 12:112–115.

Heitschmidt, R. K., J. R. Conner, S. K. Canon, W. E. Pinchak, J. W. Walker, and S. L. Dowhower. 1990. Cow/calf production and economic returns from yearlong continuous, deferred rotation and rotational grazing treatments. J. Prod. Agr. 3:92–99.

Heitschmidt, R. K., S. L. Dowhower, R. A. Gordon, and D. L. Price. 1985. Response of vegetation to livestock grazing at the Texas Experimental Ranch. Texas Agr. Exp. Sta. B-1515. College Station. USA.

Heitschmidt, R. K., S. L. Dowhower, and J. W. Walker. 1987a. 14-vs. 42-paddock rotational grazing: above-ground biomass dynamics, forage production and harvest efficiency. J. Range Manage. 40:216–223.

Heitschmidt, R. K., S. L. Dowhower, and J. W. Walker. 1987b. 14-vs. 42-paddock rotational grazing: forage quality. J. Range Manage. 40:315–317.

Heitschmidt, R. K., S. L. Dowhower, and J. W. Walker. 1987c. Some effects of a rotational grazing treatment on quantity and quality of available forage and amount of ground litter. J. Range Manage. 40:318–321.

Heitschmidt, R. K., D. L. Price, R. A. Gordon, and J. R. Frasure. 1982. Short duration grazing at the Texas experimental ranch: effects on above-ground net primary production and seasonal growth dynamics. J. Range Manage. 35:367–372.

Heitschmidt, R. K., R. D. Schultz, and C. J. Scifres. 1986. Herbaceous biomass dynamics and net primary production following chemical control of honey mesquite. J. Range Manage. 39:67–71.

Heitschmidt, R. K., and J. W. Walker. 1982. Short duration grazing and the Savory grazing method in perspective. Rangelands 5:145–150.

Henke, S. E., S. Demarais, and J. A. Pfister. 1988. Digestive capacity and diets of white-tailed deer and exotic ruminants. J. Wildl. Manage. 52:595–598.

Herbel, C. H. 1974. A review of research related to development of grazing systems on native ranges of the western United States, p. 138–149. *In*: K. W. Kreitlow and R. H. Hart (Coord.), Plant morphogensis for scientific management of range resources. USDA Misc. Pub. 1271.

Herbel, C. H. 1979. Utilization of grass- and shrublands of the southwestern United States, p. 161–203. *In*: B. H. Walker (ed.), Management of semi-arid ecosystems, Elsevier, New York.

Herbel, C. H., and K. L. Anderson. 1959. Response of true prairie vegetation on major Flint Hills range sites to grazing treatments. Ecol. Monogr. 29:171–186.

Herbel, C. H., F. N. Ares, and R. A. Wright. 1972. Drought effects on a semidesert grassland range. Ecology 53:1084–1093.

Hibbert, A. R. 1983. Water yield improvement potential by vegetation management on western rangelands. Water Resour. Bull. 19:375–381.

Hickey, W. C. Jr. 1961. Growth form of crested wheatgrass as affected by site and grazing. Ecology 42:173–176.

Hilder, E. J. 1964. The distribution of plant nutrients by sheep at pasture. Proc. Aust. Soc. Anim. Prod. 5:241–248.

Hillman, J. R. 1984. Apical dominance, p. 127–148. In: M. B. Wilkins (ed.), Advanced plant physiology. Pitman, Marshfield, MA. USA.

Hitchcock, A. S. 1971. Manual of the Grasses of the United States. 2nd Edition. Two Volumes. Dover Publications, Inc., New York.

Hixon, M. A. 1982. Energy maximizers and time minimizers: theory and reality. Am. Nat. 119:596–599.

Hobbs, R. J., and H. A. Mooney. 1986. Community changes following shrub invasion of grassland. Oecologia (Berlin) 70:508–513.

Hodgkinson, K. 1976. The effects of frequency and extent of defoliation, summer irrigation, and fertilizer on the production and survival of the grass *Danthonia caespitosa* (Gaud.). Aust. J. Agr. Res. 27:755–767.

Hodgkinson, K. C., and H. G. Baas Becking. 1977. Effect of defoliation on root growth of some arid zone perennial plants. Aust. J. Agr. Res. 29:31–42.

Hodgkinson, K. C., N. G. Smith, and G. E. Miles. 1972. The photosynthetic capacity of stubble leaves and their contribution to growth of the lucerne plant after high level cutting. Aust. J. Agr. Res. 23:225–238.

Hodgson, J. 1976. Influence of grazing pressure and stocking rate on herbage intake and animal performance, p. 93–103. In: J. Hodgson and D. K. Jackson (eds.), Proc. Occasional Symp. No. 8. Brit. Grassl. Soc.

Hodgson, J. 1977. Factors limiting herbage intake by the grazing animal, p.70–75. In: Proc. Int. Meet. Anim. Prod. Temperate Grassl. Dublin. Ireland

Hodgson, J. 1979. Nomenclature and definitions in grazing studies. Grass Forage Sci. 34:11–18.

Hofmann, R. R. 1989. Evolutionary steps of ecophysiological adaptation and diversification of ruminants: a comparative view of their digestive system. Oecologia (Berlin) 78:443–457.

Hofmann, R. R., and D. R. M. Stewart. 1972. Grazer or browser: A classification based on the stomach structure and feeding habits of East African ruminants. Mammalia. 36:226–240.

Hogan, J. P., R. H. Weston, and J. R. Lindsay. 1969. The digestion of pasture plants by sheep. IV. The digestion of *phalaris tuberosa* at different stages of maturity. Aust. J. Agr. Res. 20:925–940.

Holechek, J. L., R. D. Pieper, and C. H. Herbel. 1989. Range management principles and practices. Prentice Hall, Englewood Cliffs, NJ. USA.

Holechek, J. L., and T. Stephenson. 1983. Comparison of big sagebrush vegetation in northcentral New Mexico under moderately grazed and grazing excluded conditions. J. Range Manage. 36:455–456.

Holechek, J. L., and M. Vavra. 1982. Forage intake by cattle on forest and grassland ranges. J. Range Manage. 35:737–740.

Holechek, J. L., and M. Vavra. 1983. Drought effects on diet and weight gains of yearling heifers in Northeastern Oregon. J. Range Manage. 36:227–231.

Holechek, J. L., M. Vavra, and D. Arthun. 1982. Relationship between performance, intake, diet nutritive quality and fecal nutritive quality of cattle on mountain range. J. Range Manage. 35:741–744.

Holland, E. A., and J. K. Detling. 1990. Plant response to herbivory and below-ground nitrogen cycling. Ecology 71:1040–1049.

Holloway, J. W., and W. T. Butts. 1983. Patterns of forage intake, milk yield, calf growth and efficiency of Angus cow-calf pairs grazing fescue-legume or fescue pastures. Tennessee Agr. Exp. Sta. Bull. 621. Knoxville. USA.

Holmgren, A. H., and J. L. Reveal. 1966. Check List of the Vascular Plants of the Intermountain Region. U.S. Forest Service Research Paper Int.-32.

Hood, R. E., and J. M. Inglis. 1974. Behavioral responses of white-tailed deer to intensive ranching operations. J. Wildl. Manage. 38:488–498.

Horn, F. P., J. P. Telford, J. E. McCroskey, D. F. Stephens, J. V. Whiteman, and R. Totusek. 1979. Relationship of animal performance and dry matter intake to chemical constituents of grazed forage. J. Anim. Sci. 49:1051–1058.

Horner, J. M., and J. E. R. Staddon. 1987. Probalistic choice: simple invariance. Behav. Proc. 15:59–92.

Hudson, N. 1981. Soil conservation. Cornell Univ. Press. Ithaca, New York. USA.

Hume, I. D., and A. C. I. Warner. 1980. Evolution of microbial digestion in mammals, p. 665–

684. *In*: Y. Ruckebusch and P. Thivend (eds.), Digestive physiology and metabolism in ruminants. AVI, Westport, CT. USA.

Humphrey, R. R. 1949. Field comments on the range condition method of forage survey. J. Range Manage. 2:1–10.

Hungate, R. E. 1966. The rumen and its microbes. Academic Press, New York.

Hungate, R. E., G. D. Phillips, A. McGregor, D. P. Hungate, and H. K. Buechner. 1959. Microbial fermentation in certain animals. Science 130:1192–1194.

Hunter, R. A., and B. D. Siebert. 1985a. Utilization of low-quality roughage by *Bos taurus* and *Bos indicus* cattle. I. Rumen digestion. Brit. J. Nutr. 53:637–648.

Hunter, R. A., and B. D. Siebert. 1985b. Utilization of low-quality roughage by *Bos taurus* and *Bos indicus* cattle. II. The effect of rumen-degradable nitrogen and sulphur on voluntary food intake and rumen characteristics. Brit. J. Nutr. 53:649–656.

Huntly, N., and R. Inouye. 1988. Pocket gophers in ecosystems: patterns and mechanisms. BioScience 38:786–793.

Hunziker, J. H., R. A. Palacois, L. Poggio, and C. A. Naranjo. 1977. Geographic distribution, morphology, hybridization, crytogenetics and evolution, p. 10–47. *In*: T. J. Mabry, J. H. Hunziker and D. R. DiFeo (eds.), Creosotebush: the biology and chemistry of *Larrea* in New World deserts. Dowden, Hutchinson and Ross, Stroudsburg, PA. USA.

Huston, J. E. 1978. Forage utilization and nutrient requirements of the goat. J. Dairy Sci. 61:988–993.

Huston, J. E., B. S. Engdahl, and K. W. Bales. 1988. Intake and digestibility in sheep and goats fed three forages with different levels of supplemental protein. Small Rumin. Res. 1:81–92.

Huston, J. E., B. S. Rector, W. C. Ellis, and M. L. Allen. 1986. Dynamics of digestion in cattle, sheep, goats and deer. J. Anim. Sci. 62:208–215.

Huston, J. E., B. S. Rector, L. B. Merrill, and B. S. Engdahl. 1981. Nutritional value of range plants in the Edwards Plateau region of Texas. Texas Agr. Exp. Sta. Bull. B-1357, College Station. USA.

Huston, M. 1979. A general hypothesis of species diversity. Am. Nat. 113:81–101.

Hyder, D. N. 1972. Defoliation in relation to vegetative growth, p. 302–317. *In*: V. B. Youngner and C. M. McKell (eds.), The biology and utilization of grasses. Academic Press, New York.

Hyder, D. N. 1974. Morphogenesis and management of perennial grasses in the U.S., p. 89–98. *In*: Plant morphogenesis as the basis for scientific management of range resources. USDA Misc. Publ. 1271.

Ingham, E. R., D. A. Klein, and M. J. Trlica. 1985. Responses of microbial components of the rhizosphere to plant management strategies in a semiarid rangeland. Plant Soil 85:65–76.

Inglis, J. M. 1985. Wildlife management and IBMS. Texas Agr. Exp. Sta. Bull. B-1493. College Station. USA.

Inglis, J. M., B. A. Brown, C. A. McMahan, and R. E. Hood. 1986. Deer-brush relationships on the Rio Grande Plain, Texas. Kleberg Studies Nat. Res., Texas Agr. Exp. Sta. Bull. RM14/KS6. College Station. USA.

Ingraham, N. L., and R. A. Matthews. 1988. Fog drip as a source of groundwater recharge in northern Kenya. Water Resour. Res. 24:1406–1410.

Jaksic, F. M., and E. R. Fuentes. 1980. Why are native herbs in the Chilean matorral more abundant beneath bushes: microclimate or grazing? J. Ecol. 68:665–669.

Jameson, D. A. 1963. Responses of individual plants to harvesting. Bot. Rev. 29:532–594.

Jameson, D. A. 1967. The relationship of tree overstory and herbaceous understory vegetation. J. Range Manage. 20:247–249.

Janis, C. 1976. The evolutionary strategy of the Equidae and the origins of rumen and cecal digestion. Evolution 30:757–774.

Jewiss, O. R. 1972. Tillering in grasses-its significance and control. J. Brit. Grassl. Soc. 27:65–82.

Johansen, J. R. 1986. Importance of cryptogamic soil crusts to arid rangelands: implications for short duration grazing, p. 127–136. *In*: J. A. Tiedeman (ed.), Proc. Short Duration Grazing Current Issues Grazing Manage. Short-Course. Washington St. Univ., Pullman. USA.

Johnsen, T. N. 1962. One-seed juniper invasion of northern Arizona grasslands. Ecol. Monogr. 32:187–207.

Johnson, K. L. 1987. Sagebrush over time: a photographic study of rangeland change, p. 223–252. *In*: E. D. McArthur and B. L. Welch (eds.), Proc.—Symp. on the Biology of *Artemisia* and *Chrysothamnus*. USDA For. Serv. Gen. Tech. Rep. INT-200. Ogden, UT. USA.

Johnston, M. C. 1963. Past and present grasslands of southern Texas and northeastern Mexico. Ecology 44:456–466.

Jones, A. L., A. L. Goetsch, S. R. Stokes, and M. Colberg. 1988. Intake and digestion in cattle fed

warm- or cool-season grass hay with or without supplemental grain. J. Anim. Sci. 66:194–203.
Jones, M. B., B. Collett, and S. Brown. 1982. Sward growth under cutting and continuous stocking managements: sward canopy structure, tiller density and leaf turnover. Grass Forage Sci. 37:67–73.
Jones, R. M., and J. J. Mott. 1980. Population dynamics in grazed pastures. Trop. Grassl. 14:218–224.
Jordan, W. A., E. E. Lister, J. M. Mauthy, and J. E. Comeau. 1973. Voluntary roughage intake by nonpregnant and pregnant or lactating beef cows. Can. J. Anim. Sci. 53:733–738.
Judkins, M. B., J. D. Wallace, M. L. Galyean, L. J. Krysl, and E. E. Parker. 1987. Passage rates, rumen fermentation, and weight change in protein supplemented grazing cattle. J. Range Manage. 40:100–104.
Kako, Y., and H. Toyoda. 1981. Soil conservation of sloping grassland. V. The infiltration capacity of water on sloping grassland. Bull. Nat. Grassl. Res. Inst. 18:127–136.
Kartchner, R. J. 1980. Effects of protein and energy supplementation of cows grazing native winter range forage on intake and digestibility. J. Anim. Sci. 51:432–438.
Kasperbauer, M. J., and D. L. Karlen. 1986. Light-mediated bioregulation of tillering and photosynthate partitioning in wheat. Physiol. Plant 66:159–163.
Kay, R. D. 1981. Farm management—planning, control and implementation. McGraw-Hill, New York.
Kelly, R. D., and B. H. Walker. 1976. The effects of different forms of land use on the ecology of a semi-arid region of southeastern Rhodesia J. Ecol. 64:553–576.
Kemper, W. D., and R. C. Rosenau. 1986. Aggregate stability and size distribution, p. 425–461. In: A. Klute (ed.), Methods of soil analysis, Part 1, 2nd ed. Agron. Monogr. No. 9. Am. Soc. Agron. Madison, WI. USA.
Kempton, T. J., and R. A. Leng. 1979. Protein nutrition of growing lambs. 1. Responses in growth and rumen function to supplementation of a low-protein-cellulosic diet with either urea, casein or formaldehyde-treated casein. Brit. J. Nutr. 42:289–302.
Kenney, P. A., and J. L. Black. 1984. Factors affecting diet selection by sheep. I. Potential intake rate and acceptability of feed. Aust. J. Agr. Res. 35:551–563.
Kie, J. G., and J. W. Thomas. 1988. Rangeland vegetation as wildlife habitat, p. 585–607. In: P. T. Tueller (ed.), Handbook of vegetation science. Kluwer, Boston.
King, R. T. 1966. Wildlife and man. N.Y. Conserv. 20:8–11.
Kinucan, R. J. 1987. Influence of soil seed bank, seed rain inhibition competition and site disturbance on successional processes within three long-term grazing regimes on the Edwards Plateau, Texas. Ph. D. Diss., Texas A&M University, College Station. USA.
Kirby, D. R., M. F. Pessin, and G. K. Clambey. 1986. Disappearance of forage under short-duration and season-long grazing. J. Range Manage. 39:496–500.
Klein, D. A., B. A. Frederick, M. Biondini, and M. J. Trlica. 1988. Rhizosphere microorganism effects on soluble amino acids, sugars and organic acids in the root zone of *Agropyron cristatum*, *A. smithii* and *Bouteloua graciis*. Plant Soil 110:19–25.
Klopatek, J. M., and P. G. Risser. 1982. Energy analysis of Oklahoma rangelands and improved pastures. J. Range Manage. 35:637–643.
Knoll, G., and H. H. Hopkins. 1959. The effects of grazing and trampling upon certain soil properties. Trans. Kansas Acad. Sci. 62:221–231. USA.
Koniak, S., and R. L. Everett. 1982. Seed reserves in soils of successional stages of pinyon woodlands. Am. Midl. Nat. 108:295–303.
Koontz, H., and C. O'Donnell. 1972. Principles of management: an analysis of managerial functions. McGraw-Hill, New York.
Kothmann, M. M. 1980. Integrating livestock needs to the grazing system, p. 65–83. In: K. C. McDaniel and C. D. Allison (eds.), Proc. Grazing Manage. Syst. for S. W. Range. Symp. New Mexico St. Univ., Las Cruces. USA.
Kramer, A. 1979. The specific behavior and dispersion of two sympatric deer species. J. Wildl. Manage. 37:288–300.
Krebs, J. R., and J. B. Davies. 1978. Behavioral ecology: an evolutionary approach. Sinauer, Sunderland, MA. USA.
Krebs, J. R., and N. B. Davies. 1984. Behavioural ecology. 2nd ed., Blackwell, Oxford. ENG.
Krysl, L. J., M. E. Branine, M. L. Galyean, R. E. Estell, and W. C. Hoefler. 1987a. Influence of cottonseed meal supplementation on voluntary intake, ruminal and cecal fermentation, digesta kinetics and serum insulin and growth hormone in mature ewes fed prairie hay. J. Anim. Sci. 64:1178–1188.

Krysl, L. J., M. L. Galyean, M. B. Judkins, M. E. Branine, and R. E. Estell. 1987b. Digestive physiology of steers grazing fertilized and nonfertilized blue grama rangeland. J. Range Manage. 40:493–501.
Laidlaw, A. S., and A. M.M. Berrie. 1974. The influence of expanding leaves and the reproductive stem apex on apical dominance in *Lolium multiflorum*. Ann. Appl. Biol. 78:75–82.
Langer, P. 1984. Anatomical and nutritional adaptations in wild herbivores, p. 185–203. In: F. M. C. Gilchrist and R. I. Mackie (eds.), Herbivore nutrition in the subtropics and tropics. Science Press, Craighall. S. AFR.
Langer, R. H. M. 1956. Growth and nutrition of timothy (*Phleum pratense*). I. The life history of individual tillers. Ann. Appl. Biol. 44:166–187.
Langer, R. H. M. 1963. Tillering in herbage grasses. Herb. Abst. 33:141–148.
Langer, R. H. M. 1972. How grasses grow. Edward Arnold, London.
Lauenroth, W. K., and W. A. Laycock (eds.). 1989. Secondary succession and the evaluation of rangeland condition. Westview Press, Boulder, CO. USA
Launchbaugh, J. L. 1955. Vegetation changes in the San Antonio prairie associated with grazing, retirement from grazing and abandonment from cultivation. Ecol. Monogr. 25:39–57.
Launchbaugh, J. L. 1986. Intensive-early season stocking, p. 52–66. In: P. E. Reece and J. T. Nichols (eds.), Proc.: ranch manage. symp. Nebraska Coop. Ext. Serv., Lincoln. USA.
Launchbaugh, J. L., C. E. Owensby, J. R. Brethour, and E. F. Smith. 1983. Intensive-early stocking studies on Kansas ranges. Kansas Agr. Exp. Sta. Prog. Rep. 441, Manhattan. USA.
Launchbaugh, J. L., C. E. Owensby, F. L. Schwartz, and L. R. Corah. 1978. Grazing management to meet nutritional and functional needs of livestock, p. 541–546. In: D. N. Hyder (ed.), Proc. 1st Int. Range. Congr. Soc. Range Manage., Denver, CO. USA.
Launchbaugh, K., J. W. Stuth, and J. W. Holloway. 1990. Influence of range site on diet selection and nutrient intake of cattle. J. Range Manage: 43:109–115.
Lawlor, D. W. 1987. Photosynthesis: metabolism, control and physiology. Longman, Essex. ENG.
Laycock, W. A. 1967. How heavy grazing and protection affect sagebrush-grass ranges. J. Range Manage. 29:206–213.
Lehmann, V. W. 1984. Bobwhites in the Rio Grande Plain of Texas. Texas A&M Univ. Press, College Station. USA.
Leith, H. 1978. Primary productivity in ecosystems: Comparative analysis of global patterns, p. 300–321. In: H. F. H. Leith (ed.), Patterns of primary production in the biosphere. Dowden, Hutchinson and Ross, Stroudberg, PA. USA.
Leopold, A. C. 1949. The control of tillering in grasses by auxin. Am. J. Bot. 36:437–440.
Lewis, J. K. 1969. Primary producers in grassland ecosystems, p. 91–187. In: G. M. Van Dyne (ed.), The ecosystem concept in natural resource management. Academic Press, New York.
Lidicker, W. Z. 1988. The synergistic effects of reductionist and holistic approaches in animal ecology. Oikos 53:279–280.
Litton, C. W. 1977. Food habits of the Rio Grande turkey in the Permian Basin of Texas. Texas Dep. Parks Wildl. Bull. 18. Austin. USA.
Lonsdale, M., and R. Braithwaite. 1988. The shrub that conquered the bush. New Scient. 15:52–55.
Loope, W. L., and G. F. Gifford. 1972. Influence of soil microfloral crust on select properties of soils under pinyon-juniper in southeastern Utah. J. Soil Water Conserv. 27:164–167.
Loucks, O. L., M. L. Plumb-Mentjes, and D. Rogers. 1985. Gap processes and large-scale disturbances in sand prairies, p. 71–83. In: S. T. A. Picket and P. S. White (eds.), The ecology of natural disturbance and patch dynamics. Academic Press, New York.
Ludwig, J. A., G. L. Cunningham, and P. D. Whitson. 1988. Distribution of annual plants in North American deserts. J. Arid Environ. 15:221–227.
Lull, H. W. 1959. Soil compaction on forest and rangelands. USDA Misc. Publ. 768.
Lull, H. W. 1964. Ecological and silviculture aspects, p. 6.1–6.30. In: V. T. Chow (ed.), Handbook of applied hydrology. McGraw-Hill, New York.
Lusby, K. S., D. F. Stephens, and R. Totusek. 1976. Influence of breed and level of winter supplement on forage intake of range cows. J. Anim. Sci. 43:543–548.
Lynch, J. M., and E. Bragg. 1985. Microorganisms and soil aggregate stability. Adv. Soil Sci. 2:133–171.
Macfadyen, A. 1964. Energy flow in ecosystems and its exploitation by grazing, p. 3–20. In: D. J. Crisp (ed.), Grazing in terrestrial and marine environments. Blackwell, Oxford. ENG.
Mack, R. N., and J. N. Thompson. 1982. Evolution in steppe with few large, hooved mammals. Am. Nat. 119:757–773.

MacMahon, J. A. 1980. Ecosystems over time: succession and other types of change, p. 27–58. *In*: R. Waring (ed.), Forests: fresh perspectives from ecosystem analyses. Oregon St. Univ. Press, Corvallis. USA.

MacMahon, J. A., D. L. Phillips, J. V. Robinson, and D. J. Schimpf. 1978. Levels of biological organization: an organism centered approach. BioScience 28:700–704.

Madany, M. H., and N. E. West. 1983. Livestock grazing-fire regime interactions within montane forests of Zion National Park, Utah. Ecology 64:661–667.

Malechek, J. C. 1984. Impacts of grazing intensity and specialized grazing systems on livestock response, p. 1129–1158. *In*: Developing strategies for rangeland management. Westview Press, Boulder, CO. USA.

Malechek, J. C., and D. D. Dwyer. 1983. Short duration grazing doubles your livestock? Utah Sci. 44:32–37.

Mangel, M., and C. W. Clark. 1986. Towards a unified foraging theory. Ecology 67:1127–1138.

Marshall, J. K. 1973. Drought, land use and soil erosion, p. 55–77. *In*: J. V. Lovett (ed.), The environmental, economic and social significance of drought. Angus and Robertson, London.

Martin, S.C. 1978. The Santa Rita grazing system, p. 573–575. *In*: D. N. Hyder (ed.), Proc. 1st Int. Range. Congr. Soc. Range Manage., Denver, CO. USA.

Maslow, A. H. 1954. Motivation and personality. Harper and Row, New York.

Matches, A. G. 1966. Influence of intact tillers and height of stubble on growth responses of tall fescue (*Festuca arundinacea* Schreb.). Crop Sci. 6:484–487.

Mattson, W. J., Jr. 1980. Herbivory in relation to plant nitrogen content. Ann. Rev. Ecol. Syst. 11:119–161.

May, L. H. 1960. The utilization of carbohydrate reserves in pasture plants after defoliation. Herb. Abst. 30:239–245.

McCartor, M. M., and F. M. Rouquette, Jr. 1977. Grazing pressures and animal performance from pearl millet. Agron. J. 69:983–987.

McCown, R. L. 1982. The climatic potential for beef cattle production in tropical Australia: Part IV—Variation in seasonal and annual productivity. Agr. Syst. 8:3–15.

McGinty, W. A., F. E. Smeins, and L. B. Merrill. 1978. Infiltration and sediment production of Edwards Plateau rangelands as affected by soil characteristics and grazing management. J. Range Manage. 32:33–37.

McInnis, M. L., and M. Vavra. 1987. Dietary relationships among feral horses, cattle and pronghorn in Southeastern Oregon. J. Range Manage. 40:60–66.

McIntyre, G. I. 1967. Environmental control of bud and rhizome development in the seedlings of *Agropyron repens* L. Beauv. Can. J. Bot. 45:1315–1326.

McKown, C. D., J. W. Stuth, J. W. Walker, and R. K. Heitschmidt. 1990. Nutrient intake of cattle in rotational and continuous grazing treatments. J. Range Mange (In press).

McLeod, M. N., and D. J. Minson. 1988. Large particle breakdown by cattle eating ryegrass and alfalfa. J. Anim. Sci. 66:992–999.

McMahan, C. A. 1964. Comparative food habits of deer and three classes of livestock. J. Wildl. Manage. 28:789–808.

McMahan, C. A. 1966. Suitability of grazing enclosures for deer and wildlife research on the Kerr Wildlife Management Area, Texas. J. Wildl. Manage. 30:151–165.

McMeekan, C. P., and M. J. Walsh. 1963. The inter-relationships of grazing method and stocking rate in the efficiency of pasture utilization by dairy cows. J. Agr. Sci. 61:147–163.

McNaughton, S. J. 1977. Diversity and stability of ecological communities: a comment on the role of empiricism in ecology. Am. Nat. 111:515–525.

McNaughton, S. J. 1978. Serengeti ungulates: Feeding selectivity influences the effectiveness of plant defense guilds. Science 199:806–807.

McNaughton, S. J. 1979. Grazing as an optimization process: Grass-ungulate relationships in the Serengeti. Am. Nat. 113:691–703.

McNaughton, S. J. 1983a. Serengeti grassland ecology: the role of composite environmental factors and contingency in community organization. Ecol. Monogr. 53:291–320.

McNaughton, S. J. 1983b. Compensatory plant growth as a response to herbivory. Oikos 40:329–336.

McNaughton, S. J. 1984. Grazing lawns: animals in herds, plant form and co-evolution. Am. Nat. 24:863–886.

McNaughton, S. J. 1987. Adaptation of herbivores to seasonal changes in nutrient supply, p. 391–408. *In*: J. B. Hacker and J. H. Ternought (eds.), The nutrition of herbivores. Academic Press, New York.

McNaughton, S. J. 1988. Mineral nutrition and spatial concentrations of African ungulates. Nature 334:343–345.

McNaughton, S. J., M. B. Coughenhour, and L. L. Wallace. 1982. Interactive processes in grassland ecosystems, p. 167–193. *In*: J. R. Estes, R. J. Tyrl and J. N. Brunken (eds.), Grasses and grasslands: systematics and ecology. Univ. Oklahoma Press, Norman. USA.

McNaughton, S. J., J. L. Tarrants, M. M. McNaughton, and R. H. Davis. 1985. Silica as a defense against herbivory and a growth promotor in African grasses. Ecology 66:528–535.

McNaughton, S. J., L. L. Wallace, and M. B. Coughenour. 1983. Plant adaptation in an ecosystem context: effects of defoliation, nitrogen, and water on growth of an African C_4 sedge. Ecology 64:307–318.

McPherson, G. R., and H. A. Wright. 1987. Factors affecting reproductive maturity of redberry juniper (*Juniperus pinchotti*). For. Ecol. Manage. 21:191–196.

McPherson, G. R., H. A. Wright, and D. B. Wester. 1988. Patterns of shrub invasion in semiarid Texas rangelands. Am. Midl. Nat. 120:391–397.

Merrill, L. B. 1954. A variation of deferred rotation grazing for use under southwest range conditions. J. Range Manage. 7:152–154.

Merrill, L. B., J. G. Teer, and O. C. Wallmo. 1957. Reaction of deer populations to grazing practices, p. 10–12. *In*: Texas Agr. Exp. Sta. PR 3. College Station. USA.

Meyer, R. E., and R. W. Bovey. 1982. Establishment of honey mesquite and huisache on a native pasture. J. Range Manage. 35:548–550.

Milchunas, D. G., O. E. Sala, and W. K. Lauenroth. 1988. A generalized model of the effects of grazing by large herbivores on grassland community structure. Am. Nat. 132:87–106.

Mitchell, J. M. 1980. History and mechanisms of climate, p. 31–42. *In*: H. Oeschger, B. Messerli and M. Svilar (eds.), Das klima-analysen und modelle, geschichte und zukunft, Springer-Verlag, Berlin.

Mooney, H. A., and S. L. Gulmon. 1982. Constraints on leaf structure and function in reference to herbivory. BioScience 32:198–206.

Morgan, J. K. 1971. Ecology of the Morgan Creek and East Fork of the Salmon River bighorn sheep herds and management of bighorn sheep in Idaho. M.S. Thesis. Utah St. Univ., Logan. USA.

Morley, F. H. W. 1966. Stability and productivity of pastures. N.Z. Soc. Anim. Prod. Proc. 26:8–21.

Morley, F. H. W. 1981. Management of grazing systems, p. 379–400. *In*: F. H. W. Morley (ed.), Grazing animals. Elsevier, New York.

Mueggler, W. F. 1972. Influence of competition on the response of bluebunch wheatgrass to clipping. J. Range Manage. 25:88–92.

Mueggler, W. F. 1975. Rate and pattern of vigor recovery in Idaho fescue and bluebunch wheatgrass. J. Range Manage. 28:198–204.

Mueller, R. J., and J. H. Richards. 1986. Morphological analysis of tillering in *Agropyron spicatum* and *Agropyron desertorum*. Ann. Bot. 58:911–921.

National Atmospheric Deposition Program (IR-7)/National Trends Network. 1987. Annual report. Nat. Resourc. Ecol. Lab. Colorado St. Univ., Ft. Collins. USA.

National Research Council. 1978. Nutrient requirements of dairy cattle, 5th ed. National Academy Press, Washington DC.

National Research Council. 1981a. Effect of environment on nutrient requirements of domestic animals. National Academy Press, Washington, DC.

National Research Council. 1981b. Nutrient requirements of domestic animals, No. 15. Nutrient requirements of goats: Angora, dairy, and meat goats in temperate and tropical countries. National Academy Press, Washington, DC.

National Research Council. 1984. Nutrient requirements of beef cattle, 6th ed. National Academy Press, Washington DC.

National Research Council. 1985a. Nutrient requirements of sheep, 6th ed. National Academy Press, Washington DC.

National Research Council. 1985b. Ruminant nitrogen usage. National Academy Press, Washington DC.

Neilson, R. P. 1986. High resolution climatic analysis and southwest biogeography. Science 232:27–34.

Neilson, R. P. 1987. Biotic regionalization and climatic controls in western North American. Vegetatio 70:135–147.

Neilson, R. P., and L. H. Wullstein. 1985. Comparative drought physiology and biogeography of *Quercus gambelii* and *Quercus turbinella*. Am. Midl. Nat. 114:259–271.

Nelson, J. R. 1984. A modelling approach to large herbivore competition, p. 491–525. *In*: Developing strategies for rangeland management. Westview Press, Boulder, CO. USA.

Niering, W. A., and R. H. Goodwin. 1974. Creation of relatively stable shrublands with herbicides: arresting "succession" on rights-of-way and pastureland. Ecology 55:784–795.

Noble, I. R. 1986. The dynamics of range ecosystems, p.3–5. *In*: P. J. Joss, P. W. Lynch and O. B. Williams (eds.), Rangelands: a resource under siege. Aust. Acad. Sci., Canberra.

Noble, J. C., A. D. Bell, and J. L. Harper. 1979. The population biology of plants with clonal growth I. The morphology and structural demography of *Carex arenaria*. J. Ecol. 67:983–1008.

Norton, B. E., and P. S. Johnson. 1983. Pattern of defoliation by cattle grazing crested wheatgrass pastures, p. 462–464. *In*: J. A. Smith and V. W. Hayes (eds.), Proc. 14th Int. Grassl. Congr. Westview Press, Boulder, CO. USA.

Norton, G. A., and J. D. Munford. 1984. Decision-making in pest control. Adv. Appl. Biol. 8:87–119.

Novellie, P. A. 1978. Comparison of the foraging strategies of blesbok and springbok on Transvaal highveld. S. Afr. J. Wildl. Res. 8:137–144.

Nowak, R. S., and M. M. Caldwell. 1984. A test of compensatory photosynthesis in the field: implications for herbivory tolerance. Oecologia (Berlin) 61:311–318.

Noy-Meir, I., M. Gutman, and Y. Kaplan. 1989. Responses of Mediterranean grassland plants to grazing and protection. J. Ecol. 77:290–310.

Noy-Meir, I., and B. H. Walker. 1986. Stability and resilience in rangelands, p. 21–25. *In*: P. J. Joss, P. W. Lynch and O. B. Williams (eds.), Rangelands: a resource under seige. Aust. Acad. Sci., Canberra.

Odum, E. P. 1971. Fundamentals of ecology, 3rd ed. W. B. Saunders, Philadelphia.

Office of Technology Assessment. 1982. Impacts of technology on U.S. cropland and rangeland productivity. Rep. USA Congress.

Olson, B. E., and J. H. Richards. 1988a. Spatial arrangement of tiller replacement in *Agropyron desertorum* following grazing. Oecologia (Berlin) 76:7–10.

Olson, B. E., and J. H. Richards. 1988b. Annual replacement of the tillers of *Agropyron desertorum* following grazing. Oecologia (Berlin) 76:1–6.

Olson, B. E., and J. H. Richards. 1988c. Tussock regrowth after grazing: intercalary meristem and axillary bud activity of *Agropyron desertorum*. Oikos 51:374–382.

Ong, C. K. 1978. The physiology of tiller death in grasses. 1. The influence of tiller age, size and position. J. Brit. Grassl. Soc. 33:197–203.

Ong, C. K., and C. Marshall. 1979. The growth and survival of severely-shaded tillers in *Lolium perenne* L. Ann. Bot. 43:147–155.

Orodhu, A. B., M. J. Trlica, and C. D. Bonham. 1990. Long-term heavy-grazing effects on soil and vegetation in the four corners region. S. W. Nat. (in press).

Osuji, P. O. 1974. The physiology of eating and the energy expenditure of the ruminant at pasture. J. Range Manage. 27:437–443.

Otterman, J. 1977. Anthropogenic impact on the albedo of the earth. Clim. Change. 1:2.

Owen-Smith, N., and S. Cooper. 1987. Palatability of woody plants to browsing ruminants in a South African savanna. Ecology 68:319–331.

Owen-Smith, N., and P. Novellie. 1982. What should a clever ungulate eat? Am. Nat. 119:151–178.

Paige, K. N., and T. G. Whitham 1987. Overcompensation in response to mammalian herbivory: The advantage of being eaten. Am. Nat. 129:407–416.

Parsons, A. J., E. L. Leafe, B. Collett, P. D. Penning, and J. Lewis. 1983. The physiology of grass production under grazing. II. Photosynthesis, crop growth and animal intake of continuously grazed swards. J. Appl. Ecol. 20:127–139.

Parton, W. J., and P. G. Risser. 1980. Impact of management practices on the tallgrass prairie. Oecologia (Berlin) 52:37–375.

Passioura, J. B. 1979. Accountability, philosophy and plant physiology. Search 10:347–350.

Peak, J. M., and P. D. Dalke (eds.). 1982. Proc. Wildlife-Livestock Relationships Symp. Idaho For., Wildl. Range Exp. Sta., Moscow. USA.

Penfound, W. T. 1964. The relation of grazing to plant succession in the tallgrass prairie. J. Range Manage. 17:256–260.

Penman, H. L. 1948. Natural evaporation from open water, bare soil and grass. Roy. Soc., Proc. (London). Ser. A. 193:120–145.

Petersen, J. L., D. N. Ueckert, R. L. Potter, and J. E. Huston. 1987. Ecotypic variation in selected fourwing saltbush populations in Texas. J. Range Manage. 40:361–366.

Petersen, M. K., D.C. Clanton, and Robert Britton. 1985. Influence of protein degradability in range supplements on abomasal nitrogen flow, nitrogen balance and nutrient digestibility. J. Anim. Sci. 60:1324–1329.

Peterson, R. G., H. L. Lucas, and G. O. Mott. 1965. Relationship between rate of stocking and per animal and per acre performance on pasture. Agron. J. 57:27–30.

Pfister, J. A., and J. C. Malechek. 1986. The voluntary forage intake and nutrition of goats and sheep in the semi-arid tropics of Northeastern Brazil. J. Anim. Sci. 63:1078–1086.

Phillips, I. D. J. 1975. Apical dominance. Ann. Rev. Plant Physiol. 26:341–367.

Phillipson, A. T., and R. W. Ash. 1965. Physiological mechanism affecting the flow of digesta in ruminants, p. 97–107. In: R. W. Dougherty, R. S. Allen, W. Burroughs, N. L. Jacobson and A. D. McGilliard (eds.), Physiology of digestion in ruminants. Butterworths, Washington DC.

Pieper, R. D. 1980. Impacts of grazing systems on livestock, p. 133–151. In: K. C. McDaniel and C. D. Allison (eds.), Proc. Grazing Manage. Syst. for S. W. Range. Symp. New Mexico St. Univ., Las Cruces. USA.

Pieper R. D., G. B. Donart, E. E. Parker, and J. D. Wallace. 1978. Livestock and vegetation response to continuous and 4-pasture, 1-herd grazing systems in New Mexico, p. 560–562. In: D. N. Hyder (ed.), Proc. 1st Int. Range. Congr. Soc. Range Manage. Denver, CO. USA.

Pieper, R. D., and R. K. Heitschmidt. 1988. Is short-duration grazing the answer? J. Soil Water Conserv. 43:133–137.

Pimentel, D., and M. Burgess. 1980. Energy inputs in corn production, p. 67–84. In: D. Pimentel (ed.), Handbook of energy utilization in agriculture. CRC Press, Boca Raton, FL. USA.

Pimentel, D., P. A. Oltenacu, M. C. Nesheim, J. Krummel, M.S. Allen, and S. Chick. 1980. The potential for grass-fed livestock: resource constraints. Science 207:843–848.

Pimm, S. L. 1988. Energy flow and trophic structure, p. 263–278. In: L. R. Pomeroy and J. J. Alberts (eds.), Concepts of ecosystem ecology. Springer-Verlag, New York.

Pinchak, W. E., S. K. Canon, R. K. Heitschmidt, and S. L. Dowhower. 1990. Effect of long-term, yearlong grazing at moderate and heavy rates of stocking on diet selection and forage intake dynamics. J. Range Manage. (in press).

Pitelka, L. F., and J. W. Ashman. 1985. Physiology and integration of ramets in clonal plants, p. 399–435. In: J. B. G. Jackson, L. W. Buss and R. E. Cook (eds.), The population biology and evolution of clonal organisms. Yale Univ. Press, New Haven, CT. USA.

Pluhar, J. J., R. W. Knight, and R. K. Heitschmidt. 1987. Infiltration rates and sediment production as influenced by grazing systems in the Texas Rolling Plains. J. Range Manage. 40:428–431.

Pond, F. W. 1960. Vigor of Idaho Fescue in relation to different grazing intensities. J. Range Manage. 13:28–30.

Pond, K. R., W. C. Ellis, C. E. Lascano, and D. E. Akin. 1987. Fragmentation and flow of grazed coastal Bermudagrass through the digestive tract of cattle. J. Anim. Sci. 65:609–618.

Poppi, D. P., D. J. Minson, and J. H. Ternouth. 1980. Studies of cattle and sheep eating leaf and stem fractions of grasses. I. The voluntary intake, digestibility and retention time in the reticulo-rumen. Aust. J. Agr. Res. 32:99–108.

Poppi, D. P., D. J. Minson, and J. H. Ternouth. 1981. Studies of cattle and sheep eating leaf and stem fractions of grasses. III. The retention time in the rumen of large feed particles. Aust. J. Agr. Res. 32:123–137.

Potvin, M. A., and A. T. Harrison. 1984. Vegetation and litter changes of a Nebraska sandhills prairie protected from grazing. J. Range Manage. 37:55–58.

Powell, D. J., D.C. Clanton, and J. T. Nichols. 1982. Effect of range condition on the diet and performance of steers grazing native sandhills range in Nebraska. J. Range Manage. 35:96–99.

Provenza, F. D., and D. F. Balph. 1987a. Diet learning by domestic ruminants: theory evidence and practical implications. Appl. Anim. Behav. Sci. 18:211–232.

Provenza, F. D., and D. F. Balph. 1987b. The development of dietary choice in livestock on rangelands and its implications for management, p. 2356–2368. In: D. R. Mertens (ed.), Proc. Forage Selection and Intake by Grazing Rumin. Symp. Am. Soc. Anim. Sci., Champaign, IL. USA.

Provenza, F. D., and J. C. Malechek. 1983. Tannin allocation in blackbrush (*Caleogyne ramosissima*). Biochem. Syst. Ecol. 11:233–238.

Pyke, G. H., H. R. Pulliam, and E. C. Charnov. 1977. Optimal foraging: a selective review of theory and tests. Quart. Rev. Biol. 55:137–154.

Rabotnov, T. A. 1974. Differences between fluctuations and successions, p. 19–24. In: R. Knapp (ed.), Vegetation dynamics. Dr. Junk, The Hague. NETHERLANDS.

Reardon, P. O., L. B. Merrill, and C. A. Taylor, Jr. 1978. White-tailed deer preferences and hunter success under various grazing systems. J. Range Manage. 31:40–42.

Rechenthin, C. A. 1956. Elementary morphology of grass growth and how it affects utilization. J. Range Manage. 9:167–170.

Rector, B. S., and J. E. Houston. 1982. Composition of diets selected by livestock in combination, p. 320. (Abstr.) Ann. Meet. Am. Soc. Anim. Sci., Champaign, IL. USA.

Reiner, R. J., F. C. Bryant, R. D. Farfan, and B. F. Craddock. 1987. Forage intake of alpacas grazing Andean rangeland in Peru. J. Anim. Sci. 64:868–871.

Rhoades, E. D., L. F. Locke, H. M. Taylor, and E. H. McIlvain. 1964. Water intake on a sandy range as affected by 20 years of differential cattle stocking rates. J. Range Manage. 17:185–190.

Rhodes, D. F. 1979. Evolution of plant chemical defense against herbivores, p. 3–54. In: G. A. Rosenthal and D. H. Janzen, (eds.), Herbivores: their interaction with secondary plant metabolites. Academic Press, New York.

Rhodes, D. F. 1985. Offensive-defensive interactions between herbivores and plants: their relevance in herbivore population dynamics and ecological theory. Am. Nat. 125:205–238.

Rhodes, D. F., and R. G. Cates. 1976. Toward a general theory of plant antiherbivore chemistry. Rec. Adv. Phytochem. 10:168–213.

Rich, R. L., and H. G. Reynolds. 1963. Grazing in relation to erosion on some chaparral watersheds of central Arizona. J. Range Manage. 16:322–326.

Richards, J. H. 1984. Root growth response to defoliation in two *Agropyron* bunchgrasses: field observations with an improved root periscope. Oecologia (Berlin) 64:21–25.

Richards, J. H., and M. M. Caldwell. 1985. Soluble carbohydrates, concurrent photosynthesis and efficiency in regrowth following defoliation: A field study with *Agropyron* species. J. Appl. Ecol. 22:907–920.

Richards, J. H., R. J. Mueller, and J. J. Mott. 1988. Tillering in tussock grasses in relation to defoliation and apical bud removal. Ann. Bot. 62:173–179.

Ridsdill Smith, T. J. 1977. Effects of root-feeding by scarabaeid larvae on growth of perennial ryegrass plants. J. Appl. Ecol. 14:73–80.

Riechers, R. K., J. R. Conner, and R. K. Heitschmidt. 1989. Economic consequences of alternative stocking rate adjustment tactics: A simulation approach. J. Range Manage. 42:165–171.

Ripple, C. D., J. Rubin, and T. E. A. van Hylckama. 1972. Estimating steady-state evaporation rates from bare soils under conditions of high water table. US Geol. Surv. Water-Supply Paper 2019-A.

Rittenhouse, L. R., C. D. Clanton, and C. L. Streeter. 1970. Intake and digestibility of winter range forage by cattle with and without supplementation. J. Anim. Sci. 31:1215–1221.

Robbins, L. 1932. The nature and significance of economic science. Macmillan, London.

Robertson, J. H. 1971. Changes on a grass-shrub range in Nevada ungrazed for 30 years. J. Range Manage. 24:397–400.

Robinson, R. R., and R. B. Alderfer. 1952. Runoff from permanent pastures in Pennsylvania. Agron. J. 44:459–462.

Robson, M. J. 1968. The changing tiller population of spaced plants of S.170 tall fescue (*Festuca arundinacea*). J. Appl. Ecol. 5:575–590.

Rode, L. M., G. H. Coulter, G. J. Mears, and J. E. Lawson. 1986. Biological constraints to ruminant production. Can. J. Anim. Sci. 66:859–875.

Rosene, W. 1969. The bobwhite quail: its life and management. Rutgers Univ. Press, New Brunswick, NJ. USA.

Rosenzweig, M. L. 1981. A theory of habitat selection. Ecology 62:327–355.

Rosiere, R. E., J. D. Wallace, and R. D. Pieper. 1980. Forage intake in two year old cows and heifers grazing blue grama summer range. J. Range Manage. 33:71–73.

Rowe, J. S. 1961. The level-of-integration concept and ecology. Ecology 42:420–427.

Ruess, R. W. 1984. Nutrient movement and grazing: experimental effects of clipping and nitrogen source on nutrient uptake in *Kyllinga nervosa*. Oikos 43:183–188.

Ruyle, G. B., and D. D. Dwyer. 1985. Feeding stations of sheep as an indicator of diminished forage supply. J. Anim. Sci. 61:349–353.

Ryle, G. J., and C. E. Powell. 1975. Defoliation and regrowth in the graminaceous plant: the role of current assimilate. Ann. Bot. 39:297–310.

Sala, O. E., W. J. Parton, L. A. Joyce, and W. K. Lauenroth. 1988. Primary production of the central grassland regions of the United States. Ecology 69:40–45.

Salihi, D. O., and B. E. Norton. 1987. Survival of perennial grass seedlings under intensive grazing in semi-arid rangelands. J. Appl. Ecol. 24:145–151.

Sambo, E. Y. 1983. Leaf extension rate in temperate pasture grasses in relation to assimilate pool in the extension zone. J. Exp. Bot. 34:1281–1290.
Samuelson, P. A. 1964. Economics: an introductory analysis. McGraw-Hill, New York.
Satterlund, D. R. 1972. Wildland watershed management. Ronald Press, New York.
Savory, A. 1978. A holistic approach to ranch management using short duration grazing, p. 555–557. In: Proc. 1st Int. Range. Congr. Soc. Range Manage., Denver, CO. USA.
Savory, A. 1979. Range management principles underlying short duration grazing, p. 375–379. In: Beef Cattle Science Handbook. Agriservices Found., Clovis, CA. USA.
Savory, A. 1983. The Savory grazing method or holistic resource management. Rangelands 5:155–159.
Savory, A., and S. D. Parsons. 1980. The Savory grazing method. Rangelands 2:234–237.
Scales, G. H., A. H. Denham, C. L. Streeter, and G. M. Ward. 1974. Winter supplementation of beef calves on sandhill range. J. Anim. Sci. 38:442–448.
Scanlan, J. C. 1988. Spatial and temporal vegetation patterns in a subtropical Prosopis savanna woodland, Texas. Ph. D. Diss., Texas A&M Univ., College Station. USA.
Scarnecchia, D. L., and M. M. Kothmann. 1982. A dynamic approach to grazing management terminology. J. Range Manage. 35:262–264.
Schemnitz, S.D (ed.). 1980. Wildfife management techniques manual, 4th ed. Wildl. Soc., Washington, DC.
Schimel, D. S., W. J. Parton, F. J. Adamsen, R. G. Woodmansee, R. L. Senft, and M. A. Stillwell. 1986. The role of cattle in the volatile loss of nitrogen from a shortgrass steppe. Biogeochemistry 2:39–52.
Schimel, D., M. A. Stillwell, and R. G. Woodmansee. 1985. Biogeochemistry of C, N and P in a soil catena of the shortgrass steppe. Ecology 66:276–282.
Schladwieler, P. 1974. Ecology of shiras moose in Montana. Mimeo. Montana Dep. Fish Game, Helena. USA.
Schoener, T. W. 1969. Models of optimal size for solitary predators. Am. Nat. 103:277–313.
Schoener, T. W. 1971. Theory of feeding strategies. Ann. Rev. Ecol. Syst. 2:369–403.
Schoener, T. W. 1983. Simple models of optimal feeding-territory size: a reconciliation. Am. Nat. 121:608–629.
Schofield, C. J., and E. H. Bucher. 1986. Industrial contributions to desertification in South America. Tree 1:78–80.
Schuster, J. L. 1964. Root development of native plants under three grazing intensities. Ecology 45:63–70.
Scifres, C. J. 1980. Brush management: principles and practices for Texas and the Southwest. Texas A&M Univ. Press, College Station. USA.
Scifres, C. J. 1987. Decision-analysis approach to brush management planning: ramification for integrated range resource management. J. Range Manage. 41:482–498.
Scifres, C. J., W. T. Hamilton, J. R. Conner, J. M. Inglis, G. A. Rasmussen, R. P. Smith, J. W. Stuth, and T. G. Welch. 1985. Development and implementation of integrated brush management systems (IBMS) for South Texas. Texas Agr. Exp. Sta. B–1493. College Station. USA.
Scifres, C. J., W. T. Hamilton, J. M. Inglis, and J. R. Conner. 1983. Development of integrated brush management systems (IBMS): decision-making processes, p. 97–104. In: K. W. McDaniel (ed.), Proc. Brush Manage. Symp. Texas Tech Press, Lubbock. USA.
Scifres, C. J., W. T. Hamilton, B. H. Koerth, R. C. Flinn, and R. A. Crane. 1988. Bionomics of patterned herbicide application for wildlife habitat enhancement. J. Range Manage. 41:317–321.
Scifres, C. J. J. L. Mutz, R. E. Whitson, and D. L. Drawe. 1982. Interrelationships of huisache canopy cover with range forage on the coastal prairie. J. Range Manage. 35:558–562.
Scott, J. A., N. R. French, and J. W. Leetham. 1979. Patterns of consumption in grasslands, p. 89–105. In: N. R. French (ed.), Perspectives in grassland ecology. Springer-Verlag, New York.
Seastedt, T. R. 1985. Canopy interception of nitrogen in bulk precipitation by annually burned and unburned tallgrass prairie. Oecologia (Berlin) 66:88–92.
Seastedt, T. R., R. A. Ramundo, and D.C. Hayes. 1988. Maximization of densities of soil animals by foliage herbivory: empirical evidence, graphical and conceptual models. Oikos 51:243–248.
Senft, R. L. 1989. Hierarchical foraging models: Effects of stocking and landscape composition on simulated resource use by cattle. Ecol. Modeling 46:283–303.
Senft, R. L., M. B. Coughenour, D. W. Bailey, L. R. Rittenhouse, O. E. Sala, and D. M. Swift. 1987. Large herbivore foraging and ecological hierarchies. Bioscience 37:789–799.
Senft, R. L., L. R. Rittenhouse, and R. G. Woodmansee. 1985. Factors influencing patterns of cattle grazing behavior on shortgrass steppe. J. Range Manage. 38:82–87.

Seoane, J. R., M. Cote, and S. A. Visser. 1982. The relationship between voluntary intake and the physical properties of forages. Can. J. Anim. Sci. 62:473–480.

Sharman, B.C. 1945. Leaf and bud initiation in the gramineae. Bot. Gaz. 106:269–289.

Sharp, A. L., L. J. Bond, J. W. Neuberger, A. R. Kuhlman, and J. K. Lewis. 1964. Runoff as affected by intensity of grazing on rangeland. J. Soil Water Conserv. 19:103–106.

Sheehy, D. P. 1988. Grazing relationships of elk, deer and cattle on seasonal rangeland in northeastern Oregon. Ph. D. Diss. Oregon St. Univ. Corvallis. USA.

Shiflet, T. N. 1973. Range sites and range soils in the United States, p. 26–33. In: Arid shrublands—Proc. 3rd Workshop of the United States/Australia Rangelands Panel. Soc. Range Manage., Denver, CO. USA.

Shiflet, T. N., and H. F. Heady. 1971. Specialized grazing systems: their place in range management. USDA-SCS TP-152.

Simon, J. C., and G. Lemaire. 1987. Tillering and leaf area index in grasses in the vegetative phase. Grass Forage Sci. 72:373–380.

Simons, A. B., and G. C. Marten. 1971. Relationships of indole alkaloids to palatability of *Phalaris arundinacea* L. Agron. J. 63:915–919.

Sims, P. L., and J. S. Singh. 1978a. The structure and function of ten western North American grasslands. II. Intra-seasonal dynamics in primary producer compartments. J. Ecology 66:547–572.

Sims, P. L., and J. S. Singh. 1978b. The structure and function of ten western North American grasslands. III. Net primary production, turnover and efficiency of energy capture and water use. J. Ecol. 66:573–597.

Sims, P. L., R. E. Sosebee, and D. M. Engle. 1982. Plant and vegetation responses to grazing management, p. 4–31. In: D. D. Briske and M. M. Kothmann (eds.), Proc. Nat. Conf. Grazing Manage. Tech. Texas A&M Univ., College Station. USA.

Sinclair, A. R. E. 1975. The resource limitations of trophic levels in tropical grassland ecosystems. J. Anim. Ecol. 44:497–520.

Singh, J. S., and M. C. Joshi. 1979. Ecology of the semi-arid regions of India with emphasis on land-use, p. 243–273. In: B. H. Walker (ed.), Management of semi-arid ecosystems. Elsevier, New York.

Skovlin, J. 1987. Southern Africa's experience with intensive short duration grazing. Rangelands 4:162–167.

Skovlin, J. M., R. W. Harris, G. S. Strickler, and G. A. Garrison. 1976. Effects of cattle grazing methods on ponderosa pine-bunchgrass range in the Pacific northwest. USDA For. Serv. Tech. Bull. 1531.

Smeins, F. E. 1984. Origin of the brush problem—a geographical and ecological perspective of contemporary distributions, p. 5–16. In: K. W. McDaniel (ed.), Proc. Brush Manage. Symp. Texas Tech Press, Lubbock. USA.

Smeins, F. E., and L. B. Merrill. 1988. Long-term change in a semi-arid grassland, p. 101–114. In: B. B. Amos and F. R. Gehlbach (eds.), Edwards Plateau vegetation. Baylor Univ. Press, Waco, TX. USA.

Smith, D. A., and E. M. Schmutz. 1975. Vegetative changes on protected versus grazed desert grassland ranges in Arizona. J. Range Manage. 28:453–457.

Smith, D. D., and W. H. Wischmeir. 1957. Factors affecting sheet and rill erosion. Trans. Am. Geophys. Union. 38:889–896.

Smith, E. L. 1988a. Successional concepts in relation to range condition assessment, p. 113–133. In: P. T. Tueller (ed.), Vegetation science applications for rangeland analysis and management. Kluwer, Dordrecht. NETHERLANDS.

Smith, M.S. 1984. Behavioral ecology of sheep in the Australian arid zone. Ph. D. Diss., Aust. Nat. Univ., Canberra.

Smith, M.S. 1988b. Modeling: three approaches to predicting how herbivore impact is distributed in rangelands. New Mex. Agr. Exp. Sta. Reg. Res. Rep. 628. Las Cruces. USA.

Smith, R. L. 1974. Ecology and field biology. Harper & Row, New York.

Smoliak, S., J. F. Dormaar, and A. Johnston. 1972. Long-term grazing effects on *Stipa-Bouteloua* prairie soils. J. Range Manage. 25:246–250.

Snayden, R. W. 1981. The ecology of grazing pastures, p. 13–31. In: F. H. W. Morley (ed.), Grazing animals. Elsevier, New York.

Society for Range Management. 1974. A glossary of terms used in range management, 2nd ed. Soc. Range Manage., Denver, CO. USA.

Society for Range Management. 1983. Guidelines and terminology for range inventories and monitoring. Soc. Range Manage., Denver, CO. USA.

Society for Range Management. 1989. A glossary of terms used in range management, 3rd ed. Soc. Range Manage., Denver, CO. USA.

Solis, J. C., F. M. Byers, G. T. Schelling, C. R. Long, and L. W. Greene. 1988. Maintenance requirements and energetic efficiency of cows of different breed types. J. Anim. Sci. 66:764–773.

Sosebee, R. E. 1976. Hydrology: the state of the science evapotranspiration, p. 95–104. In: H. F. Heady, D. H. Falkenborg and J. P. Riley (eds.), Watershed management of range and forest lands. Utah Water Res. Lab. Logan. USA.

Southwood, T. R. E. 1985. Interactions of plants and animals: patterns and processes. Oikos 44:5–11.

Sowell, B. F., B. H. Koerth and F. C. Bryant. 1985. Seasonal nutrient estimates of mule deer diets in the Texas Panhandle. J. Range Manage. 38:163–167.

Squires, V. 1981. Livestock management in the arid zone. Inkata Press, Melbourne.

Squires, V. R. 1982. Behavior of free-ranging livestock on native grasslands and shrublands. Trop. Grassl. 16:161–170.

Stanton, N. L. 1988. The underground in grasslands. Ann. Rev. Ecol. Syst. 19:573–589.

Stebbins, G. L. 1981. Coevolution of grasses and herbivores. Ann. Missouri Bot. Gard. 68:75–86.

Stillwell, M. A., and R. G. Woodmansee. 1981. Chemical transformations of urea-nitrogen and movement of nitrogen in a shortgrass prairie soil. Soil Sci. Soc. Am. J. 45:893–898.

Stobbs, T. H. 1973. The effect of plant structure on the intake of tropical pastures. II. Differences in sward structure, nutritive value, and bite size of animals grazing *Setaria anceps* cv Kazungula swards. Aust. J. Agr. Res. 24:824–829.

Stuart-Hill, G. C., and M. T. Mentis. 1982. Coevolution of African grasses and large herbivores. Proc. Grassl. Soc. S. Afr. 17:122–128.

Stuart-Hill, G. C., N. N. Tainton, and H. J. Barnard. 1987. The influence of an *Acacia karroo* tree on grass production in its vicinity. J. Grassl. Soc. S. Afr. 4:83–88.

Stubbendieck, J., S. L. Hatch, and K. J. Hirsch. 1986. North American Range Plants. 3rd Edition. Univ. of Nebraska Press. Lincoln and London.

Stuth, J. W., J. R. Brown, P. D. Olson, M. R. Araujo, and H. D. Aljoe. 1987. Effects of stocking rate on critical plant-animal interactions in a rotational grazed *Schizachyrium—Paspalum* savanna, p. 115–139. In: F. P. Horn, J. Hodgson, J. J. Mott and R. N. Brougham (eds.), Grazing-lands research at the plant-animal interface. Winrock Int., Morrilton, AR. USA.

Stuth, J. W., J. R. Conner, W. T. Hamilton, D. A. Riegel, B. G. Lyons, B. R. Myrick, and M. J. Couch. 1990. RSPM—A resource planning model for integrated resource management. J. Biogeography: 17:531–540.

Stuth, J. W., and S. Searcy. 1987. A new electronic approach to monitoring ingestive behavior of cattle, p. 81–82. In: J. B. Hacker and J. H. Ternouth (eds.), The nutrition of herbivores. Academic Press, Sydney.

Svejcar, T. J. 1989. Animal performance and diet quality as influenced by burning on tallgrass prairie. J. Range Manage. 42:11–15.

Svejcar, T. J., and S. Christiansen. 1987. Grazing effects of water relations of caucasian bluestem. J. Range Mange. 40:15–18.

Tainton, N. M., P. J. Edwards, and M. T. Mentis. 1980. A revised method for assessing veld condition. Proc. Grassl. Soc. S. Afr. 15:37–42.

Taylor, C. A. 1985. Multispecies grazing research overview (Texas), p. 65–83. In: F. H. Baker and R. K. Jones (eds.), Proc. Conf. on Multispecies Grazing. Winrock Int., Morrilton, AR. USA.

Taylor, C. A. 1989. Short duration grazing: experiences from the Edwards Plateau region in Texas. J. Soil Water Conserv. 44:297–302.

Thimann, K. V., and F. Skoog. 1933. Studies on the growth hormone of plants. 3. The inhibiting action of the growth substance on bud development. Proc. US Nat. Acad. Sci. 19:714–716.

Thomas, R. J., A. B. Logan, A. D. Ironside, and G. R. Bolton. 1988. Transformation and fate of sheep urine-N applied to an upland U. K. pasture at different times during the growing season. Plant Soil 107:173–181.

Thurow, T. L., W. H. Blackburn, and C. A. Taylor, Jr. 1986. Hydrologic characteristics of vegetation types as affected by livestock grazing systems, Edwards Plateau, Texas. J. Range Manage. 39:505–509.

Thurow, T. L., W. H. Blackburn, and C. A. Taylor, Jr. 1987. Rainfall interception losses by midgrass, shortgrass, and live oak mottes. J. Range Manage. 40:455–460.

Thurow, T. L., W. H. Blackburn, and C. A. Taylor, Jr. 1988a. Infiltration and interrill erosion responses to selected livestock grazing strategies, Edwards Plateau, Texas. J. Range Manage. 41:296–302.

Thurow, T. L., W. H. Blackburn, and C. A. Taylor, Jr. 1988b. Some vegetation responses to selected livestock grazing strategies, Edwards Plateau, Texas. J. Range Manage. 41:108–114.
Tiedemann, A. R., D. A. Higgins, T. M. Quigley, H. R. Sanderson, and D. B. Marx. 1987. Responses of fecal coliform in streamwater to four grazing strategies. J. Range Manage. 40:322–329.
Tisdale, E. W., and M. Hironaka. 1981. The sagebrush-grass region: a review of the ecological literature. Idaho For., Wildl. Range Exp. Sta., Bull. No. 33. Moscow. USA.
Tomlinson, P. B. 1974. Vegetative morphology and meristem dependence—the foundation of productivity in seagrasses. Aquaculture 4:107–130.
Torell, L. A. 1984. Economic optimum stocking rates and retreatment schedules for crested wheatgrass stands. Ph. D. Diss., Utah St. Univ., Logan. USA.
Tothill, J. C., and J. J. Mott. 1985. Ecology and management of the world's savannas. Aust. Acad. Sci., Canberra.
Tripathi, R. S., and J. L. Harper. 1973. The comparative biology of *Agropyron repens* (L.) Beauv. and *A. caninum* (L.) Beauv. I. The growth of mixed populations established from tillers and from seeds. J. Ecol. 61:353–368.
Troughton, A. 1981. Length of life of grass roots. Grass Forage Sci. 36:117–120.
Ueckert, D. N. 1979. Impact of a white grub (*Phyllophasga crinita*) on a shortgrass community and evaluation of selected rehabilitation practices. J. Range Mange. 32:445–448.
Urness, P. J. 1982. Livestock as tools for managing big game winter range in the intermountain west, p. 20–31. *In*: J. M. Peak and P. D. Dalke (eds.), Wildlife-Livestock Relationships Symp. Proc. Idaho For., Wildl. Range Exp. Sta., Moscow. USA.
Urness, P. J., D. D. Austin, and L. C. Fierro. 1983. Nutritional value of crested wheatgrass for wintering mule deer. J. Range Manage. 36:225–226.
Valentine, K. A. 1947. Distance from water as a factor in grazing capacity of rangeland. J. Range Manage. 45:749–754.
Vali, G., M. Christensen, R. W. Fresh, E. L. Galyan, L. R. Maki, and R. C. Schnell. 1976. Biogenic ice nuclei: Part II. Bacterial sources. J. Atmos. Sci. 33:1565–1570.
Van Haveren, B. P. 1983. Soil bulk density as influenced by grazing intensity and soil type on a shortgrass prairie site. J. Range Manage. 36:586–588.
Van Soest, P. J. 1967. Development of a comprehensive system of feed analyses and its application to forages. J. Anim. Sci. 26:119–128.
Van Soest, P. J. 1982. Nutritional ecology of the ruminant. O&B, Corvallis, OR. USA.
Van Soest, P. J., R. H. Wine, and L. A. Moore. 1966. Estimation of the true digestibility of forage by the in vitro digestion of cell walls, p. 438–441. *In*: A. G.G. Hill (ed.), Proc. 10th Int. Grassl. Congr. Valtioneuvofton Kirjapaino, Helsinke. FINLAND.
Van Tassell, L. W., R. K. Heitschmidt, and J. R. Conner. 1987. Modeling variation in range calf growth under conditions of environmental uncertainty. J. Range Manage. 40:310–314.
Van Vegten, J. A. 1983. Thornbush invasion in a savanna ecosystem in eastern Botswana. Vegetatio 56:3–7.
Vavra, M., R. W. Rice, and R. E. Bement. 1973. Chemical composition of the diet, intake and gain of yearling cattle at different stocking intensities. J. Anim. Sci. 36:411–414.
Veihmeyer, F. J. 1964. Evapotranspiration, p. 1–38. *In*: V. T. Chow (ed.), Handbook of applied hydrology. McGraw-Hill, New York.
Ventura, M., J. E. Moore, O. C. Ruelke, and D. E. Franke. 1975. Effect of maturity and protein supplementation on voluntary intake and nutrient digestibility of pangola digitgrass hays. J. Anim. Sci. 40:769–774.
Vine, D. A. 1983. Sward structure changes within a perennial ryegrass sward: leaf appearance and death. Grass Forage Sci. 38:231–242.
Virtanen, A. I. 1968. On nitrogen metabolism in milking cows. Ann. Acad. Sci. Fennicae. Ser. A. II Chemica. 141. Suomalainen Tiedeakatemia, Helsinke. FINLAND.
Voigt, J. W., and J. E. Weaver. 1951. Range condition classes of native midwestern pasture: an ecological analysis. Ecol. Monogr. 21:39–60.
Volenec, J. J., and C. J. Nelson. 1984. Carbohydrate metabolism in leaf meristem of tall fescue. I. Relationship to genetically altered leaf elongation rates. Plant Physiol. 74:590–594.
Walker, B. H. 1988. Autecology, synecology, climate and livestock as agents of rangeland dynamics. Australian Rangelands Journal. 10:69–75.
Walker, B. H., R. H. Emslie, R. N. Owen-Smith, and R. J. Scholes. 1987. To cull or not to cull: lessons from a southern African drought. J. Appl. Ecol. 24:381–401.
Walker, B. H., D. Ludwig, C. S. Holling, and R. M. Peterman. 1981. Stability of semi-arid savannah grazing systems. J. Ecol. 69:473–498.

Walker, B. H., D. A. Matthews, and P. J. Dye. 1986a. Management of grazing systems—existing versus an event oriented approach. S. Afr. J. Sci. 82:172–180.

Walker, J., J. A. Robertson, L. K. Penridge, and P. J. H. Sharpe. 1986b. Herbage response to tree thinning in a *Eucalyptus crebra* woodland. Aust. J. Ecol. 11:135–140.

Walker, J. W., and R. K. Heitschmidt. 1986. Effect of various grazing treatments on type and density of cattle trails. J. Range Manage. 39:428–431.

Walker, J. W., and R. K. Heitschmidt. 1989. Some effects of a rotational grazing treatment on cattle grazing behavior. J. Range Manage. 42:337–342.

Walker, J. W., R. K. Heitschmidt, E. A. DeMoraes, M. M. Kothmann, and S. L. Dowhower. 1989b. Quality and botanical composition of cattle diets under rotational and continuous grazing treatments. J. Range Manage. 42:239–242.

Walker, J. W., and R. K. Heitschmidt, and S. L. Dowhower. 1989c. Some effects of a rotational grazing treatment on cattle preference for plant communities. J. Range Manage. 42:143–148.

Walker, J. W., J. W. Stuth, and R. K. Heitschmidt. 1989a. A simulation approach for investigating field data from grazing trials. Agr. Syst. 30:301–316.

Wallace, L. L., S. J. McNaughton, and M. B. Coughenour. 1984. Compensatory photosynthetic responses of three African graminoids to different fertilization, watering, and clipping regimes. Bot. Gaz. 145:151–156.

Wallace, L. L., S. J. McNaughton, and M. B. Coughenour. 1985. Effects of clipping and four levels of nitrogen on the gas exchange, growth, and productivity of two east African graminoids. Am. J. Bot. 72:222–230.

Walter, J. 1984. Rangeland revolutionary: an interview with Allan Savory. J. Soil Water Conserv. 29:235–240.

Wanyoike, M. M., and W. Holmes. 1981. The effects of winter nutrition on the subsequent live weight performance and intake of herbage by beef cattle. J. Agr. Sci. 97:221–226.

Warren, S. D., W. H. Blackburn, and C. A. Taylor, Jr. 1986a. Effects of season and stage of rotation cycle on hydrologic condition of rangeland under intensive rotation grazing. J. Range Manage. 39:486–491.

Warren, S. D., M. B. Neville, W. H. Blackburn, and N. E. Garza. 1986b. Soil response to trampling under intensive rotation grazing. Soil Sci. Soc. Am. J. 50:1336–1341.

Warren, S. D., T. L. Thurow, W. H. Blackburn, and N. E. Garza. 1986c. The influence of livestock trampling under intensive rotation grazing on soil hydrologic characteristics. J. Range Manage. 39:491–495.

Warrington, B. G., F. M. Byers, G. T. Schelling, D. W. Forrest, J. F. Baker, and L. W. Greene. 1988. Gestation nutrition, tissue exchange and maintenance requirements of heifers. J. Anim. Sci. 66:774–782.

Watson, M. A., and B. B. Casper. 1984. Morphogenetic constraints on patterns of carbon distribution in plants. Ann. Rev. Ecol. Syst. 15:233–258.

Weaver, J. E., and E. Zink. 1946. Length of life of roots of ten species of perennial range and pasture grasses. Plant Physiol. 21:201–217.

Webster, J. R. 1979. Hierarchical organization of ecosystems, p. 119–129. *In*: E. Halfon (ed.), Theoretical system ecology, Academic Press, New York.

Weeda, W. C. 1967. The effects of cattle dung patches on pasture growth, botanical composition and pasture utilization. N.Z.J. Agr. Res. 10:150–159.

Weinmann, H. 1948. Underground development and reserves of grasses. A review. J. Brit. Grassl. Soc. 3:115–140.

Welch, D. 1985. Studies in the grazing of heather moorland in northeast Scotland. IV. Seed dispersal and plant establishment in dung. J. Appl. Ecol. 22:461–472.

Welker, J. M., D. D. Briske, and R. W. Weaver. 1987. Nitrogen-15 partitioning within a three generation tiller sequence of the bunchgrass *Schizachyrium scoparium*: response to selective defoliation. Oecologia (Berlin) 24:330–334.

Welsh, R. G., and R. F. Beck. 1976. Some ecological relationships between creosotebush and bush muhly. J. Range Manage. 29:472–475.

Weltz, M. A. 1987. Observed and estimated (Erhym-II Model) water budgets for south Texas rangelands. PhD. Diss., Texas A&M Univ., College Station. USA.

Weltzin, J. F., and M. B. Coughenour. 1990. *Acacia tortilis* tree canopy influence on understory vegetation and soil nutrients in an arid Kenyan savanna. J. Vegetation Sci. (in press).

West, N. E., F. D. Provenza, P. S. Johnson, and K. M. Owens. 1984. Vegetation change after 13 years of livestock grazing exclusion on sagebrush semi-arid-desert in West Central Utah. J. Range Manage. 37:262–264.

West, N. E., K. H. Rea, and R. O. Harniss. 1979. Plant demographic studies in sagebrush-grass communities of southeastern Idaho. Ecology 60:376–388.
Wester, D. B., and H. A. Wright. 1987. Ordination of vegetation change in Guadalupe Mountain, New Mexico, USA. Vegetatio 72:27–33.
Western, D., and C. van Praet. 1973. Cyclical changes in the habitat and climate of an East African ecosystem. Nature 241:104–106.
Westoby, M. 1974. An analysis of diet selection by large generalist herbivores. Am. Nat. 108:290–304.
Westoby, M., B. Walker, and I. Noy-Meir. 1989. Opportunistic management for rangelands not at equilibrium. J. Range Manage. 42:266–274.
Weston, R. H. 1982. Animal factors affecting feed intake, p. 183–198. In: J. B. Hacker (ed.), Nutritional limits to animal production from pastures. Commonwealth Agr. Bur. Farnham Royal, Slough. UK.
Weston, R. H., and J. P. Hogan. 1967. The transfer of nitrogen from the blood to the rumen in sheep. Aust. J. Biol. Sci. 20:967–973.
Weston, R. H., and D. P. Poppi. 1987. Comparative aspects of food intake, p. 133–162. In: J. B. Hacker and J. H. Ternouth (eds.), The nutrition of herbivores. Academic Press, New York.
Wheeler, J. L., and R. D. Mochrie. 1981. Forage evaluation, concepts and techniques. CSIRO. Melbourne.
Whicker. A. D., and J. K. Detling. 1988. Ecological consequences of prairie dog disturbances. BioScience 38:778–785.
White, J. 1979. The plant as a metapopulation. Ann. Rev. Ecol. Syst.10:109–145.
White, J. 1980. Demographic factors in populations of plants, p. 21–48. In: O. T. Solbrig (ed.), Demography and evolution in plant populations. Univ. Calif. Press, Berkley. USA.
White, L. 1973. Carbohydrate reserves of grasses: a review. J. Range Manage. 26:13–18.
White, L. D., T. R. Troxel, J. G. Pena, and D. E. Guynn. 1988. Total ranch management-meeting goals, p. 597–603. In: L. S. Pope (ed.), Beef cattle science handbook. Vol. 21. Spillman Press. Sacramento, CA. USA.
Whitman, T. G. 1980. The theory of habitat selection: examined and extended using Pemphigus aphids. Am. Nat. 115:449–466.
Whitman, W. C. 1971. Influence of grazing on the microclimate of mixed-grass prairie, p. 207–218. In: K. M. Kreitlow and R. H. Hart (eds.), Plant morphogenesis as the basis for scientific management of range resources. USDA Agr. Res. Serv. Misc. Publ. No. 1271.
Whitmore, J. S. 1971. South Africa's water budget. S. Afr. J. Sci. 42:147–153.
Whitson, R. E., R. K. Heitschmidt, M. M. Kothmann, and G. K. Lundgren. 1982. The impact of grazing systems on the magnitude and stability of ranch income in the Rolling Plains of Texas. J. Range Manage. 35:526–532.
Whittaker, R. H. 1972. Communities and ecosystems. Macmillan, New York.
Wight, J. R., and A. L. Black. 1972. Energy fixation and precipitation-use efficiency in a fertilized rangeland ecosystem of the northern Great Plains. J. Range Manage. 25:376–380.
Wight, J. R., and J. W. Skiles. (eds.). 1987. SPUR: Simulation of production and utilization of rangelands. Documentation and user guide. USDA Agr. Res. Serv., ARS 63.
Wilcox, D. G. 1982. The importance of flexibility in ranch management strategies, p. 15–26. In: L. D. White and L. R. Hoermann (eds.), Proc. 1982 Int. Ranchers Roundup. Texas Agr. Ext. Serv. College Station. USA.
Wilkinson, S. R., and R. W. Lowrey. 1973. Cycling in mineral nutrients in pasture ecosystems, p. 247–315. In: G. W. Butler and R. W. Bailey (eds.), Chemistry and biochemistry of herbage. Vol. 2, Academic Press, New York.
Willatt, S. T., and D. M. Pullar. 1983. Changes in soil physical properties under grazed pastures. Aust. J. Soil Res. 22:343–348.
Willemoes, J. G., J. Beltrano, and E. R. Montaldi. 1987. Stolon differentiation in *Cynodon dactylon* (L.) Pers. mediated by phytochrome. Environ. Exp. Bot. 27:15–20.
Williams, K., R. J. Hobbs, and S. P. Hamburg. 1987. Invasion of an annual grassland in northern California by *Baccharis pilularis* spp. *consanguinea*. Oecologia (Berlin) 72:461–465.
Williams, O. B. 1969. Studies in the ecology of the riverine plain. V. Plant density response of species in a *Danthonia caespitosa* grassland to 16 years of grazing by merino sheep. Aust. J. Bot. 17:255–268.
Williams, R. F., and R. H. M. Langer. 1975. Growth and development of the wheat tiller. II. The dynamics of tiller growth. Aust. J. Bot. 23:745–759.
Williams, W. A. 1966. Range improvement as related to net productivity, energy flow, and foliage configuration. J. Range Manage. 19:29–34.

Wilson, A. D. 1989. The development of systems of assessing the condition of rangeland in Australia, p. 77–102. *In*: W. K. Lauenroth and W. A. Laycock (eds.), Secondary succession and the valuation of rangeland condition. Spring-Verlag, New York. (in press).

Wilson, A. D., G. N. Harrington, and I. F. Beale. 1984. Grazing management, p. 129–139. *In*: G. N. Harrington, A. D. Wilson and M. D. Young (eds.), Management of Australia's rangelands. CSIRO. Melbourne.

Wilson, A. M., and D. D. Briske. 1979. Seminal and adventitious root growth of blue grama seedlings on the central plains. J. Range Manage. 32:209–213.

Wilson, A. M., D. N. Hyder, and D. D. Briske. 1976. Drought resistance characteristics of blue grama seedlings. Agron. J. 68:479–484.

Wolf, D. D., and D. J. Perry. 1982. Short-term growth response of tall fescue to changes in soil water potential and to defoliation. Crop Sci. 22:996–999.

Wood, M. K., and W. H. Blackburn. 1981. Grazing systems: Their influence on infiltration rates in the Rolling Plains of Texas. J. Range Manage. 34:331–335.

Wood, M. K., and W. H. Blackburn. 1984. Vegetation and soil responses to cattle grazing systems in the Texas Rolling Plains. J. Range Manage. 37:303–308.

Woodmansee, R. G. 1978. Additions and losses of nitrogen in grassland ecosystems. BioScience 28:448–453.

Woodmansee, R. G. 1988. Ecosystem processes and global change, p. 11–27. *In*: T. Rosswell, R. G. Woodmansee and P. G. Risser (eds.), Scales and global change. John Wiley, New York.

Woodmansee, R. G., and F. J. Adamsen. 1983. Biogeochemical cycles and ecological hierarchies, p. 11–27. *In*: R. R. Lowrance, R. L. Todd, L. E. Asmussen and R. A. Leonard (eds.), Nutrient cycling in agricultural ecosystems. Georgia Agr. Exp. Sta., Athens. USA.

Woodmansee, R. G., J. L. Dodd, R. A. Bowman, F. E. Clark, and C. E. Dickinson. 1978. Nitrogen budget of a shortgrass prairie ecosystem. Oecologia (Berlin) 34:363–376.

Woodmansee, R. G., I. Vallis, and J. J. Mott. 1981. Grassland nitrogen. *In*: F. E. Clark and T. Rosswall (eds.), Terrestrial nitrogen cycles. Ecol. Bull. 33:443–462.

Wooton, E. O. 1908. The range problem in New Mexico. New Mexico Agricultural Experiment Station Bulletin 66, 46 p.

Wraith, J. M., D. A. Johnson, R. J. Hanks, and D. V. Sisson. 1987. Soil and plant water relations in a crested wheatgrass pasture: response to spring grazing by cattle. Oecologia (Berlin) 73:573–578.

Wright, R. G., and G. M. Van Dyne. 1976. Environmental factors influencing semidesert grassland perennial grass demography. S. W. Nat. 21:259–274.

Young, B. A. 1986. Food intake of cattle in cold climates, p. 328-340. *In*: F. N. Owens (ed.), Proc. Feed Intake by Beef Cattle Symp. Oklahoma St. Univ., Stillwater. USA.

Young, B. A. 1987a. Thermal factors influencing energy requirements of livestock, p. 37–38. *In*: Proc. Grazing Livestock Nutr. Conf. Univ. Wyoming, Laramie. USA.

Young, J. A., and R. A. Evans. 1981. Demography and fire history of a western juniper stand. J. Range Manage. 34:501–506.

Young, T. P. 1987b. Increased thorn length in *Acacia depranolobium*—an induced response to browsing. Oecologia (Berlin) 71:436–438.

Youngner, V. B. 1972. Physiology of defoliation and regrowth, p. 292–303. *In*: V. B. Youngner and C. M. McKell (eds.), The biology and utilization of grasses. Academic Press, New York.

Zarrough, K. M., C. J. Nelson, and D. A. Sleper. 1984. Interrelationships between rates of leaf appearance and tillering in selected tall fescue populations. Crop Sci. 24:565–569.

INDEX

adaptation 109, 111
aggregate stability 81–83, 149, 151 (*see also* soil aggregates)
animal unit (AU) 161–162
animal requirements, relative to (*see also* nutrient requirements)
   food, water, cover, and space 67–81, 180–183
   livestock-wildlife interactions 183–189, 198
apical dominance 99
apical meristem 86
assimilation efficiency, as
   a measure of ecological efficiency 21
   a factor affecting secondary production 21 (*see also* livestock production)
autotrophs 12
axillary buds 87

biological inertia 114
bite size and rate 73, 77
browse
   as component of diet(s) 49, 75, 78–79, 81, 193
   nutritive value 48
bunchgrasses
   developmental morphology 79, 87–88
   impact on hydrological condition 146, 150, 154
caespitose growth form 79 (*see also* bunchgrass)
carbohydrates
   allocation patterns 28, 97
   functional role in plants 28, 98
   nutritive value 28, 29
   plant reserves 97
carrying capacity 68, 192–193 (*see also* stocking rate)
catena 111
cellulose 29, 30, 31 (*see also* carbohydrates)
climate, effects on
   carrying capacity 69
   ecological succession, condition, and trend 112, 126

ecosystem processes and energy flow 22, 109
economic risk 194–196
foraging behavior 69
hydrologic condition 148, 154
nutritive value of forage 45
primary production 14, 22, 109, 112, 164 (*see also* forage production)
secondary production 164 (*see also* livestock production)
supplemental feeding tactics 57
wildlife 180–183
community 67, 70, 102, 107, 109
compensatory growth 19, 55, 101 (*see also* grazing optimization hypothesis)
compensatory intake 56
compensatory photosynthesis 96
competition 80, 102, 109, 112, 116, 119, 129
continuous grazing 162, 165, 172–177, 195
cover
   as an animal requirement 182 (*see also* animal requirements) types of wildlife cover 182, 188
critical stocking rate 163
cryptogamic crusts 152

decision-making 183–189
decision support systems 183
decomposers 13
deep drainage 142–143, 147
deferred rotation grazing 172–177 (*see also* grazing systems)
desertification 156
diet selection
   animal attributes affecting 36, 80
   impacts on energy flow 17, 22
   landscape features affecting 65–75, 81
   plant attributes affecting 75–79, 93

dietary overlap, effects on
   harvest efficiency 71, 167
   livestock production 167–168, 193 (*see also* multi-species grazing)
   livestock-wildlife interactions 180–181, 184–186
dietary plasticity 80
digestion (*see also* fermentation)
   flow dynamics 30, 31, 34
   non-ruminant vs. ruminant 30
   post-gastric vs. pre-gastric 30
discount factor
   determination of 218
   rationale for using 158, 218
disturbance theory
   intermediate disturbance hypothesis 121
   scale and magnitude 112
diversity 194
drought 113

early intensive stocking 169, 175, 194
ecological
   efficiencies 13, 17–18
   stability and resilience 123, 134
   thresholds 130, 133, 135
economic goals and risk 191, 204–205
economics of grazing management, as affected by
   grazing systems 195, 216
   kinds of animals 80–81, 193
   number of animals 192 (*see also* stocking rate)
   ownership goals 196, 204–205 (*see also* land ownership and management strategies)
   production/animal 193
   production/unit area 193
   spatial distribution of animals 67–70, 194
   temporal distribution of animals 69, 194
ecosystem, as an ecological unit 109

256 Index

emergent properties 109
functional components 13, 109
hierarchical organization 83, 103, 107, 109
structural components 12, 109
edge effect 188
electivity index 70, 75, 77 (*see also* preference)
emergent properties 110
energy, as a nutrient
   functional role and prioritized use 13, 21, 39
   metabolism 39
energy flow
   ecological constraints affecting 14, 21
   the ecological dilemma of 16
   factors affecting magnitude and rate of 13, 21, 110
   impact of management on 20, 106
energy, solar, capturing efficiency, factors affecting
   abiotic variables 14
   composition and density of plants 22
   harvest efficiency 18, 21
energy subsidies 24
enterprise budgets and optimization 212–217 (*see also* planning process)
erosion
   and desertification 156–158
   natural or geologic vs. accelerated 153
   sequential process of 153–157
   water 154–155 (*see also* sediment production)
   wind 155
evapotranspiration, factors affecting
   climatic condition 147
   plant attributes 147, 152–153
   soil attributes 117, 147, 152–153
extensive management strategies 23, 192

feedback, as
   in ecological systems 81, 156–158
   a managerial process 220
feeding station 73, 76 (*see also* foraging behavior and diet selection)
fermentation, as a digestive process
   functional aspects 30
   pre-gastric vs. post-gastric 30
fire 128, 133

floral induction 87
food chains
   detrital 13
   grazing 13
forage
   character of 46, 75, 79
   functional classification 47
   selectivity classification 35, 75
forage availability, as affected by
   forage demand 162–170
   forage production 22, 162–170
   grazing intensity 16, 162–170
forage availability, effects on
   diet selection and forage behavior 80–81
   grazing pressure 161–177
   livestock production 20, 163–167
   nutrient intake 163–167
forage demand, as affected by
   kinds and class of animals 80–81, 167–168, 186
   numbers of animals 163–167, 186 (*see also* stocking rate)
forage intake *see* nutrient intake
forage production (*see also* primary productivity)
   as influenced by species composition 22, 175
   effect on livestock production 22, 27, 163–177
   effects of climate on 22, 163–166
forage quality *see* nutritive value
foraging behavior (*see also* diet selection)
   decision making hierarchy 65–66
   impact of thermal balance on 69
   impact of water foci on 68, 151
   role of preference 75 (*see also* preference)
foraging strategies
   as affected by animal attributes 35, 80–81
   classification of grazers based on 35, 180
forbs
   as component of diet 8–81
   nutritive value 48

goals and goal setting 204–206 (*see also* economic goals)
   cultural influences 191
   effect of land ownership 196 (*see also* land ownership and management strategies)
   impact of psychological needs on 191, 203
   impact of technology on 191
grazing avoidance 93
   biochemical mechanisms 77, 79, 95 (*see also* secondary compounds)
   impact on community structure 81, 102, 107, 121
   mechanical mechanisms 75, 94
grazing behavior 65, 83 (*see also* diet selection and foraging behavior)
grazing foci 67–68
grazing food chain 13
grazing intensity, effects on
   diet selection 80–81
   ecological efficiencies 17, 21, 161
   ecological succession 102, 109, 115, 132, 166,
   economic returns 193, 195
   energy flow 16, 106
   grazing pressure 115, 161–177
   harvest efficiency 21, 161
   hydrologic condition 116, 148–153, 155–158
   livestock production 16, 21, 161–177, 193
   microenvironment 109, 116, 144–145, 156
   nutrient cycling 18, 117
   nutritive intake 163–170
   plant growth and development 96, 109
   plant species composition 102, 109, 119, 121, 132, 166
   primary production 106, 109, 114, 136, 164
   soil properties 109, 144–145, 151–153, 155–156
   species replacement processes 102
   wildlife production 184–186
grazing management
   the ecological dilemma of 150, 161
   fundamental principles of 109, 163–170, 191
grazing optimization hypothesis 19, 65, 101
grazing plan 206–210
grazing pressure (*see also* grazing intensity)
   livestock vs. forage 170
   tactical management of 172 (*see also* managerial grazing tactics)
grazing resistance 93
   associated costs 101

avoidance vs. tolerance 93, 121
  inducible defenses 95
  optimal defense theory 95
grazing sites
  factors affecting use 65–73
  preferential use classification 72
grazing systems
  effects on economic returns 195, 216
  effects on livestock production 170–177, 195
  types of 172, 195
grazing time, factors affecting 70–71
grazing tolerance 96, 112, 121
  morphological mechanisms 96

habitat 70, 180–183
harvest efficiency, as a factor affecting (*see also* managerial grazing tactics)
  ecological efficiencies 16–17, 21
  secondary production 21 (*see also* livestock production)
herbivory *see* grazing
heterotrophs 12
hierarchical theory 110, 191
high intensity-low frequency grazing 172–177
homeothermy 69
hydrologic condition 150
hydrologic cycle, as affected by
  deep drainage 148–149, 153
  evapotranspiration 147, 152–153
  grazing intensity 148–153
  interception 143–144, 148–149
  surface runoff 146–147, 152
  types and amounts of inflow 142–143
hysteresis 135

increaser species 104, 123
infiltration, factors affecting
  aerial and ground cover 145–146, 150–151
  soil attributes and topographic features 144–146, 151–152
  storm characteristics 146–147, 152
intensive management strategies 23
intercalary meristems 86
interception, factors affecting
  storm characteristics 143–144

surface cover and structural attributes 143–144, 148–149
interflow 143, 146
internal rate of return 205, 217–218
invader species 123
investment
  analytical methodology 210–219
  diversification and economic risk 207
land ownership and management strategies
  effects on goal setting 196, 204–205 (*see also* goals and goal setting)
  private vs. public 196–197
landscape 67, 109–111
leaf
  demography 92
  developmental morphology 79, 86
  longevity 92
  nutritive value 28
  replacement potential 96
limiting factors 180
linear programming 210
livestock production, as a measure of
  ecological efficiencies 17, 21
  economic risk 194
livestock production, factors affecting (*see also* secondary productivity)
  climatic variation 22, 163–167
  ecological condition 23, 163–167
  grazing system 170–177
  individual animal performance 20, 163–167
  kinds and classes of animals 167–168 (*see also* multi-species grazing)
  number of animals 20, 163–167 (*see also* stocking rate)
  quantity and quality of available forage 163–177
  spatial distribution of animals 68, 168–169
  temporal distribution of animals 69, 169–170
livestock-wildlife interactions, as affected by
  kinds of animals 183–186
  number of animals 186
  spatial distribution of animals 68, 186–189
  temporal distribution of animals 69, 186–189
livestock-wildlife planning model 183–189

management, functions of
  monitoring 205
  operation 204
  planning 204, 206 (*see also* planning process)
managerial expertise
  impact of perception on 203–204
  psychological basis 202
  role of cognitive and judgmental skills 203
managerial grazing tactics
  as affected by grazing intensity 172
  as affected by grazing system 172–173
  effects on livestock production 173–177
  high production (HPG) vs. high utilization (HUG) 172
meristematic growth limitations 96, 98
metabolizable energy 15, 40
midgrass
  developmental morphology 87–88
  impact on hydrologic condition 145–146, 150
mineralization 18
minerals
  functional role 15, 18
  macro- vs. micro-minerals 43
multi-species grazing, effects on
  grazing pressure 167–168, 185–186
  livestock production 167–168

net present value 217–218
nutrient cycling
  ecological process and significance 15, 109–110, 115, 117
  factors affecting rate of 18
nutrient intake, as affected by
  abiotic factors 19, 54
  animal attributes 49, 53
  forage availability and presentation 56
  forage quality 19, 49, 54
  foraging behavior 35, 71
  supplementation 57 (*see also* supplemental feeding) 57
nutrient requirements, relative to
  energy 15, 39
  minerals 43
  protein 42
  vitamins 43
nutritive value of forage, determination of
  Detergent Fiber System 45
  Proximate Analysis System 44

nutritive value of forage, factors affecting 118
  abiotic variables 47
  nutrient requirements 29
  plant attributes 28, 45, 94, 169

operational planning 220
opportunity costs 210
overgrazing 22, 153, 155–158, 166
ownership costs 210

palatability 75, 93 (*see also* preference) 75, 93
partial budgeting 219
phytomer 86
planning hierarchy
  operational 220
  strategic 208
  tactical 219
planning process
  alternative selection 207
  goal setting 184, 206
  identification and analyses of alternatives 207
  resource assessment 206, 208
plant species composition, effects on
  diet selection and nutrient intake 75, 77, 163–170, 180–181
  economic returns 193
  foraging behavior 75
  hydrologic condition 145–146, 150–151
  livestock-wildlife interactions 183–186
  primary production 106
  secondary production 22
plant species composition, factors affecting
  abiotic factors 23, 120, 122, 166
  competitive interactions 102, 119, 124, 126, 134
  grazing intensity 102, 131, 163–170 (*see also* stocking rate)
  selective grazing 22, 81, 102, 163–170 (*see also* diet selection)
population 102, 105, 110, 123
precipitation *see* water
predator avoidance tactics 69, 182
preference, effects on
  diet selection and nutrient intake 75, 77, 162–163, 167, 180–181, 184–186
  ecological efficiencies 17, 21–22

ecological succession 81, 102
foraging behavior 35, 72, 75, 77
plant species composition 23, 81, 102
secondary production 167–168
primary productivity, as
  affected by ecological constraints 14, 21, 120, 122
  affected by herbivory 16, 19, 120, 136
  a measure of ecological efficiency 16–17, 21
protein
  functional role and prioritized use 32, 33
  metabolism 31, 32, 33
  sources of 31

range condition 131
  and wildlife habitat 184–189
  effects on livestock production 22, 163
  and hydrological condition 150
  impact of grazing on 105, 131, 163–167
range improvement 163
range site 67, 131, 164
range trend *see* range condition
residual return 214–215
resilience, and risk management 195
rest rotation grazing 172–177
retrogressive succession 113, 131
rhizomes 88
risk management 137, 221
roots
  adventitious 89
  response to defoliation 99
  seminal 89
rumen
  flow dynamics 31, 34 (*see also* digestion)
  functional role 30
  retention time (RT) 31, 35, 36
rumination 31
runoff, factors affecting
  grazing intensity 151
  plant species composition and cover 150, 152
  soil attributes and topographic features 144, 151–152
  storm characteristics 146–147, 152

season long grazing 169, 194–195 (*see also* continuous grazing)
secondary compounds in plants
  as grazing deterrents 77, 79, 95
  qualitative vs. quantitative 95
secondary productivity, as (*see also* livestock production)
  affected by ecological constraints 14, 21
  affected by harvest efficiency 16, 20, 22
  affected by primary productivity 14
  a measure of ecological efficiency 18
sediment production, factors affecting (*see also* erosion)
  aerial and ground cover 153–156 (*see also* interception)
  soil attributes and topographic features 153–155
  storm characteristics 154
  water infiltration rate 154 (*see also* runoff)
seedbank 128, 134–135
selection ratio 80–81 (*see also* preference)
selective grazing, impacts on (*see also* diet selection)
  ecological efficiencies 22
  ecological succession and plant species composition 22, 81–83, 102
  energy flow 22, 106
short duration grazing 172–177
shortgrasses, developmental morphology 87–88
shrubs 77, 78, 80 (*see also* browse and woody plants)
social goals 191
  relationship to economic goals 192, 207 (*see also* economic goals and risk)
  relative to hierarchy of human needs 191
soil aggregates
  factors affecting 144–145, 150–152
  impacts on water infiltration, runoff, and erosion 144–145, 150–152
  structural architecture 144
soil erosion *see* erosion
soil organisms 13, 116–117
soil porosity *see* soil aggregates
soil structure 144–145
solar energy 13–14, 17
stability *see* ecological stability
stocking density 161–162, 171
stocking rate, effects on

diet selection and foraging behavior 68, 163–167
economic returns and risk 192
energy flow and ecological efficiency 16–17, 106
forage demand, forage availability, and grazing pressure 161–167
hydrological condition 148–153
livestock production 21, 163–167
livestock-wildlife interactions 186–189
nutrient intake 163–167
quantity and quality of available forage 163–167
stolons 88
strategic planning 208
succession 131 (*see also* retrogressive succession, ecological thresholds, ecological stability and resilience, and hysteresis)
factors affecting magnitude, rate, and direction of 102, 166
impact of climate and management on 22, 102
and range condition and trend 102, 131, 150
supplemental feeding
as affected by quantity and quality of available forage 57
effects on nutrient intake, digestion, and animal production 57
sustained yield 109
systems 16, 20, 110, 148, 221–223

tactical planning 219
temperature, effects on foraging behavior 69
terminal infiltration rate *see* infiltration
thermal neutral zone 69
threshold 130, 133
tiller
demography and associated processes 90
developmental morphology 87 (*see also* axillary buds)
hierarchical arrangement 86, 103
longevity 92
recruitment 90
trophic levels 14 (*see also* food chains)
tussock growth form *see* bunchgrasses

undergrazing 22, 166

vegetative cover, effects on
erosion 153–155
infiltration 144–146, 149–151
interception 143–144, 148–149

runoff 146, 152
vitamins
functional role as nutrient 32
metabolism 32
voluntary intake 71 (*see also* nutrient intake)

water
chemical and physical properties of 141–142
as an essential nutrient 141
water balance 143
water cycle *see* hydrologic cycle
water source, effects on
animal distribution and foraging behavior 67–69
livestock-wildlife interactions 181–182, 187–189
plant distribution 95, 102, 109
wildlife
behavioral attributes 179–183
grazing tactics 180–181
habitat requirements 180–183 (*see also* animal requirements)
management goals 184
planning model 183–189
wind erosion 149, 155 (*see also* erosion)
woody plants 77, 78, 80, 124, 128

SF 85.G73 1991
22629610

Grazing management